Hans-Dieter Höltje, Wolfgang Sippl, Didier Rognan, and Gerd Folkers

Molecular Modeling

Basic Principles and Applications

Third, Revised and Expanded Edition

WITHDRAWN

WILEY-VCH

WILEY-VCH Verlag GmbH & Co. KGaA

The Authors

Prof. Dr. Dr. h.c. Hans-Dieter Höltje
Institute of Pharmaceutical Chemistry
Heinrich-Heine-Universität
Universitätsstr. 1
40225 Düsseldorf
Germany

Prof. Dr. Wolfgang Sippl
Institute of Pharmacy
Martin-Luther-Universität Halle-Wittenberg
Wolfgang-Langenbeck-Str. 4
06120 Halle/Saale
Germany

Prof. Dr. Didier Rognan
CNRS UMR 7175-LC1
Inst. Gilbert Laustriat
Boulevard Sébastien Brant
67412 Illkirch
France

Prof. Dr. Gerd Folkers
Collegium Helveticum
ETH-Zentrum /STW
Schmelzbergstr. 25
8092 Zürich
Switzerland

Library of Congress Card No.: applied for

British Library Cataloguing-in-Publication Data
A catalogue record for this book is available from the British Library.

Bibliographic information published by the Deutsche Nationalbibliothek
Die Deutsche Nationalbibliothek lists this publication in the Deutsche Nationalbibliografie; detailed bibliographic data are available on the Internet at <http://dnb.d-nb.de>.

© 2008 WILEY-VCH Verlag GmbH & Co. KGaA, Weinheim

Composition Laserwords Private Ltd, Chennai, India
Printing Betz-Druck GmbH, Darmstadt
Bookbinding Litgas & Dopf GmbH, Heppenheim
Cover Design Schulz Grafik Design, Fußgönheim

Printed in the Federal Republic of Germany
Printed on acid-free paper

ISBN: 978-3-527-31568-0

Hans-Dieter Höltje, Wolfgang Sippl,
Didier Rognan, and Gerd Folkers

Molecular Modeling

Related Titles

Schneider, G., Baringhaus, K.-H.

Molecular Design

Concepts and Applications for Beginners

2008
ISBN: 978-3-527-31432-4

Tramontano, A.

Protein Structure Prediction

Concepts and Applications

2006
ISBN: 978-3-527-31167-5

Baxevanis, A. D., Ouellette, B. F. F.
(eds.)

Bioinformatics
A Practical Guide to the Analysis of Genes
and Proteins

2004
ISBN: 978-0-471-47878-2

Eidhammer, I., Jonassen, I.,
Taylor, W. R.

Protein Bioinformatics
An Algorithmic Approach to Sequence and
Structure Analysis

2004
ISBN: 978-0-470-84839-5

Comba, P., Hambley, T. W.,
Martin, B.

**Molecular Modeling of
Inorganic Compounds**

2008
ISBN: 978-3-527-31799-8

Contents

Molecular Modeling. Basic Principles and Applications. 3ʳᵈ Edition
H.-D. Höltje, W. Sippl, D. Rognan, and G. Folkers
Copyright © 2008 Wiley-VCH Verlag GmbH & Co. KGaA, Weinheim
ISBN: 978-3-527-31568-0

Preface to the Third Edition

> *. . . And when after the long search, some new fragment of truth has been captured, it comes not as something discovered but as something revealed. The answer is so unexpected, and yet so simple and aesthetically satisfying, that it carries instant conviction, and that wonderful never-to-be forgotten moment comes when one says to oneself 'Of course that's it'. . .*

These words of Lawrence Bragg are most apt in describing the outstanding potential of the molecular modeling method for enhancing the comprehension of complicated interactions between ligands and their targets, which are so typical for many fields of research in biosciences. I, myself, have always considered this attribute of molecular modeling methods to be the most profitable and intriguing.

Generated by the rapid development of hardware and software tools, the application of molecular modeling techniques has yielded much advancement in the four years since the second edition of this book was published. It is therefore that we, the authors and the publisher, decided that it is time for revision and amplification of our book.

Hence the following revisions have been made in this edition:

- An additional chapter on Chemogenomics has been added.
- The protein modeling case study was replaced by a very actual work on nucleohormone receptor modeling.
- All other chapters have been carefully revised and expanded.

We also decided to include the titles of all cited papers in order to facilitate the reader in their search for additional information.

We hope that novices in the field will also find this volume adequate to assist their work.

On behalf of my coauthors, I acknowledge the valuable input of all coworkers and express our gratitude to the publisher, Wiley-VCH, for their continued cooperation.

October 2007
Düsseldorf   **Hans-Dieter Höltje**

Molecular Modeling. Basic Principles and Applications. 3rd Edition
H.-D. Höltje, W. Sippl, D. Rognan, and G. Folkers
Copyright © 2008 Wiley-VCH Verlag GmbH & Co. KGaA, Weinheim
ISBN: 978-3-527-31568-0

1
Introduction

'Dear Venus that beneath the gliding stars...'. Lukrez (Titus Lucretius Carus, +55 B.C.) starts his most famous poem *De Rerum Natura* with the wish to the Goddess of love to reconcile the war God Mars, who in this time, when the Roman Empire was at its zenith, ruled the world.

Explanation was the vision of Lukrez. His aim was in odd opposition to his invocatory wish to the Goddess of love: the liberation of people from their fear of God, from the dark power of unpredictable nature.

The explanation of mechanism to the common man is the measure with which Lukrez would take away the fear, the fear of the Gods and their priests, the fear of the might of nature and the power of the stars.

Lightning, fire and light, wine and olive oil were, perhaps, the simple things of daily experience that people needed and which people were afraid of and so these were dear to him:

> '... again, light passes through the horn
> of the lantern's side, while rain is dashed away.
> And why? - unless those bodies of light should be
> finer than those of water's genial showers.
> We see how quickly through a colander
> the wines will flow; how, on the other hand
> the sluggish olive oil delays: no doubt
> because 'tis wrought of elements more large,
> or else more crook'd and intertangled ...'

The atom theory of Demokrit led Lukrez to the description of the quality of light, water and wine. For this derivation of structure-quality relationships, he used models. The fundamental building stones of Lukretian models look a little like our atoms, called *primordials* by Lukrez, elementary individuals that were not cleavable any further. Those elementary building stones could associate with each other. Lukrez even presupposed recognition and interaction. He provided his building stones with mechanic tools that guaranteed recognition and interaction. The most important of this conceptually tools to hold fast are the complementary structure (sic!) and the barked hook. With these primordials, Lukrez built his world.

Molecular Modeling. Basic Principles and Applications. 3rd Edition
H.-D. Höltje, W. Sippl, D. Rognan, and G. Folkers
Copyright © 2008 Wiley-VCH Verlag GmbH & Co. KGaA, Weinheim
ISBN: 978-3-527-31568-0

How well the modeling fits is shown in his explanation of the fluidity of wine and oil. A comparison of the Corey, Pauling, Koltun (CPK) models of the fatty acid and water molecules amazes because of its similarity with the 2000-year-old image of Lukrez.

1.1
Modern History of Molecular Modeling

The roots from which the methods of the modern molecular modeling developed go back to the beginning of the twentieth century. The first successful representation of molecular structures was closely related to the fast ascent of nuclear physics.

Crystallography was the decisive line of development of molecular modeling. The complexity of the crystal structures increased very fast, but their solution was connected with a huge arithmetic exercise and an inadequate two-dimensional (2D) paper representation. To get a three-dimensional (3D) impression of the crystal structure, usage of molecular kits was the only alternative.

In this context, the Dreiding models became famous. All the accumulated knowledge of structural chemistry of the time was contained in these models. Prefabricated modular elements, for instance, different nitrogen atoms with the correct number of bonds and angles corresponding to their hybridization state, or aromatic moieties, rendered it possible to build up near exact 3D models of the crystal structures. Thus they permitted molecular modeling. Dimensions were linearly translated from the Angstrom area. Steric hindrances of substituents, hydrogen bond interactions, and so on, were fairly well represented by the models.

The Stuart–Briegleb or CPK models provided a similar quality of modeling, which was, however, less accurate and space filling.

Watson and Crick described their initial fumbling with the molecular kits and self-constructed building parts in their efforts to model the base pairing and the DNA helix in the end.

Molecular modeling was not an a priori part of computer science. Could the computer be equipped with an additional dimension in molecular modeling/molecular design?

There were synergetic advances in computer science. Processors with increasing speed could carry out the necessary computational steps in shorter times. Thus, it became possible to handle proteins containing thousands of atoms. The molecular graphics technology looked for a further quantum leap bound to the same development of these fast processors. For the first time in the 1970s, the pseudo 3D description of a molecule, color coded and rotatable, was possible on the computer screen. 'Virtual Dreiding models' had been created. Without computer technology, the flood of data emerging from a complex structure like that of a protein would have exceeded the saturation limits of human efficiency. Proteins would not have been measurable with methods like X-ray structure analysis and nuclear magnetic resonance without the corresponding computer technology, which has first made these methods what they are today.

There was, however, a second surge of investigation and development, without which today's computer-assisted molecular design would be unthinkable. Since the 1930s, nuclear physics needed not only analytical, but also systemic thinking. (In fact, this was important for the construction of the atom bomb.) Thereafter, calculations included a sense of mathematical modeling for the computation of physical states and their predictions.

In the 1940s, the *computers* in Los Alamos, in the true sense of the word, were made of army recruits. Gathered in large groups, each one had to solve a certain calculation step, but it was always the same step for a certain man. This was the advent of a revolution in the development of the computer. The Monte Carlo simulation, which originated in that period, was applied to the prediction of physical states of gas particles. The first applications of mechanical analogies on molecular systems originated during that period. The force fields were born, and were optimized in course of time till they reached today's unbelievable efficiency.

Mathematical approximation techniques that have rendered possible the quantum chemical calculation of systems larger than the hydrogen atom have now made possible even "quantum dynamic" simulations of ligands in the active site of enzymes.

1.2
Do Today's Molecular Modeling Methods Only Make Pictures of the Lukretian World or Do They Make Anything More?

Actually this is a question of quality of use. The methods used could be naive or intelligent. The results will speak for themselves. Naive usage should not be condemned. It is imperative for the quality of the use, however, that a sufficiently critical view is taken, while considering the results. This means that the user realizes his naive use of the methods. Now, the researcher is conscious of the restricted nature of the methods, and knows how to judge on the basis of the results. Here, this critical position results, even in case of very simplistic approaches, in further knowledge of correlations of structure and properties.

Very often, however, this critical attitude is absent, and this might be due to the nature of modern, commercial modeling systems. These programs always give you a result. Its evaluation is in the hands of the user. The programs tend to calculate stubbornly every absurd application and present a result that could not only be a number, but could also be a graph. This is a further instrument of seduction for an uncritical use of the algorithms. However, the merits of molecular graphics is undisputed on account of its essential contribution to the development of other analytic methods, for example, nuclear magnetic resonance spectroscopy and X-ray analysis of proteins.

The tendency to perfect data presentation is the other side of the coin.

There are many examples for this. Let us consider one of them.

The visualization of the isoelectric potentials is one of the most valuable techniques for the comparison of molecular attributes. Very often, a positive and a negative potential of certain energy is used for the description of structures. The presentation of the potentials is based upon a charge calculation and is, perhaps, used to find a suitable alignment of a training set of biologically active molecules. The latter can be realized on different quality levels. There are, for example, algorithms that perform well in calculations for simple carbohydrates, but are incapable of handling aromatic structures.

Unfortunately, these algorithms do not always signal their incapability if an aromatic system has to be calculated. A result is obtained, an isopotential surface is calculated, and a graph is created. With that, one tries to derive structure-activity relationships. The second trap comes next.

The training set that is selected represents of course a drastic reduction of the parameter space. You may hope to receive a possible representative distribution of the attributes by careful selection. You are never sure, however. Thus, the correlations originate from coincidental reciprocal effects of two mistakes, which can be traced back to uncritical selection of methods and data sets.

1.3
What are Models Used For?

Models in science have different uses and can be classified accordingly.

They serve, first of all, to *simplify*, that is explain by limitation of the analysis, the phenomena that are believed to be important.

The second type of models serve as *didactical illustrations* of very complicated circumstances, which are not accessible otherwise. Here, you take into account the explanation that the models do not show the complete reality.

Mechanical analogies are a third kind of models. They are useful because of the fact that laws of classical mechanics are completely defined, like for example Hook's law. Model building of this kind plays a decisive role in the development of uniform theories. It is their special feature that it is not presupposed that the models reflect reality, but that first of all, a structural similarity of two different fields is assumed. This is, for example the presumption that the behavior of bonds in a molecule is partly similar to the action springs, which are described by Hook's law.

These mechanical analogy models have very successfully helped in enlarging theories because the validity of a theory can, in many cases, be scrutinized experimentally. But the most important fact is that one can make predictions on new phenomena.

These models are also often called *empirical.* Force fields belong to this class. The advantage of empirical models is that their parameters are optimized on the basis of reality. The 'mechanization' does not give us explicit information from the nonmechanical contributions, but an empirical correction can eliminate the convolutions due to the nonmechanical contributions. This is why empirical models are often very close to reality.

Finally, the fourth application of models is in the area of *mathematical modeling.* These models help in the simulation of processes, as for instance, the kinetic simulation of a chemical reaction step in an enzyme. By suitable choice of parameters, kinetic simulations of the real processes can be performed.

1.4
Molecular Modeling Uses all Four Kinds for Model Building

Didactical models are used for the combined representation of structure and molecular properties. This is, for instance, for small molecules, the graphical representation of the results from quantum chemical calculations or the representation of the mobility of flexible ligandlike peptides.

In case of proteins the structure itself is already a complex problem. Interactions of ligand and protein can also be studied with these didactical models. It is already clear, that the different types of models overlap in their functions. Mechanical analogies, as well as reductions aiming at simplification of essential parts of the objects under study, are typical applications in molecular modeling.

1.5
The Final Step Is *Design*

Design is perhaps the most essential element of modeling. Molecular modeling creates a virtual world, which is connected with reality by one of the four model types. Within this world, which exists in the computer, extrapolations can be made because, contrary to the real world, we create a completely deterministic universe. On the basis of an analytical description of the system, it becomes possible to design the inhibitors in advance of the synthesis, and to test them in a virtual computer experiment.

With that step of design, the circular course of a scientific study actually closes. That study does not remain (like it often happens) on a simple analytical description of a system, which has been set up like a clockwork, but comes back by reassembling the parts of the system. Molecular design makes us realize that a system could be more than the sum of its parts. This is effective especially for biological systems with which drug design is confronted by preference.

The design step itself is not as straightforward as it would be desirable, even in the virtual world. The situation is similar to the one that Gulliver encounters during his visit to the academy of Lagado.

He learns that there is a machine, which will have written sometime, every important scientific book of the world once, by systematic combination of letters and words. Jonathan Swift's wonderful science fiction of the eighteenth century brings us to the crux of the problem: the time span of human life is not large enough to test all possibilities. There has to be an intelligent algorithm to get at the correct solutions. This is, in the above example, the author of the book. By this, Swift meant that there had to be somebody who introduces an additional criterion of quality. This is based on knowledge, experience and the ability to reject combinations of words and sentences: the human–machine network. Actually, Swift introduces such a criterion in the person of the professor, who instructs his students, who serve the machines and decide after every experiment upon the result – that is, should the combination of words be included in the book.

Unfortunately, the experimenter himself is not defined qualitatively in Swift's novel. This is the irony in Gulliver's travels. Hence, the result is not only dependent on an error-free function of the machine, but also on the quality of its user! (Figure 1.1).

The same problem is presented to us in the artificial world of modeling. Systematic exploration of properties is possible only for small numbers. Because of the combinatorics, the system 'explodes' after a few steps. Flexibility studies on peptides is one such example. The change from four torsion angles to five or six increases the number of possible conformations from some thousands to several billions.

For the design of a ligand, the situation becomes more complex. It demands a most intelligent restraining by suitable experiments, intuition or knowledge. Even here the quality of the human–machine network plays a decisive role. Fully automatic design systems seem to be like a Swift prediction machine in Gulliver's travels to the academy of Lagado.

1.6
Scope of the Book

The objective of this book is to provide support for the beginner. The recognition of principal concepts and their limitations is important. More importantly it is the complete presentation of all available algorithms, programs and data banks. Like in all areas associated with computer techniques, the technical development in this area is also exponential. Nearly everyday new algorithms are offered on the network for comparison of protein sequences, or for searching of new data banks, and so on. The user has no means of judging the quality, other than to use the programs and to explore their limitations.

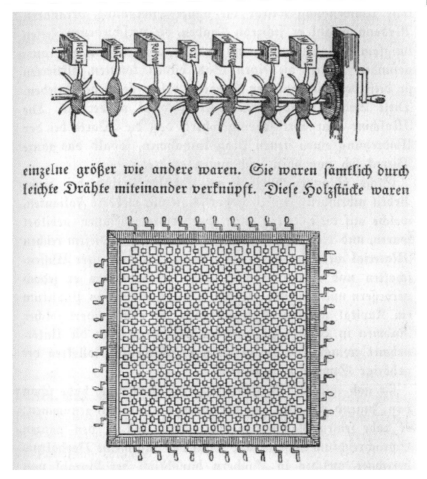

einzelne größer wie andere waren. Sie waren sämtlich durch leichte Drähte miteinander verknüpft. Diese Holzstücke waren

Figure 1.1 J. J. Grandville's imagination of the 'book writing machine' in Gulliver's visit of Lagado.

He has to know, therefore, that energy minimizing in vacuo does not make sense in any case for the analysis of the interaction geometry of a ligand.

He has also to know that a protein cannot be simply folded up from a linear polypeptide chain.

He has to realize that there is an alternative or multiple binding mode: inhibitors that bind to enzymes show alternative binding geometries in the active site even within a set of analogs. Very small changes in the molecular structure could provoke another orientation of the ligands in the active site. It is not necessarily true that a structure-orientated overlapping would be better than an intuitive one or even one that is oriented by steric or electrostatic surface properties.

In essence, today's modeling goes far beyond the example of Lukrez. Modeling does not stay on the level of analytic description of properties or correlations. It not only creates 'colored pictures', but also introduces one to systemic thinking. It even demands systemic thinking to avoid very simple applications and to have in mind the limitations of the methods.

It is here that we want to give support. By describing our own experience with molecular design using two examples, one for the 'small molecules', (ligands) and another for 'big molecules', (proteins), we seek to encourage the beginner toward a critical engagement in this direction.

2
Small Molecules

2.1
Generation of 3D Coordinates

When starting on a molecular modeling study, the first step is to generate
a model of the molecule in the computer by defining the relative positions
of the atoms in space using a set of Cartesian coordinates. A reasonable and
reliable starting geometry essentially determines the quality of the subsequent
investigations. It can be obtained from several sources. The basic methods for
generating 3D molecular structures are as follows:
  (1) use of X-ray crystallographic databases;
  (2) compilation from fragment libraries with standard
        geometries; and
  (3) 2D to 3D conversion using automated approaches.

2.1.1
Crystal Data

First we focus on the use of X-ray data for molecular building. The most
important database for crystallographic information of small molecules is the
Cambridge Crystallographic Database [1]. The Cambridge Crystallographic
Database contains experimentally derived atomic coordinates for organic and
inorganic compounds up to a size of about 500 atoms and is continuously
updated. The Cambridge Crystallographic Data Centre leases the database as
well as software for searching the database and for analyzing the results. The
output of the database search is a simple, readable file containing the 3D
structural information on the molecule of interest. This data file can be read
by most of the commercial molecular modeling packages (e.g Refs. 2–4).
    The atomic coordinates listed in the database are converted automatically
to Cartesian coordinates when reading the file into the modeling program.
Subsequently, the structure can be displayed using molecular graphics and
studied in its 3D shape.

Molecular Modeling. Basic Principles and Applications. 3rd Edition
H.-D. Höltje, W. Sippl, D. Rognan, and G. Folkers
Copyright © 2008 Wiley-VCH Verlag GmbH & Co. KGaA, Weinheim
ISBN: 978-3-527-31568-0

In general, small molecule X-ray structures are very well resolved, but there is no guaranty for the accuracy of the data. The localization of hydrogen atoms is always a problem because they are difficult to observe using X-ray crystallography. The principle of the X-ray method is the scattering of the X rays by the electron cloud around an atom. Because hydrogen atoms have one electron only, their influence on X-ray scattering is low and they are normally disregarded in structural determination. But of course, positions of hydrogen can be appointed on the basis of knowledge obtained on standard bond lengths and bond angles. According to this procedure, all bond lengths involving hydrogen atoms are usually not very specific. Before using the information from the X-ray database, it is therefore advisable to check the atomic coordinates, bond lengths and bond angles for internal consistency. In particular, the following points should be clarified before starting on any work with an X-ray structure:

(1) whether the atom types are correct;
(2) whether the bond lengths and bond angles are reasonable;
(3) whether the bond orders are correct; and
(4) in case of chiral molecules, whether the data correspond to the correct enantiomer.

After taking care of these details, the molecule can be saved in a molecular data file. The organization, extension name, format and the information contained in the file are program dependent.

It should be kept in mind that the crystalline state geometry of a molecule is subject to the influence of crystal packing forces. Therefore, bond lengths and bond angles can differ from theoretical standard values. Furthermore, the solid state structure corresponds to only one of perhaps many low energy conformations accessible to a flexible molecule and is always affected by the neighbor molecules in the crystal unity cell and sometimes also influenced by solvent molecules in the crystal. Other energetically allowed conformations must be explored by a conformational analysis to eventually reveal conformations of biological relevance. Also, knowledge of the most stable conformation called *the global energy minimum structure* is important for the evaluation of probabilities for conformers with higher energy content. Procedures for this purpose are described in Section 2.2.

2.1.2
Fragment Libraries

The second very common building method is the construction of molecules from preexisting fragment libraries. This is the method of choice when there is no access to crystallographic databases or if X-ray data for the desired structures are not available. Almost all commercial molecular modeling programs nowadays do offer the possibility to construct molecules using fragment libraries.

Fragment libraries can be utilized like an electronic 3D structure tool kit, which is easy to handle. Because of the preoptimized standard geometries of all entries in the fragment pool, the resulting 3D structures already have an acceptable geometry. Mostly only torsion angles have to be cleared to avoid atom overlapping or close van der Waals contacts. Problems may arise with fused ring systems because of the different ways in which saturated rings can be joined to each other. To solve this problem it is recommendable, wherever possible, to refer to X-ray data or to experimental data of comparable ring systems to select the correct ring connection.

Each atom in any arbitrary structure does carry characteristic features that are defined by the so-called atom type. Properties distinguishing different atoms in molecular modeling terms are for example hybridization, volume, and so on. The corresponding parameters define the particular atom type. All atomic parameters taken collectively represent the atomistic part of a force field. On preexisting fragments selected from libraries, the atom types are already defined and, in general, are correct. In many cases, however, the decision regarding which atom type will be appropriate is not so easy to take. We discuss this problem with the example of N-acetylpiperidine.

When N-acetylpiperidine is generated from the fragment library using a piperidine ring and an acetyl residue, the nitrogen atom in the piperidine is defined as sp^3 nitrogen atom type with tetrahedral geometry. But when this nitrogen is connected with the acetyl residue, it can also be considered as an amide nitrogen atom demanding planar trigonal sp^2 geometry. In such a case, the correct decision can only be made by either comparing the geometry obtained from the building routine with X-ray data or performing a quantum mechanical calculation for the structural element of interest to get a reliable geometry. Figure 2.1.1 shows the results of a semiempirical and an ab initio calculation in comparison with force field geometries and the crystal structure of N-acetylpiperidine-2-carboxylic-acid [5].

Whereas the sp^3 nitrogen atom of the force field structure bears a tetrahedral geometry, the crystal structure and the quantum chemically calculated geometries indicate an almost planar nitrogen atom. To avoid errors in subsequent calculations, the nitrogen atom has to be assigned an atom type with planar geometry.

Another problem occurring when building substituted saturated ring systems is the correct conformation of the cycle because it may be influenced by the substituents. Cyclohexane is one of the cyclic molecules studied in most detail in organic chemistry. The different possible conformations and the energy barriers separating them have been subject of many investigations [6, 7]. There is no doubt that the chair form is the most stable conformation of this molecule. For mono-substituted cyclohexane this still holds true. The preferred position of any substituent is found to be the equatorial one. The energy difference determined between the equatorial and axial

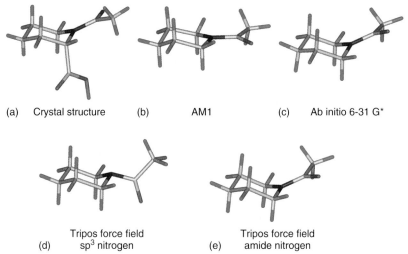

(a)   Crystal structure   (b)   AM1   (c)   Ab initio 6-31 G*

Tripos force field   Tripos force field
(d)   sp³ nitrogen   (e)   amide nitrogen

Figure 2.1.1 The geometry of the amide group in *N*-acetylpiperidine decisively depends on the method used as well as the atom types used for optimization. For comparison, the crystal structure of piperidine-2-carboxylic acid is shown in (a). The color code used here is carbon = white, oxygen = red, nitrogen = blue, hydrogen = cyan and this is used throughout this book.

position is not very distinct for small substituents but is increasing for larger groups [8]. Therefore it is necessary and advisable to always check the results of structure building from fragment libraries with experimental data in comparable situations.

2.1.3
Conversion of 2D Structural Data into 3D Form

An alternative way for generating 3D molecular structures is to start from 1D or 2D representations of molecules and to convert this information into a 3D form. Numerous software programs are available to handle structural information on molecules in different ways. All of these programs have one task in common: to save the structural data in a file. Therefore, a variety of software suppliers have developed their own file format. As many different file formats have been developed in the past, the need for a standard chemical structure format has been increasingly felt. Two of the numerous formats have achieved unanimous acceptance. The Molecular Design Limited (MDL) Molfile format (and the related (structure-data) SD format) developed in the 1980s by MDL [9] and the SMILES (Simplified Molecular Input Line Entry System) notation developed by David Weininger in 1986 [10, 11] represent de facto standard file formats for molecular structures. Besides the MDL Molfile and SMILES notation format, other file formats are often used in modeling systems. The Tripos Mol2 format [2] is the standard format for representing 3D structures of small molecules (see next chapter), whereas the Protein Data Bank (PDB) file format [12, 13] is primarily used for storing 3D structure information on proteins and other macromolecules (see Section 4.1).

In SMILES notations, the chemical structure information is highly compressed and simplified. Chemical structures are described as simple 1D line notations [10]. SMILES notations have found widespread acceptance as a universal chemical nomenclature for the representation and exchange of chemical structure information, independently of the software program. Several predefined rules are used by this language to convert a 3D structure into a character string. More details about SMILES can be found in the literature [10, 11] and at http://www.daylight.com/daylight/smiles. The compact textual coding in SMILES makes it possible to store a large number of structural information and permits a fast exchange between different programs or computers.

An extension of the MDL Molfile is the SD file format. It contains structure and data (properties) for one or more molecules, which makes it especially convenient to handle large number of molecules. Many organizations and suppliers that offer chemical compounds store their structures in this 2D format. In order to generate 3D structures out of SD files or SMILES notations, 3D structure generators such as CONCORD [14, 15], CORINA [16] or Omega [17] are used. These programs offer the possibility to automatically generate 3D structures. They use a rule and data based system that automatically generates 3D atomic coordinates from the constitution of a molecule as expressed by a connection table (SD file) or linear code (SMILES), and which is powerful and reliable to convert large databases of several hundreds of thousand or even millions of compounds.

For structure generation, CONCORD uses a very detailed table of bond lengths. In addition to information like atomic number, hybridization and bond type, the program regards the 'environment' of the atoms included in the bond before assigning the bond lengths. This precise selection of bond lengths is especially important for the construction of ring systems. Deviations from correct bond lengths may have a dramatic effect on the resulting ring conformation.

When starting the 2D conversion, the program identifies the so-called smallest set of small rings. Subsequently, a logical analysis is performed for each particular ring system. On the basis of ring adjacency and ring constraints, these logical rules decide how the rings will be constructed. In addition, rough conformation of each ring system is determined taking into consideration planarity or stereochemical constraints.

If fusion atoms of multicyclic systems are not specified, CONCORD creates the isomer with the lowest energy content. After constructing and connecting the ring systems, the program modifies the gross conformations to remove the internal strain by distributing the strain symmetrically over all atoms in the ring. This procedure leads to cyclic structures with sufficiently relaxed geometries.

The next step in structure generation is to add the acyclic substructures. Bond lengths and bond angles are again taken from predefined tables. To avoid close van der Waals contacts in the built structure, the torsion angles are modified to obtain energetically acceptable conformations. Besides computational speed, the main advantage of CONCORD is that the entire topology of the growing molecule is considered at each step. As a result of this, CONCORD yields 3D structures of good quality in lesser computational time. This is an important criterion when large databases of 2D information are to be converted into 3D space.

CORINA works similar to CONCORD. The starting point in creating ring systems is analogous to CONCORD. But in the subsequent step CORINA uses a different approach to connect the ring systems. The rings are fused and the energies of possible ring conformations are calculated using a crude force field. If the actual choice of a particular ring connection is detected to be energetically unfavorable, a new attempt is made using other energetically possible conformations of the rings. The generation of ring structures is followed by a geometry-optimization step.

Similar to CONCORD, the acyclic substructures are constructed after the ring system is completed. The chains added to the rings are usually in fully extended conformations. This of course leads to geometries needing further refinement. The torsion angles are rotated until the first conformation is reached, which relieves close contacts. As a result of this rough conformational search, the program indeed yields acceptable structures.

It is important to note that the resulting conformations, only as a matter of chance, correspond either to a conformation in the crystal environment or to a low energy conformation. The structure finally obtained therefore has to be subjected to a conformational analysis to detect all possible low energy conformations.

All of the reviewed programs are effective alternatives in structure generation. They are fast, robust and provide good conversion rates (99.5%) tested on the conversion of 250 000 2D structures of the National Cancer Institute Open Database [18]. This database is freely available and contains a huge number of small organic compounds and drug molecules tested for cancer activity at the National Cancer Institute.

Since more and more suppliers and organizations are offering compound databases where the chemical structures are stored in SD files or SMILES, powerful 3D structure generator programs are needed to convert the large compound databases in reasonable time. Today, another way to get the corresponding 3D structures is to download the 3D structures already converted – for example from the ZINC (Zinc is not commercial) web site (http://blaster.docking.org/zinc/) [19]. The group of Brian Shoichet has carefully converted 4.6 million molecules from available compound databases into 3D structures.

For more information about structure representation, structure generators and exchange between different programs, the reader is referred to the literature [20, 21].

References

1. Olga Kennard, F.R.S. Cambridge Structural Database, Cambridge Crystallographic Data Centre, http://www.ccdc.cam.ac.uk.
2. SYBYL, Tripos Associates, St. Louis, http://www.tripos.com.
3. INSIGHT/DISCOVER, Accelrys Inc., San Diego, http://www.accelrys.com.
4. MOE, Chemical Computing Group, Montreal, http://www.chemcomp.com.
5. Rae, I.D., Raston, C.L., and White, A.H. (1980) Crystal and molecular-structure of (+)-(e)-n-acetylpiperidine-2-carboxylic acid. *Australian Journal of Chemistry*, **33**, 215.
6. Bucourt, R. (1974) The torsion angle concept in conformational analysis, in *Topics in Stereochemistry* (eds E.L., Eliel and N.L., Allinger) John Wiley & Sons, New York, Vol. 8, pp. 159–224.
7. Shopee, C.W. (1946) Steroids and the Walden inversion. Part II. Derivatives of 5-cholestene and 5-androstene. *Journal of the Chemical Society*, 1147–51.
8. Hirsch, J.A. (1967) Tables of conformational energies, in *Topics in Stereochemistry*, (eds E.L., Eliel and N.L., Allinger) John Wiley & Sons, New York, Vol. 1, pp. 199–222.
9. MDL Informations Systems, http://www.mdli.com.
10. Weininger, D. (1988) SMILES, a chemical language and information-system.1. Introduction to methodology and encoding rules. *Journal of Chemical Information and Computer Sciences*, **28**, 31–36.
11. Weininger, D. (1990) SMILES. 3. DEPICT – graphical depiction of chemical structures. *Journal of Chemical Information and Computer Sciences*, **30**, 237–43.
12. Bernstein, F.C., Koetzle, T.F., Williams, G.J.B. *et al.* (1977) Protein data bank – computer-based archival file for macromolecular structures. *Journal of Molecular Biology*, **112**, 535–42.
13. PDB Format Description, http://www.rcsb.org/pdb.
14. Pearlman, R.S. (1987) Rapid generation of high quality approximate 3D molecular structures. *Chemical Design Automation News*, **2**, 1–7.
15. Pearlman, R.S. (1993) 3D molecular structures: generation and use in 3D searching, in *3D QSAR in Drug Design* (ed. H., Kubinyi), Escom Science Publishers, Leiden, pp. 41–79.
16. Gasteiger, J., Rudolph, C., and Sadowski, J. (1990) Automatic generation of 3D-atomic coordinates for organic molecules. *Tetrahedron Computer Methods*, **3**, 537–47.
17. Omega, Version 2.0, Openeyes Scientific Software, Santa Fe, http://www.eyesopen.com.
18. National Cancer Institute, National Institute of Health (Development Therapeutics Program) http://dtp.nci.nih.gov/docs/3d_data.
19. Irwin, J.J. and Shoichet, B.K. (2005) ZINC – A free database of commercially available compounds for virtual screening. *Journal of Chemical Information and Modeling*, **45**, 177–82.
20. Engel, T. (2003) Representation of chemical compounds, in *Chemoinformatics* (eds J., Gasteiger and T., Engel), Wiley-VCH, Weinheim, pp. 15–168.
21. Engel, T. (2006) Basic overview of chemoinformatics. *Journal of Chemical Information and Modeling*, **46**, 2267–77.

2.2
Computational Tools for Geometry Optimization

2.2.1
Force Fields

Molecular structures generated using the procedures described in the Section 2.1 should always be geometry-optimized to find the individual energy minimum state. This is normally done by applying a molecular mechanics method. The expression 'molecular mechanics' is used to define a widely accepted computational method used to calculate molecular geometries and energies.

Unlike quantum mechanical (QM) approaches, the electrons and nuclei of the atoms are not explicitly included in the calculations. Molecular mechanics considers the atomic composition of a molecule to be a collection of masses interacting with each other via harmonic forces. As a result of this simplification, molecular mechanics is a relatively fast computational method practicable for small molecules as well as for larger molecules and even oligomolecular systems.

In the framework of the molecular mechanics method, the atoms in molecules are treated as rubber balls of different sizes (atom types) joined together by springs of varying length (bonds). Hooks law is used for calculating the potential energy of the atomic ensemble. In the course of a calculation, the total energy is minimized with respect to atomic coordinates, where

$$E_{tot} = E_{str} + E_{bend} + E_{tors} + E_{vdw} + E_{elec} + \cdots \tag{1}$$

with

E_{tot} = total energy of the molecule
E_{str} = bond stretching energy term
E_{bend} = angle bending energy term
E_{tors} = torsional energy term
E_{vdw} = van der Waals energy term
E_{elec} = electrostatic energy term.

Molecular mechanics enables the calculation of the total steric energy of a molecule in terms of deviations from reference 'unstrained' bond lengths, angles and torsions plus nonbonded interactions. A collection of these unstrained values together with what may be termed *force constants* (but in reality are empirically derived fit parameters) is known as the *force field*. The first term in Equation 1 describes the energy change as a bond stretches and contracts from its ideal unstrained length. It is assumed that the interatomic forces are harmonic, so the bond stretching energy term can be described by a simple quadratic function given in Equation 2:

$$E_{str} = \tfrac{1}{2}k_b(b - b_0)^2 \tag{2}$$

Molecular Modeling. Basic Principles and Applications. 3rd Edition
H.-D. Höltje, W. Sippl, D. Rognan, and G. Folkers
Copyright © 2008 Wiley-VCH Verlag GmbH & Co. KGaA, Weinheim
ISBN: 978-3-527-31568-0

where

k_b = bond stretching force constant
b_0 = unstrained bond length
b = actual bond length.

In more refined force fields, a cubic term [1], a quartic function [2–4] or a Morse function [5] has been included.

Also, for angle bending mostly a simple harmonic, springlike representation is used. The expression describing the angle bending term is shown in Equation 3.

$$E_{\text{bend}} = \tfrac{1}{2} k_\theta (\theta - \theta_0)^2 \qquad (3)$$

where

k_θ = angle bending force constant
θ_0 = equilibrium value for θ
θ = actual value for θ.

A common expression for the dihedral potential energy term is a cosine series (Equation 4):

$$E_{\text{tors}} = \tfrac{1}{2} k_\varphi (1 + \cos(n\varphi - \varphi_0)) \qquad (4)$$

where

k_φ = torsional barrier
φ = actual torsional angle
n = periodicity (number of energy minima within one full cycle)
φ_0 = reference torsional angle (the value usually is $0°$ for a cosine function with an energy maximum at $0°$ or $180°$ for a sine function with an energy minimum at $0°$)

The van der Waals interactions between atoms that are not connected directly are usually represented by a Lennard-Jones potential [6] (Equation 5):

$$E_{\text{vdw}} = \sum \frac{A_{ij}}{r_{ij}^{12}} - \frac{B_{ij}}{r_{ij}^{6}} \qquad (5)$$

where

A_{ij} = repulsive term coefficient
B_{ij} = attractive term coefficient
r_{ij} = distance between the atoms i and j

This is one form of the Lennard-Jones potential but there exist several modifications of this term that are used in the different force fields. An additional function is used to describe the electrostatic forces. In general, it makes use of the Coulomb interaction term (Equation 6):

$$E_{\text{elec}} = \frac{1}{\varepsilon} \cdot \frac{Q_1 \cdot Q_2}{r}$$

(6)

where

ε = dielectric constant
Q_1, Q_2 = atomic charges of interacting atoms
r = interatomic distance

Charges may be calculated using the methods described in Section 2.4.1.1 or are implemented in some of the force fields [2–4] as empirically derived parameter sets.

Some force fields also include cross terms, out of plane terms, hydrogen bonding terms, and so on, and use more differentiated potential energy functions to describe the system. As force fields are varying in their functional form, not all can be discussed here in detail, but they have been subject of excellent reviews [7, 8].

The basic idea of molecular mechanics is that the bonds have 'natural' lengths and angles. The equilibrium values of these bond lengths and bond angles and the corresponding force constants used in the potential energy function are defined in the force field and will be denoted as force field parameters. Each deviation from these standard values will result in increasing total energy of the molecule. So the total energy is a measure of intramolecular strain relative to a hypothetical molecule with ideal geometry. By itself, the total energy has no physical meaning.

The objective of a good and generally usable force field is to describe as many different classes of molecules as possible with reasonable accuracy. The reliability of the molecular mechanics calculation is dependent on the potential energy functions and the quality of the parameters incorporated in these functions. So it is easy to understand that a calculation of high quality cannot be performed if parameters for important geometrical elements are missing. To avoid this situation, it is necessary to choose a suitable force field for a particular investigation.

Several force fields have been developed to examine a wide range of organic compounds and small molecules [1–4, 9, 10], while other programs contain force fields primarily for proteins and other biomolecules [11–13]. Inadequate availability of experimental data can considerably inhibit the development of improved energy functions for more accurate calculations. This has led to the development of the so-named class II force fields such as the consistent force field (CFF) [11] and the Merck molecular force field (MMFF) [10], which are both based primarily on QM calculations of the energy surface. The purpose of MMFF is to handle all functional groups of interest in pharmaceutical design, including small molecules and macromolecular structures. The current version, MMFF94, has been implemented in various programs and commercial software packages (e.g. SYBYL, MOE or Cerius2).

For all force fields it must be stated that they can only be applied for a particular problem if the necessary parameters are included. If parameters for particular atom types, bond types, bond lengths, bond angles or torsional angles are missing, it is unavoidable to add the missing data to the force field [14, 15].

2.2.2
Geometry Optimization

As already mentioned, it is almost certain that the generated 3D model of a given molecule does not have ideal geometry, and it is therefore necessary that a geometry optimization should be performed subsequently. In the course of the minimization procedure, the molecular structure will be relaxed. As can be deduced from the example presented in Figure 2.2.1 and Table 2.2.1, the internal strain in structures obtained from crystal data is mainly influenced by small deviations from the 'ideal' bond lengths. Therefore all the above corresponding energy terms (bond stretching term, angle bending term) are altered in the course of a force field optimization. Despite the remarkable change in energy content, torsional angles are affected only to a lesser extent. This is a clear indication to the well-known observation that in crystals, almost exclusively, low-energy conformations are found. It should also be realized that crystal structures are by no means 'bad' geometries. As can be easily deduced from Figure 2.2.1, the distortion of the crystal structure is only very subtle when compared to the relaxed geometry of the force field structure in terms of geometry differences. This fact can be interpreted also in the sense that large variations in geometry are not to be expected when different well-parameterized force fields are applied. In the case considered here, the individual but real crystal packing of ramiprilate is compared to the well-known general Tripos force field [9].

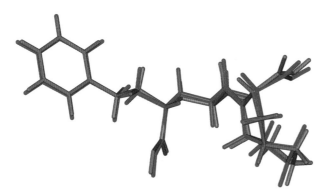

Figure 2.2.1 Superposition of the crystal structure (red) and force field (Tripos force field) optimized geometry (green) of the angiotensin-converting-enzyme inhibitor ramiprilate.

Table 2.2.1 Force field energy terms (Tripos force field) for the ramiprilate molecule before and after geometry optimization.

Energy terms	Energy (kcal mol^{-1})
Crystal structure	
Bond stretching energy	179.514
Angle bending energy	15.693
Torsional energy	17.230
Out of plane bending energy	0.043
1-4 van der Waals energy	18.538
van der Waals energy	−3.839
Total energy	227.179
Optimized structure	
Bond stretching energy	0.982
Angle bending energy	10.372
Torsional energy	14.335
Out of plane bending energy	0.011
1-4 van der Waals energy	4.791
van der Waals energy	−7.822
Total energy	22.669

STO 3G	3-21 G	6-31 G

Figure 2.2.2 The figure shows the final geometries of 2,6-diazaspiro[3.3]hept-2-yl-formamide after geometry optimization using different basis sets. The example clearly indicates the dependence of the resulting geometry on the applied basis set. The minimal basis set STO-3G and the 3-21G basis set yield very different geometries. The inclusion of the d orbitals (6-31G*) in the calculations leads to a structure reflecting the polarization effects and the ring tension more precisely as the resulting geometry of the amide nitrogen atom lies in between tetrahedral and trigonal planar hybridization states.

Before starting a geometry optimization, bad van der Waals contacts should be removed because the minimum energy geometry at the end of the optimization will depend on the starting geometry [7].

Several advantages like speed, sufficient accuracy and the broad applicability on small molecules as well as on large systems have established the force field geometry optimization as the most important standard method. Because of the complexity and the demanding computational costs, QM methods should be reserved for special problems, which is discussed later.

We will now focus on some common energy-minimization procedures used by molecular mechanics. It is important to note that the minimization algorithms only find local minima on the potential energy surface, but not implicitly the global energy minimum.

2.2.3
Energy-minimizing Procedures

The energy-minimization methods can be divided into two classes: the first derivative techniques like steepest descent, conjugate gradient and Powell and the second derivative methods like the Newton–Raphson and related algorithms.

2.2.3.1 Steepest Descent Minimizer

The steepest descent minimizer uses the numerically calculated first derivative of the energy function to approach the energy minimum. The energy is calculated for the initial geometry and then again when one of the atoms has been moved in a small increment in one of the directions of the coordinate system. This process will be repeated for all atoms, which are finally moved to new positions downhill the energy surface [7]. The procedure will stop if the predetermined minimum condition is fulfilled. The optimization process is slow near the minimum, so the steepest descent method is often used for structures far from the minimum. It is the method most likely to generate low-energy structures of poorly refined crystallographic data or to relax graphically built molecules. In most cases, the steepest descent minimization is used as a first rough and introductory run followed by a subsequent minimization using a more advanced algorithm like conjugate gradients.

2.2.3.2 Conjugate-Gradient Method

The conjugate-gradient method accumulates the information about the function from one iteration to the next. With this proceeding, the reverse of the progress made in an earlier iteration can be avoided. For each minimization step, the gradient is calculated and used as additional information for computing the new direction vector of the minimization procedure. Thus each successive step continually refines the direction toward the minimum. The computational effort and the storage requirements are greater than those for steepest descent but conjugate gradients is the method of choice for larger systems. The higher total computational expense and the longer time per iteration are more than compensated for by the more efficient convergence to the minimum achieved by conjugate gradients.

The Powell method is very similar to conjugate gradients. It is faster in finding convergence and is suitable for a variety of problems. But one should be careful when using this method because torsion angles may sometimes be modified to a dramatic extent. So the Powell method is not practicable for energy minimization after a conformational analysis because the located low-energy conformations will be altered in an undesired manner. It is advisable to perform a conjugate-gradient minimization in this situation.

2.2.3.3 Newton–Raphson Minimizer

The Newton–Raphson minimizer as a second derivative method uses, in addition to the gradient, the curvature of the function to identify the search direction. The second derivative is also applied to predict where the function passes through a minimum. The efficiency of the Newton–Raphson method increases as convergence is approached. The computational effort and the storage requirements for calculating larger systems are disadvantages of this method. For structures with high strain, the minimization process can become instable, so the application of this algorithm is mostly limited to problems where rapid convergence from a preoptimized geometry to an extremely precise minimum is required. For some more detailed information about the optimization methods, see Refs 16, 17.

It can be summarized that the choice of the minimization method depends on two factors – the size of the system and the current state of the optimization. For structures far from minimum, as a general rule, the steepest descent method is often the best minimizer for the first 10–100 iterations. The minimization can be completed to convergence with conjugate gradients or a Newton–Raphson minimizer. To handle systems that are too large for storing and calculating a second derivative matrix, the conjugate-gradient minimizer is the only practicable method. The minimization procedure will continue until convergence has been achieved.

There are several ways in molecular minimization to define convergence criteria. In nongradient minimizers like steepest descent, only the increments in the energy and/or the coordinates can be taken to judge the goodness of the actual geometry of the molecular system. In all gradient minimizers however, atomic gradients are used for this purpose. The best procedure in this respect is to calculate the root mean square gradients of the forces on each atom of a molecule. It is advisable to always check, in addition, the maximum derivative in order to detect unfavorable regions in the geometry. There is no doubt about the quality of a minimum geometry if all derivatives are less than a given value. The specific value chosen, for example for the maximum derivative, depends on the objective of the minimization. If a simple relaxation of a strained molecule is desired, a rough convergence criterion like a maximum derivative of 0.1 kcal mol^{-1} Å^{-1} is sufficient, whereas for other cases, convergence to a maximum derivative less than 0.001 kcal mol^{-1} Å^{-1} is required to find a final minimum.

The choice of the convergence criteria should be a balance between attaining reasonable accuracy in determining the minimum structure and avoiding unnecessary computations when no further progress can be realized [17].

2.2.4
Use of Charges, Solvation Effects

Molecular mechanics calculations are usually carried out under vacuum conditions ($\varepsilon = 1$). For nonpolar hydrocarbons, the effect of the explicit inclusion of solvent as compared with gas phase calculations is negligible. The investigation of molecules containing charges and dipoles however requires the consideration of solvent effects [7], otherwise conformations mainly influenced by strong electrostatic interactions would be overestimated. The force field will try to maximize the attractive electrostatic interaction, resulting in energetically strongly preferred but unrealistic low-energy conformations of the molecule. This can be prevented by using the corresponding solvent dielectric constant [18]. For example, in water ε amounts to $\varepsilon = 80$. Contrary to macromolecules, the electrostatic field of small molecules is considered to be homogenous, and therefore the use of a uniform dielectric constant, in principle, is allowed. Experimentally determined dielectric constants for a large number of solvents may be found in literature and can be applied for a correct treatment of the Coulomb term of solvated molecules.

A very simple but effective way to treat the problem of charges and solvation in the course of a molecular mechanics optimization is to perform the calculation without taking charges into account. This very often yields acceptable results and is especially recommended if the results of a conformational analysis are to be minimized because usage of charges may markedly alter the conformation by electrostatic interactions. Consideration of charges is always necessary if hydrogen bonding phenomena are to be described.

The strength of the electrostatic interaction decreases with r^{-1}. Therefore, in some force fields the dielectric constant can be chosen to be distance dependent in order to simulate the effect of displacement of solvent molecules in the course of the approach of a ligand molecule to a macromolecular surface. This is of particular value if a conformational analysis is part of a pharmacophore search.

Whenever possible, experimental data should be used for testing results from theoretical calculations. Above all, nuclear magnetic resonance (NMR) data have become a valuable tool in this respect. Since most of the available NMR data have been obtained in chloroform or similar organic solvents, the explicit inclusion of the corresponding dielectric constant in the Coulomb term of a force field leads to an improved agreement with experimental results.

Consideration of the dielectric constant is one possibility to simulate solvent effects. An alternative way is to create a solvent box around the molecule containing discrete solvent molecules. The additional computational effort and the limitations in regard to the limited number of solvents that can be used in most of the available force fields are severe disadvantages of this method.

2.2.4.1 Solvent as a Statistical Continuum

To determine the most probable ligand conformation in an aqueous system, it is important to consider how the solvent affects the behavior of the system. A solution to this problem would be to model the solvent molecules explicitly as an integral part of the system. However, for larger molecules (e.g. for protein–ligand complexes) this is computationally expensive. Therefore, approaches have been developed where the solvent is considered as a continuous medium surrounding the solute, providing the solvation effects for comparatively little computational effort. These methods usually include various formulations to account for solvent-mediated charge–charge interactions and some surface area terms to account for hydrophobic interactions, van der Waals interactions, and so forth. Two major continuum models have been applied to drug design: one that uses the Poisson–Boltzmann equation (see Section 4.6.1) to account for charge–charge interactions (including the surface area terms, it is called PB–SA)[19, 20], and the other that uses the generalized Born approximation to account for the charge interactions (GB–SA) [21–23]. The PB–SA model solves the Poisson–Boltzmann equation to compute the electrostatic fields in the studied system. However, for irregularly shaped solutes, such as proteins and nucleic acids, there is no analytical solution to the Poisson–Boltzmann equation. Thus, numerical algorithms are necessary and, even with advanced computational resources, this is still a computationally expensive process. The GB–SA models try to use simplified methods for calculating solvation with significantly less computational effort, but with comparable accuracy compared to that of PB–SA models. One of the first GB–SA methods, suggested by Still *et al.* [21] was shown to compare well with the PB–SA approach in predicting solvation free energies of small molecules.

Continuum methods have shown some success in ligand–protein docking [24, 25] and in the prediction of protein–ligand free energy of binding. A variety of modeling programs contain these methods to consider solvation effects within energy minimization, conformational analysis or protein–ligand docking [25–28].

2.2.5
Quantum Mechanical Methods

Quantum mechanical methods must also be discussed, at least in short, because they are very valuable additional tools in computational chemistry. In general, properties like molecular geometry and relative conformational energies can be calculated with high accuracy for a broad variety of structures by a well-parameterized general force field. However, if force field parameters for a certain structure are not available, quantum chemical methods can be used for geometry optimization. In addition, the calculation of transition states or reaction paths as well as the determination of geometries influenced by polarization or unusual electron distribution in a molecule is the domain of QM calculations. The disadvantages relative to other methods are the computational costs and the limitation to rather small molecules. So the use of QM methods should be reserved for the treatment of special problems. The objective in this context is not to discuss the QM methods from a theoretical perspective but to give some practical hints for the application of semiempirical or ab initio programs. The reader's interest are drawn to many books and reviews on this subject to gain more insight into the theoretical aspects of these methods [29–32].

2.2.5.1 Ab initio Methods

Unlike molecular mechanics and semiempirical molecular orbital methods, the ab initio quantum chemistry is capable of reproducing experimental data without using empirical parameters. Therefore, the application of ab initio calculations is especially favored in situations in which little or no experimental information are available.

The quality of an ab initio calculation is dependent on the basis set used for the calculation [33, 34]. A wrong choice of the basis set can render the results of extremely time consuming calculations meaningless. The decision as to which basis set should be taken is related to the objective of the calculation and the molecules to be studied. It should be kept in mind that even a high basis set is not always a guarantee for agreement with experimental data [35].

Only the most commonly applied basis sets are discussed here. The STO-3G basis set has been frequently used in the past and is the smallest basis set that can be chosen. STO-3G is an abbreviation for Slater type orbitals simulated by three Gaussian functions each. This minimal basis set has the smallest number of atomic orbitals that are necessary to accommodate all electrons of the atoms in their ground state assuming spherical symmetry of the atoms.

In the more recent ab initio calculations, the split-valence basis sets have become quite popular. In the split-valence basis sets, the valence orbital shells are represented by an inner and outer set. In this way more flexibility in describing the residence of the electrons has been attained [36]. The split-valence basis sets represent a progress over the STO-3G basis set, and the 3-21G, 4-31G and 6-31G basis sets are widely used in ab initio calculations. They differ only in the number of primitive Gaussians used in expanding the inner shell and first contracted valence functions [35]. For example, in 4-31G the core orbitals consist of four Gaussian functions, and the inner and outer valence orbitals consist of three Gaussian functions and one Gaussian function, respectively.

The next level of improvement is the introduction of polarization basis sets. To all nonhydrogen atoms, d orbitals are added to allow p orbitals to shift away from the position of the nucleus, leading to a deformation (polarization) of the resulting orbitals. This adjustment is particularly important for compounds containing small rings [36]. The polarization basis sets are marked by a star like 6-31G* for example. This basis uses six primitive Gaussians for the core orbitals, a 3:1 split for the s- and p-valence orbitals and a single set of six d functions (indicated by the asterisk).

For a more detailed description of the basis sets, the reader is directed to books and reviews on this subject [32, 35].

Unfortunately, there is no general rule for choosing an adequate basis set. The level of calculation depends on the desired accuracy and the molecular properties of interest. A geometry optimization of a simple molecule with moderate size can reasonably be performed using a 3-21G basis set. For other problems, however, this degree of sophistication is not sufficient. If the geometry of the molecule is influenced by polarization effects, electron delocalization or hyperconjugative effects, a 6-31G* or higher basis set is necessary to include the d orbitals as already mentioned (Figure 2.2.2).

In spite of the rapid development in computer technology, high-level ab initio calculations cannot still always be performed. A common way to overcome the computing time problem is to use a 3-21G basis set to optimize the geometric variables and then compute the wavefunction on the 6-31G* level. This procedure is termed 6-31G*//3-21G calculation.

The use of higher basis sets does not automatically improve the accuracy of the calculated molecular properties of interest. In order to find a suitable level of calculation, it is necessary to calibrate the method against experiment or test the basis sets empirically to yield acceptable results.

2.2.5.2 Semiempirical Molecular Orbital Methods

The deep gap between molecular mechanics and the ab initio calculations is filled up by the semiempirical molecular orbital methods. They are basically QM in nature but the main difference with ab initio methods is the introduction of empirical parameters in order to reduce the high costs of computer time necessary for explicit evaluation of all integrals. One-center repulsion integrals and resonance integrals are substituted by parameters fitted as closely as possible to experimental data.

Another basic idea of the semiempirical approach is the consideration of the fact that most of the interesting molecular properties are mainly influenced only by the valence electrons of the corresponding atoms. Therefore, only the valence electrons are taken into account, leading to a further reduction in computer time.

All the semiempirical methods apply the same theoretical assumptions, and they only differ in the kind of approximations that have been made [37]. Semiempirical methods like Austin Model1 (AM1) [38] and Parametric Method3 (PM3) [39–41] provide an effective compromise between the accuracy of the results and the expense of computer time required. A calculation performed with AM1 or PM3 is able to reflect the experiment as well as an ab initio calculation using a small basis set. The advantage of semiempirical methods over ab initio calculations is not only that they are several orders of magnitude faster, but also that calculations for systems up to 200 atoms are possible only with the semiempirical methods. However, it is recommendable to check the results carefully. Like the choice of a wrong basis set in ab initio calculations, the lack of correct parameters in semiempirical studies can also lead to meaningless results. The quality of semiempirical methods in treating a wide range of molecules and the calculation of different properties have been the subject of several reviews [28–31]. It should be noted that, in general, semiempirical methods may give erroneous results for the third row elements.

2.2.5.3 Combined Quantum Mechanical (QM) and Molecular Mechanical (MM) Methods

The theoretical limitations of molecular mechanics as well as the computational intensity of QM calculations lead to the development of a hybrid model combining the individual advantages of these two approaches, which was first introduced by Warshel and Levitt in 1976 [42].

Despite the enormous progress in computer technology and theoretical methodology that have significantly increased the number of atoms that can be treated quantum mechanically, it is still not feasible to apply quantum mechanics on large biological molecules such as proteins, DNA and lipid membranes consisting of thousands of atoms. Unfortunately, the usage of quantum mechanics is indispensable to describe chemical reactions involving covalent bond breaking and formation or to calculate accurate interaction energies between drugs and their targets. On the other hand, it is often not necessary to treat the entire macromolecule and the solvent with quantum mechanics. The relevant processes that need to be described electronically accurately by quantum mechanics occur in strictly localized regions (e.g. the active site of an enzyme). These regions and ligands are treated quantum mechanically in QM–MM hybrid models, whereas the rest of the macromolecule and solvent are described with molecular mechanics. A schematic arrangement of such a QM–MM system is shown in Figure 2.2.3, which also illustrates why QM–MM methods are also called *embedding* methods: the QM region is buried (embedded) in the MM region of a QM–MM system.

In the different published and applied QM–MM programs and theories, the modifications that need to be introduced to combine a force field with

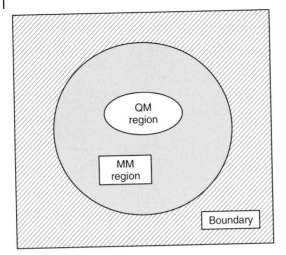

Figure 2.2.3 Partition of a QM–MM system.

an ab initio or semiempirical QM method are very similar. The basic idea is that QM and MM atoms can interact – that is, QM atoms 'see' MM atoms and vice versa. The total energy of a QM–MM system can be calculated according to Equation 7. For the calculation of the QM energy term, the MM atoms are represented as point charges and influence the QM atoms. Van der Waals interactions of QM and MM atoms are electron independent and therefore calculated by classical molecular mechanics as is the MM-energy term.

$$E_{total} = E_{QMelec} + E_{QMvdW} + E_{MM} \qquad (7)$$

where

E_{total} = total energy
E_{Qmelec} = electrostatic energy of QM atoms
E_{QMvdW} = van der Waals energy of QM atoms
E_{MM} = energy of MM atoms.

Generally, every combination of any force field and any QM program is possible. Some well-known and widely used software programs offer a QM–MM module (CHARMM, QSite, QuanteMM [43–45]).

In recent years, QM–MM potentials have been applied increasingly for the investigation of enzyme-catalyzed reactions. The description of enzyme reactions at the atomic level provides better insights and helps to understand enzyme reaction mechanisms. The first enzyme that was investigated by QM–MM calculations was lysozyme [42]. Other enzymes studied well by QM–MM approaches are triosephosphate isomerase [46], citrate synthase [47] or the pharmaceutically important β-lactamases [48].

References

1. Allinger, N.L. (1977)
Conformational-analysis. 130.
MM2 – hydrocarbon force-field
utilizing v1 and v2 torsional terms.
*Journal of the American Chemical
Society*, **99**, 8127–34.

2. Allinger, N.L., Yuh, Y.H., and Lii, J-H.
(1989) Molecular mechanics – the
MM3 force-field for hydrocarbons. 1.
*Journal of the American Chemical
Society*, **111**, 8551–66.

3. Lii, J-H. and Allinger, N.L. (1989)
Molecular mechanics – the MM3
force-field for hydrocarbons.
2. Vibrational frequencies and
thermodynamics. *Journal of the
American Chemical Society*, **111**,
8566–76.

4. Lii, J-H. and Allinger, N.L. (1989)
Molecular mechanics – the MM3
force-field for hydrocarbons. 3. The
vanderwaals potentials and crystal data
for aliphatic and
aromatic-hydrocarbons. *Journal of the
American Chemical Society*, **111**,
8576–82.

5. Morse, P.M. (1929) Diatomic
molecules according to the wave
mechanics. II. Vibrational levels.
Physical Review, **34**, 57–64.

6. Jones, J.E. (1924) On the
determination of molecular fields.
II. From the equation of state of a gas.
Proceedings of the Royal Society, **106A**,
463–77.

7. Burkert, U. and Allinger, N.L. (1982)
*Molecular Mechanics, ACS Monograph,
Vol. 177*, American Chemical Society,
Washington, DC.

8. Dinur, U. and Hagler, A.T. (1991) New
approaches to empirical force fields, in
Reviews in Computational Chemistry
(eds K.B. Lipkowitz and D.B. Boyd),
VCH, New York, Vol. 2, pp. 99–164.

9. Clark, M., Cramer III, R.D., and Van
Opdenbosch, N. (1989) Validation of
the general-purpose tripos 5.2
force-field. *Journal of Computational
Chemistry*, **10**, 982–1012.

10. Halgren, T.A. and Nachbar, R.B.
(1996) Merck molecular force field. 4.

Conformational energies and
geometries for MMFF94. *Journal of
Computational Chemistry*, **17**, 587–615.

11. Maple, J.R., Hwang, M.J., Jalkanen,
K.J. *et al.* (1998) Derivation of class II
force fields: V. Quantum force field for
amides, peptides, and related
compounds. *Journal of Computational
Chemistry*, **19**, 430–58.

12. Brooks, B.R., Bruccoleri, R.E., Olafson,
B.D. *et al.* (1983) Charmm – a
program for macromolecular energy,
minimization, and dynamics
calculations. *Journal of Computational
Chemistry*, **4**, 187–217.

13. van Gunsteren, W.F. and Berendsen,
H.J.C. (1985) Molecular dynamics
simulations: techniques and
applications to proteins, in *Molecular
Dynamics and Protein Structure* (ed.
J. Hermans), Polycrystal Books
Service, Western Springs, pp. 5–14.

14. Maple, J.R., Dinur, U., and Hagler,
A.T. (1988) Derivation of force-fields
for molecular mechanics and
dynamics from ab initio energy
surfaces. *Proceedings of the National
Academy of Sciences of the United States
of America*, **85**, 5350–54.

15. Bowen, J.P. and Allinger, N.L. (1991)
Molecular mechanics: the art and
science of parameterization, in *Reviews
in Computational Chemistry* (eds K.B.
Lipkowitz and D.B. Boyd), VCH, New
York, Vol. 2, pp. 81–97.

16. Press, W.H., Flannery, B.P.,
Teukolsky, S.A., and Vetterling, W.T.
(1988) *Numerical Recipes in C: The Art
of Scientific Computing*, Cambridge
University Press, Cambridge, p. 301.

17. Schlick, T. (1992) Optimization
methods in computational chemistry,
in *Reviews in Computational Chemistry*
(eds K.B. Lipkowitz and D.B. Boyd),
VCH, New York, Vol. 3, pp. 1–71.

18. Eliel, E.L., Allinger, N.L., Angyal, S.J.,
and Morrison, G.A. (1965)
Conformational Analysis,
Wiley-Interscience, New York.

19. Ooi, T., Oobatake, M., Nemethy, G.,
and Scheraga, H.A. (1987) Accessible

surface-areas as a measure of the thermodynamic parameters of hydration of peptides. *Proceedings of the National Academy of Sciences of the United States of America*, **84**, 3086–90.

20. Gilson, M.K., Sharp, K.A., and Honig, B. (1988) Calculating the electrostatic potential of molecules in solution – method and error assessment. *Journal of Computational Chemistry*, **9**, 327–35.

21. Still, W.C., Tempczyk, A., Hawley, R.C., and Hendrickson, T. (1990) Semianalytical treatment of solvation for molecular mechanics and dynamics. *Journal of the American Chemical Society*, **112**, 6127–29.

22. Wojciechowski, M. and Lesyng, B. (2004) Generalized born model: analysis, refinement, and applications to proteins. *Journal of Physical Chemistry B*, **108**, 18368–76.

23. Guvench, O., Weiser, J., Shenkin, P. et al. (2002) Application of the frozen atom approximation to the GB/SA continuum model for solvation free energy. *Journal of Computational Chemistry*, **23**, 214–21.

24. Apostolakis, J., Pluckthun, A., and Caflisch, A. (1998) Docking small ligands in flexible binding sites. *Journal of Computational Chemistry*, **19**, 21–37.

25. Kang, X., Shafer, R.H., and Kuntz, I.D. (2004) Calculation of ligand-nucleic acid binding free energies with the generalized-born model in DOCK. *Biopolymers*, **73**, 192–204.

26. MOE, Chemical Computing Group, Montreal. http://www.chemcomp.com.

27. Mohamadi, F., Richards, N.G.J., Guida, W.C. et al. (1990) Macromodel – an integrated software system for modeling organic and bioorganic molecules using molecular mechanics. *Journal of Computational Chemistry*, **11**, 440–67.

28. Cornell, W.D., Cieplak, P., Bayly, C.I. et al. (1995) A 2nd generation force-field for the simulation of proteins, nucleic-acids, and organic-molecules. *Journal of the*

American Chemical Society, **117**, 5179–88.

29. Pople, J.A. (1970) Molecular orbital methods in organic chemistry. *Accounts of Chemical Research*, **3**, 217–23.

30. Hehre, W.J., Radom, L., Schleyer, P.v.R., and Pople, J.A. (1986) *Ab Initio Molecular Orbital Theory*, Wiley-Interscience, New York.

31. Szabo, A. and Ostlund, N.S. (1985) *Modern Quantum Chemistry: Introduction to Advanced Electronic Structure Theory*, Revised 1st edn, McGraw-Hill, New York.

32. Clark, T. (1985) *A Handbook of Computational Chemistry: A Practical Guide to Chemical Structure and Energy Calculations*, Wiley-Interscience, New York.

33. De Frees, D.J., Levi, B.A., Pollack, S.K. et al. (1979) Effect of electron correlation on theoretical equilibrium geometries. *Journal of the American Chemical Society*, **101**, 4085–89.

34. Davidson, E.R. and Feller, D. (1986) Basis set selection for molecular calculations. *Chemical Reviews*, **86**, 681–96.

35. Feller, D. and Davidson, E.R. (1990) Basis sets for Ab initio molecular orbital calculations and intermolecular interactions, in *Reviews in Computational Chemistry* (eds K.B. Lipkowitz and D.B. Boyd), VCH, New York, Vol. 1, pp. 1–43.

36. Boyd, D.B. (1990) Aspects of molecular modeling, in *Reviews in Computational Chemistry* (eds K.B. Lipkowitz and D.B. Boyd), VCH, New York, Vol. 1, pp. 321–54.

37. Kunz, R.W. (1991) *Molecular Modelling für Anwender*, Teubner Studienbücher, Stuttgart.

38. Dewar, M.J.S., Zoebisch, E.G., Healy, E.F., and Stewart, J.J.P. (1985) The development and use of quantum-mechanical molecular-models. 76. AM1 – a new general-purpose quantum-mechanical molecular-model. *Journal of the American Chemical Society*, **107**, 3902–9.

39. Stewart, J.J.P. (1990) Semiempirical molecular orbital methods, in *Reviews in Computational Chemistry* (eds K.B. Lipkowitz and D.B. Boyd), VCH, New York, Vol. 1, pp. 45–81.

40. Stewart, J.J.P. (1989) Optimization of parameters for semiempirical methods. 1. Method. *Journal of Computational Chemistry*, **10**, 209–20.

41. Stewart, J.J.P. (1989) Optimization of parameters for semiempirical methods. 2. Applications. *Journal of Computational Chemistry*, **10**, 221–64.

42. Warshel, A. and Levitt, M. (1976) Theoretical studies of enzymic reactions – dielectric, electrostatic and steric stabilization of carbonium-ion in reaction of lysozyme. *Journal of Molecular Biology*, **103**, 227–49.

43. CHARMM, Harvard University, Cambridge. http://www.charmm.org.

44. QSite, Schrödinger Inc., Portland. http://www.schrodinger.com.

45. QuantaMM, Accelrys, http://www.accelrys.com.

46. Bash, P.A., Field, M.J., Davenport, R.C. *et al.* (1991) Computer-simulation and analysis of the reaction pathway of triosephosphate isomerase. *Biochemistry*, **30**, 5826–32.

47. Mulholland, A.J. and Richards, W.G. (1997) Acetyl-CoA enolization in citrate synthase: a quantum mechanical molecular mechanical (QM/MM) study. *Proteins*, **27**, 9–25.

48. Hermann, J.C., Hensen, C., Ridder, L. *et al.* (2005) Mechanisms of antibiotic resistance: QM/MM modeling of the acylation reaction of a class A beta-lactamase with benzylpenicillin. *Journal of the American Chemical Society*, **127**, 4454–65.

2.3
Conformational Analysis

Molecules are not rigid. The motional energy at room temperature is large enough to let all atoms in a molecule move permanently. This means that first the absolute position of atoms in a molecule and of a molecule as a whole is by no means fixed, and secondly the relative location of substituents on a single bond may vary in the course of time. Therefore, each compound containing one or several single bonds exists at each moment in many different so-called rotamers or conformers. The quantitative and qualitative composition of this mixture is permanently changing. Of course, only the low-energy conformers are found to a large extent.

A transformation from one conformation to another is primarily related to changes in torsion angles about single bonds. Only minor changes of bond lengths and angles are anticipated. The changes in molecular conformations can be regarded as movements on a multidimensional surface that describes the relationship between the potential energy and the geometry of a molecule. Each point on the potential surface represents the potential energy of a single conformation. Stable conformations of a molecule correspond to local minima on this energy surface. The relative population of a conformation depends on its statistical weight, which is influenced not only by the potential energy but also by the entropy. As a consequence, the global minimum on the potential energy surface – the conformation that contains the lowest potential energy – does not necessarily correspond to the structure with the highest statistical weight (for a detailed description, see Ref. 1).

Well-known examples for multiple conformations of molecules are the staggered and eclipsed forms of ethane, the anti-trans and gauche forms of *n*-butane or the boat and chair forms of cyclohexane. The rotation about the $C_{sp3}-C_{sp3}$ bond in the ethane molecule can be described by a sinelike curve of potential function (Figure 2.3.1). The energy minima, located at 60, 180 and 300°, correspond to the staggered form, whereas the maxima, located at 120, 240 and 360°, correspond to the eclipsed form of ethane. Because normally structures located at maxima on the potential energy function (or potential energy surface) are not viable, only the staggered form of ethane has to be taken into account when physical or chemical properties are studied. This nice and clear situation completely changes in case of larger and more flexible molecules that are existent at room temperature in several energetically accessible rotamers. For example, approximately 70% of *n*-butane existing at room temperature is in the anti-trans form and 30% is in the gauche form [2]. Thus, for a discussion on the physical behavior of this flexible aliphatic chain, both the anti-trans and the gauche conformations have to be taken into account. The same situation exists for cyclic structures like cyclohexane, where chair as well as boat forms have to be regarded.

The biological activity of a drug molecule is supposed to depend on one single unique conformation that is hidden among all the low-energy

Molecular Modeling. Basic Principles and Applications. 3rd Edition
H.-D. Höltje, W. Sippl, D. Rognan, and G. Folkers
Copyright © 2008 Wiley-VCH Verlag GmbH & Co. KGaA, Weinheim
ISBN: 978-3-527-31568-0

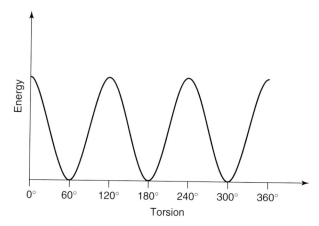

Figure 2.3.1 Sinelike potential curve of ethane shown as function between the dihedral angle and the corresponding potential energy.

conformations [3]. The search for this so-called bioactive conformation for sets of compounds is one of the major tasks in medicinal chemistry. Only the bioactive conformation can bind to the specific macromolecular environment at the active site of the receptor protein. On the basis of the information of the active conformation, one may be able to design new agents for a particular receptor system. It is widely accepted that the 'bioactive' conformation need not necessarily be identical with the lowest-energy conformation. However, on the other hand, it cannot be one that is so high in energy that it is excluded from the population of conformations in solution (for a discussion about this aspect see Ref. 4). Thus, the identification of low-energy conformations is an important part of understanding the relationship between structure and biological activity of a molecule.

Experimental techniques, like for example NMR, provide information only on one or a few conformations of a molecule. A complete overview about the conformational potential of molecules can be exclusively gained by theoretical techniques. Correspondingly, a variety of theoretical methods for conformational analysis have been developed. Many applications are reported in the literature [5–12]. The most general conformational analysis methods are those that are able to identify all minima on the potential energy surface. However, as the number of minima dramatically increases with the number of rotatable bonds, an exhaustive detection of all minima becomes a difficult and time-consuming task.

Naturally, the time expense of a conformational analysis is also directly dependent on the type of method used for energy calculation. Conformational energies can be either calculated using quantum mechanical or molecular mechanical methods. Because the quantum mechanical calculations are very time consuming, they cannot be applied in the case of large or flexible molecules. For this reason, most of the conformational search programs, as a standard, use molecular mechanics methods for the calculation of energies. In this chapter, apart from systematic search procedures, we also deal with the use of Monte Carlo and molecular dynamics techniques for conformational analyses.

2.3.1
Conformational Analysis Using Systematic Search Procedures

The systematic search [6, 7, 13] perhaps is the most natural of all different conformational analysis methods. It is performed, by varying systematically each of the torsion angles of a molecule to generate all possible conformations. If the angle increment is appropriately small, the procedure yields a complete image of the conformational space of any molecule.

The stepsize that is normally used in a systematic search is 30°. That means, during a full rotation of 360°, 12 conformations are generated. In close neighborhood to the optimal value, a smaller stepsize down to 5° may be necessary to exactly determine the minimum position of a conformation. The number of generated conformations depends not only on the stepsize, but also on the number of rotatable bonds. If 'n' is the number of rotatable bonds, then the conformation number increases with the nth order of magnitude:

$$\text{Number of conformations} = (360/\text{stepsize})^n$$

If for example a systematic conformational search is performed for a molecule with six rotatable bonds and a stepsize of 30° is used, the number of generated conformers amounts to 12^6 or 2 985 984 structures. This huge amount of data cannot be handled. Therefore it has to be reduced.

The first step in data reduction is van der Waals screening or 'bump check'. It is performed before the potential energy of the conformations is exactly calculated. The screening procedure excludes all conformations where a van der Waals volume overlap of not directly bound atoms is detected. The mathematical criterion for determining the validity of a conformation, in this respect, is simply the sum of the van der Waals radii of nonbonded atoms. The hardness of van der Waals spheres can be varied by specification of a so-called van der Waals factor. This multiplication constant controls the interpenetrability of atoms. A reduction of the van der Waals factor results in softening of contacts between nonbonded atoms, thereby increasing the number of valid conformations.

For the conformers remaining after the bump check, the potential energy is calculated using a molecular mechanics method. In general, the conformational energy is calculated after neglecting electrostatic interactions – that is, charges are not taken into consideration and the conformational analysis is performed *in vacuo*. The reasons for this procedure have been discussed in Section 2.2. If the inclusion of electrostatic interactions into a conformational analysis is justified for a special case, then the whole process becomes much more complex. Good-quality atomic charges not only depend on the connectivity but are also sensitive to the discrete spatial environment of the atoms. Therefore, atomic charges that have been calculated for the initial conformation must be constantly updated after each modification of a torsion angle. In addition, it would be necessary to mimic the effect of a solvent, which tones down the strong electrostatic interactions built up between charges *in vacuo*. Obviously, this procedure would require a large amount of additional computational time even for a small molecule. And what is even more noteworthy is that the increase in complexity of the system does not give a deeper insight into the conformational behavior of a molecule in solution, besides the fact that intramolecular interactions are diminished. The same result is obtained when charges are not considered and the analysis is performed *in vacuo*. Besides, in the active site of a receptor or enzyme, the intramolecular contacts in ligands are of minor importance.

When the conformational energies have been calculated for all conformers that survived the bump check, another possibility to reduce the number of conformations is the use of an 'energy window'. The underlying idea for applying an 'energy window' is based on the fact that conformations containing much more energy than the ones close to the minimum are found, in the conformer population, only to a neglectable quantity – that is, in our context it may be assumed that they do not have any importance for the biological activity of a particular molecule. The value for this 'energy window' depends on the size of the studied molecule as well as on the applied force field. It may vary between 5 and 15 kcal mol^{-1} [11–15].

The resulting conformations, which have passed all filter methods, should represent a complete ensemble of energetically accessible conformations for a particular molecule. But, in many cases, the number may still be too large to allow a reasonable treatment. Many of the remaining conformations are very strongly related because they, for example, only differ in a single rotor step. Obviously, these can be combined to a common family with pronounced similarity. The description of the conformational properties of a molecule does not lack comprehensiveness if we only take the minimum conformer of each conformational family into further consideration. Several methods have been developed to execute the classification into conformational families [15–17]. The parameters used for this purpose are the torsion angles. The known classification methods differ in the procedure to associate the conformers to individual families. Another method for evaluating the large amount of data accumulated in course of a systematic conformational search is the application of statistical techniques like cluster or factor analysis. For a detailed discussion of these methods, see Ref. 18.

The course of a systematic conformational analysis is demonstrated on a study performed in our group with two H$_2$-antihistaminic agents, tiotidine and ICI127032 (Figure 2.3.2) [19]. It was performed using the SEARCH module within the molecular modeling package SYBYL [16].

As rotational increment, a stepsize of 15° was chosen. Because of symmetry, only the methyl group of the cyanoguanidine system was rotated in 30° steps

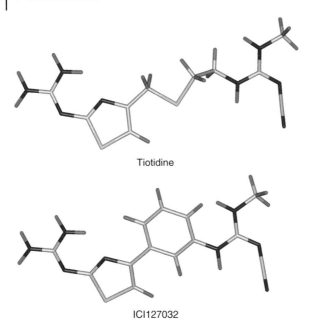

Tiotidine

ICI127032

Figure 2.3.2 Molecular formula of the histamine H_2-receptor antagonists tiotidine and ICI127032. The sulfur atoms are color coded in yellow.

between 0 and 120°. The theoretical number of conformations was reduced using the van der Waals screening from a value of 3.98×10^7 to 4.6×10^6 conformations – that is, roughly 10% of the initial number is still valid after the bump check. The application of an energy window of 15 kcal mol^{-1} leads to a further reduction of 90%. A total of 453 393 conformations were stored. Even this number cannot be handled in a reasonable way. Therefore in a next step, the conformations left were classified into families using the program IXGROS [17], which has been developed in our group. This finally yields 227 unique families that are represented by their respective minimum-energy conformations. Although the reduction from 4.6×10^7 down to 227 conformations is very impressive, one has to submit that even the rather small number left is too large. There is no chance to decide as to which of the 227 conformers is the bioactive one. But this, and only this, is the question of interest. At this point, a solution cannot be found if there do not exist rigid, or at least semirigid, congeners, which in addition, must be biologically active. It must also be proved that they bind to the same receptor site in an analogous mechanism. That is, as a rule, for finding the bioactive conformation of a flexible molecule, potent and more rigid compounds of the same series are needed. In case of the H_2-antagonists, the rigid and potent representative is ICI127032. Under consideration of the small number of low-energy conformations of the rigid matrix and repeated use of IXGROS, eight unique families survived the procedure. These remaining conformations

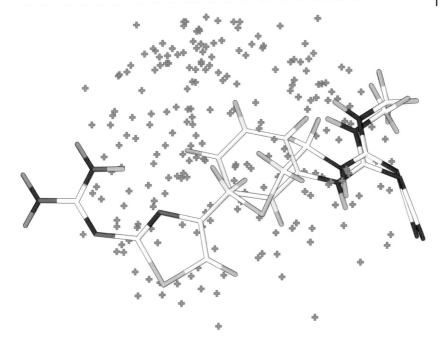

Figure 2.3.3 Representation of the results of the conformational analysis of tiotidine and ICI127032 (both displayed in a possible minimum-energy conformation). The local minimum conformation representing the different conformational families are displayed by stars symbolizing the center of the cyanoguanidine endgroup of tiotidine and ICI127032. The resulting conformations of tiotidine are indicated by green stars, while the red-colored stars mark the conformations derived for ICI127032. (The calculations have been performed using the SEARCH module within SYBYL 6.1 [16] and IXGROS [17].)

could successfully be used to determine the biological active conformation of tiotidine (Figure 2.3.3).

As discussed, it is of advantage to include rigid molecules in a conformational search for a set of flexible congeners. The rigid and biologically potent derivatives are used as a matrix for all other members of the series. Marshall [7] has extended this procedure by also including inactive rigid representatives. In doing this, the conformational space can be further restricted, and by the same token the time necessary for the search is reduced by orders of magnitude. This technique has become known as *active analog approach*.

2.3.2
Conformational Analysis Using Monte Carlo Methods

A completely different path for searching conformational space is realized in the Monte Carlo or random search. Random search techniques are of statistical nature [20]. At each stage of a Monte Carlo search, the actual conformation is randomly modified to obtain a new one.

A random search starts with an optimized structure. At each iteration in the procedure, new torsion angles [11] or new Cartesian coordinates [8, 9] are randomly assigned. The resulting conformation is minimized using molecular mechanics and the randomization process is repeated. The minimized conformation is then compared with the previously generated structures and is stored only if it represents a unique one. The random methods potentially cover all regions of conformational space, but this is true only if the process is allowed to run for a sufficiently long time. This may become extremely long because the probability to detect a new and unique conformation dramatically decreases depending on the growing number of already discovered conformers. However, even when the computation has been very long, one can never be certain that the conformational space has been completely covered. Therefore it is very important to establish a means for testing the completeness of the analysis. This can be done efficiently by performing several runs in a parallel mode, each one starting with a different initial conformation. If the results are identical or nearly identical, then completeness can be assumed. Another measure of completeness is based on the recovery rate for each low-energy conformation because the probabilistic process must reproduce it many times.

The main advantage of random search methods is that, in principle, molecules of any size can be successfully treated. In practice, however, highly flexible molecules often do not give converging results because the volume of the respective conformational space is too large. Other useful applications for Monte Carlo search methods include investigations on cyclic systems because ring systems, in general, are difficult to treat in systematic searches. The effectiveness of random search procedures is demonstrated on a practical example. Cycloheptadecane was studied using a variety of different methods including a random search method [12]. The combined results of the various procedures yielded a total of 262 different minimum conformations. None of the techniques used succeeded in finding all 262 conformers, but one of the random search analyses was, nevertheless able to detect 260 of them. It is therefore safe to comment that random search techniques are very suitable for conformational analyses of many types of molecules, but may require a large amount of computational time to ensure complete coverage of conformational space.

Another sampling technique widely applied to the problem of improved conformational searching is known as *Poling* [21]. Conformational variation is promoted through the addition of a 'poling function' to a standard molecular mechanics force field. The poling function has the effect of changing the energy surface being minimized to penalize conformational space around any previously accepted conformers. As a consequence, the method both increases conformational variation and eliminates redundancy within the limits imposed by the function. Poling has been applied within the CATALYST program [22] to search large databases of molecules.

2.3.3
Conformational Analysis Using Molecular Dynamics

The systematic conformational search procedure is a valuable tool to determine the large number of minima on the potential energy surface associated with a flexible molecule. In principle, the generation of all allowed conformations can be realized, and there is a high probability of the completeness of the conformational search. However, there are clear limitations in the applicability of this method. The multiminima problem can only be solved for rather small molecules with a limited number of rotatable bonds.

As mentioned in Section 2.3.1, the systematic conformational search of a molecule with six rotatable bonds leads to serious problems in data handling owing to the large number of generated conformers. Therefore, the investigation of flexible molecules, like for example arachidonic acid (Figure 2.3.4), which contains 15 rotors, is practically impossible. Even after applying several methods of data reduction, the systematic conformational search for this molecule yielded 500 000 different conformations. The procedure was automatically stopped by the program because of data overflow, although the conformational space was not completely sampled at this point.

Also, the conformational analysis of the same molecule by a random search procedure will be unreasonable because of the required CPU time.

Another rather difficult problem arises when saturated or partial – saturated ring system are to be treated in a systematic conformational analysis. In course of the systematic process, bonds have to be broken to produce new attainable ring conformations. Efficiency and reliability of this procedure have been the subject of several reviews [13, 14].

A very common strategy to overcome these problems is the use of molecular dynamics simulations for exploring conformational space. The aim

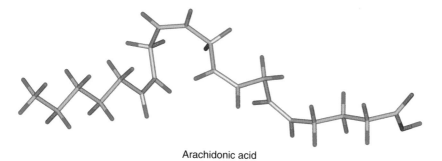

Arachidonic acid

Figure 2.3.4 One energetically allowed conformation of arachidonic acid.

of molecular dynamics is to reproduce the time-dependent motional behavior of a molecule. Molecular dynamics is based on molecular mechanics. It is assumed that the atoms in the molecule interact with each other according to the rules of the used force field (as described in Section 2.2.1). At regular time intervals, the classical equation of motion represented by Newton's second law is solved:

$$F_i(t) = m_i \, a_i(t) \tag{1}$$

where

F_i = force on atom i at time t
m_i = mass of atom i
a_i = acceleration of atom i at time t.

The gradient of the potential energy function is used to calculate the forces on the atoms, while the initial velocities on the atoms are randomly generated at the beginning of the dynamics run. On the basis of the initial atom coordinates of the system, new positions and velocities on the atoms can be calculated at time t and the atoms will be moved to these new positions. As a result of this proceeding, a new conformation is created. The cycle will then be repeated for a predefined number of time steps. The collection of energetically accessible conformations produced by this procedure is called *ensemble*.

The application of Newton's equations of motion is uniform in all the available molecular dynamics approaches; but they differ in the integration algorithms that are used. Very common methods for integrating the equations of motion are the Verlet integrator [23] and algorithms like Beeman [24] and the leapfrog scheme, [25] which are simple modifications of the Verlet algorithm. In the framework of this book, a more extended discussion of the molecular dynamics theory is not advantageous but the interested reader is urged to study more detailed reviews on this subject [26–29].

Before using molecular dynamics simulations for conformational analysis, the reader's attention is drawn to some special features of this method. Unlike the conservative geometry-optimization procedures, molecular dynamics is able to overcome energy barriers between different conformations. Therefore, it should be possible to find other local minima than those nearest in the potential energy surface. However, if the energy barrier is high or the number of degrees of freedom in the molecule is very large, then some of the potentially existing conformers of the investigated system are not possibly reached. With regard to the huge conformational space, the completeness of the conformational search during the chosen simulation time is difficult to ensure.

To enhance conformational sampling, it is a widely used tactics in molecular dynamics to apply an elevated temperature in the simulation [29]. At high temperature, the molecule is able to overcome even large energy barriers that may exist between some conformations, and therefore the chance for completeness of a conformational search does increase. It is self-evident that the choice of a particular simulation temperature and simulation time is closely dependent on the molecule of interest.

One recent and comprehensive investigation is used to demonstrate the dependence of conformational flexibility on the simulation temperature. The data and additional material were made available by courtesy of F. S. Jorgensen (Royal Danish School of Pharmacy, Copenhagen, Denmark). A molecular dynamics simulation has been performed on the experimentally well studied cyclohexane molecule using different start conformations and different simulation temperatures[1] (Figure 2.3.5).

At 400 K, the twist form of cyclohexane ($T_1 = 0$), which has been used as initial conformation, oscillates between different twist forms, whereas at 600 K, the molecule contains sufficient kinetic energy to convert to one of the chair conformations ($T_1 = 300$). Further increase in temperature up to 1000 K yielded both chair as well as twist conformations ($T_1 = 300 \rightarrow 60$) and several chair–chair interconversions could be observed. After 800 ps, one of the chair conformations ($T_1 = 60$) existed almost exclusively. In a second study, three methyl-substituted cyclohexanes (1,1-dimethylcyclohexane; 1,1,3,3-tetramethylcyclohexane and 1,1,4,4-tetramethylcyclohexane) were subjected to molecular dynamics simulations at various temperatures. The observed chair–chair interconversions at the corresponding temperatures have been compared with experimentally determined energy barriers of ring inversion [30] (Table 2.3.1). As a result of the comparison, it can be concluded that the molecular dynamics simulations are able to reflect the relative magnitude of the experimentally determined ring-inversion barriers. This example of high-temperature molecular dynamics clearly indicates the necessity to verify if the chosen simulation temperature is high enough to prevent the system from getting stuck in one particular region of conformational space.

Table 2.3.1 The table gives information on the existence of the two possible chair conformations (chair and chair') of three methyl-substituted cyclohexanes at different simulation temperatures. The data are compared to the corresponding experimentally determined ring-inversion barriers.

	600 K	800 K	1000 K	1200 K	Δ G (kcal mol^{-1})
	Chair	Chair + chair'	Chair + chair'	Chair + chair'	9.6
	Chair	Chair	Chair + chair'	Chair + chair'	10.6
	Chair	Chair	Chair	Chair + chair'	11.7

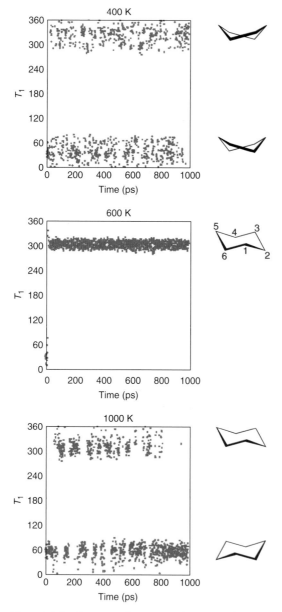

Figure 2.3.5 Variation of torsion angle T_1 (torsion angle $T_1 = C_1-C_2-C_3-C_4$) of cyclohexane in dependence of the simulation temperature. At 400 K the cyclohexane oscillates between different flexible twisted boat forms reflected by an extensive fluctuation of the observed torsion angle. Increasing the temperature up to 600 K leads to one of the possible stable minima corresponding to one chair conformation. The dynamic simulation at 1000 K yields both chair conformations as well as the already observed twist and boat conformations.

In the application of molecular dynamics, to search conformational space it is a common strategy to select conformations at regular time intervals and minimize them to the associated local minimum. This procedure has been used in several conformational analysis studies on small molecules including ring systems [14, 31]. A very impressive example in this context is the conformational analysis of the polyhydroxy analog of the sesquiterpene lactone tharpsigargin (Figure 2.3.6). This study was also performed in the laboratory of F. S. Jorgensen.

The polyhydroxy derivative has been subjected to a molecular dynamics simulation at 1200 K to get insight into the conformational behavior of the ring system.[2] The seven-member ring adopted several different conformations during the simulation, and a considerable number of ring interconversion took place. This clearly shows an extensive exploration of the conformational space.

Each of the sampled conformations has been subsequently energy minimized and compared exclusively with respect to the conformation of the seven-member ring. All conformations with a root mean square (rms) value below 0.1 Å were considered to be identical. The procedure yielded five different low-energy conformations. Fortunately NMR data [32] of tharpsigargin agree with one of the theoretically found conformations of the tricyclic ring system. This is shown in Figure 2.3.7.

In some cases, however, it is not sufficient to minimize the sampled conformations in order to reach the final minimum conformation. The intention of the high-temperature dynamics simulation is to provide the molecule with enough kinetic energy to cross energy barriers between different conformations. However, during the simulation the molecule can occupy extremely distorted geometries that sometimes cannot be relaxed by a simple minimization procedure.

If this occurs, it is recommendable to perform a high-temperature annealed molecular dynamics simulation [33]. Using this approach, all sampled

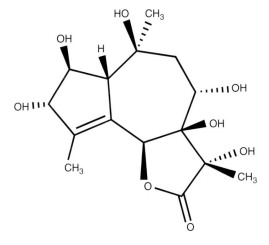

Figure 2.3.6 Molecular formula of the polyhydroxy analogue of tharpsigargin.

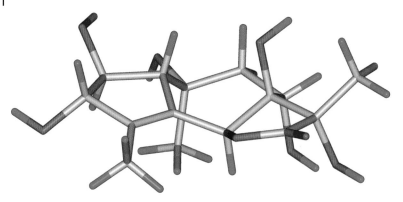

Figure 2.3.7 One of the theoretically determined conformations of the polyhydroxy analogue of tharpsigargin. The ring conformation is in accordance with the results obtained by NMR spectroscopy.

conformations of the high-temperature simulation will be subsequently optimized and then reshaken at a lower temperature, for example, 300 K to remove the internal strain of the molecule. The final optimization leads to conformations of lower energy when compared with the results of a high-temperature simulation, which is followed by a simple geometry optimization.

An additional modification of this high-temperature annealed molecular dynamics simulation is the so-called simulated annealing method [34]. In this technique, the system is cooled down at regular time intervals by decreasing the simulation temperature. As the temperature approaches 0 K, the molecule is trapped in the nearest local minimum conformation. The received geometry at the end of the annealing cycle is saved and subsequently used as starting point for further simulations at high temperature. To obtain a set of low-energy conformations, the cycle will be repeated several times. As the resulting structures should already be close to a minimum, it is not absolutely necessary to subsequently minimize the structure. The application of this method has been subject of several studies [35, 36]. Further information may be obtained from these references.

2.3.4
Which Is the Method of Choice?

With such a variety of methods available for sampling conformational space, it can be difficult to reach the right decision. Each method has its own strengths and weaknesses. Systematic searches are subject to the effects of conformational explosion, and normally they cannot be applied to cyclic molecules. Random search methods can require long runs to ensure that the conformational space has been completely sampled. In addition, duplicates have to be removed from the output.

Conformational search methods are typically validated against standard benchmark data sets. In one typical approach, several parameters are used to test both the number of conformers as well as their energies. An alternative approach is to compare generated conformations against experimentally derived X-ray structures taken from the Cambridge Crystallographic Database [37]. Sadowski *et al.* found that the CORINA program, which is used to convert 2D into 3D structures, reproduced the correct conformation for nearly half of their data set of 639 structures [38].

Now that the number of high-resolution X-ray protein–ligand complexes included in the Protein Database (PDB) [39] is increasing rapidly, a further approach for testing conformational search methods does exist. It is possible to find out whether experimentally determined bioactive conformations are present in various conformational ensembles that are generated. Multiconformer generation programs have recently been assessed in terms of how accurately generated conformers correspond to low-energy conformations. However, several studies disagree with the assumption that a bioactive conformation closely resembles the global energy minimum conformation, [40–47] and the energy threshold of protein-bound conformers is often considerably above the global energy minimum. In protein–ligand complexes resulting from X-ray analysis of a homogeneous crystal, both molecules experience a certain degree of conformational strain resulting in the bioactive conformation that may differ from that observed in crystals of the pure unbound ligand.

Boström *et al.* have recently carried out a comparison of different sampling techniques on a test set of protein–ligand complexes from the PDB [41, 42]. The authors investigated as to what extent the results of several conformational searching tools agree with the experimental results. The authors applied systematic and random search methods to a set of 32 diverse small molecules for which the receptor–ligand complex had also been obtained by X-ray crystallography [41, 42]. The comparison showed that the Low-Mode Conformational Search method within the MacroModel program [48] performs better than other algorithms. Reducing the intramolecular electrostatic interactions, either by including a solvation model or by neglecting atomic charges, favored the finding of bioactive conformation. It proved difficult to retrieve structures having more than eight rotatable bonds for all methods. According to the obtained results, a variety of ligands in energy minimum conformations do not bind to the protein.

In a recently published more comprehensive study, Langer *et al.* compared the ability of various conformer generators to reproduce bioactive confor-mations [49]. The Catalyst [22] conformational subsampling algorithm was tested in a comparative evaluation with OpenEye's conformation generator program Omega 2.0 [50]. The study was based on an enhanced test set of 778 drug molecules and pharmacologically relevant compounds extracted from the PDB. Different protocols for two common conformer generation cases were evaluated: (i) high-throughput settings for processing large databases and

(ii) high-quality settings for binding site exploration or lead structure refinement. The examination showed that the quality of conformational models is always a trade-off between the sampling depth of conformational space and the computational costs with respect to the algorithm method used. With increasing size and flexibility of the investigated compounds, larger ensembles are needed to represent the bioactive conformation in equivalent quality. In more than 80% of all investigated cases, fittings of the best-generated conformer and the bioactive conformation below root mean square deviation (RMSD) 1.50 were achieved and in 93% of all cases, the same was achieved below RMSD 2.0 by the Catalyst program. Overall, both Omega and Catalyst provide valuable solutions for conformational model generation. Whereas Omega was shown to be the better choice for the generation of high-quality models, Catalyst showed still better performance in high-throughput generation.

As a conclusion, it may be stated that a variety of methods exists nowadays which are used to sample the conformational space. The user should be careful in selecting the appropriate method, and in setting the simulation conditions to ensure the completeness of the conformational search and the validity of the results. It should be kept in mind that each approach has its strengths and its weaknesses and therefore, wherever possible, experimentally derived data should serve as verification.

End Notes

[1]Sybyl (version 6.0.3) from Tripos Associates Inc., St. Louis, USA. Energy minimizations: Tripos force field, PM3 partial charges, dielectric constant $\varepsilon = 20$ D and a convergence criteria on 0.005 kcal mol^{-1} Å^{-1}. MD simulations: 1000 ps at various temperatures with conservation of total energy, one conformation sampled ps^{-1}.

[2]Sybyl (version 6.5) from Tripos Associates Inc., St. Louis, USA. Energy minimizations: Tripos force field, PM3 partial charges, dielectric constant $\varepsilon = 20$ D and a convergence criteria on 0.005 kcal mol^{-1} Å^{-1}. MD simulations: 1000 ps at 1200 K with conservation of total energy, one conformation sampled ps^{-1}.

References

1. Scheraga, H.A. (1971) Theoretical and experimental studies of conformations of polypeptides. *Chemical Reviews*, **71**, 195–217.
2. Rademacher, P. (1987) In *Strukturen Organischer Moleküle* (ed. M. Klessinger), VCH Publishers, Weinheim, New York, p. 139.
3. Ghose, A.K., Crippen, G.M., Revankar, G.R. *et al.* (1989) Analysis of the in vitro antiviral activity of certain ribonucleosides against para-influenza virus using a novel computer-aided receptor modeling procedure. *Journal of Medicinal Chemistry*, **32**, 746–56.

4. Jörgensen, W.L. (1991) Rusting of the lock and key model for protein-ligand binding. *Science*, **254**, 954–55.

5. Howard, A.E. and Kollman, P.A. (1988) An analysis of current methodologies for conformational searching of complex-molecules. *Journal of Medicinal Chemistry*, **31**, 1669–75.

6. Smellie, A., Kahn, S.D., and Teig, S.L. (1995) Analysis of conformational coverage. 1. Validation and estimation of coverage. *Journal of Chemical Information and Computer Sciences*, **35**, 285–94.

7. Dammkoehler, R.A., Karasek, S.F., Shands, E.F.B., and Marshall, G.R. (1989) Constrained search of conformational hyperspace. *Journal of Computer-Aided Molecular Design*, **3**, 3–21.

8. Saunders, M. (1987) Stochastic exploration of molecular mechanics energy surfaces – hunting for the global minimum. *Journal of the American Chemical Society*, **109**, 3150–52.

9. Saunders, M. (1989) Stochastic search for the conformations of bicyclic hydrocarbons. *Journal of Computational Chemistry*, **10**, 203–8.

10. Ferguson, D.M. and Raber, D.J. (1989) A new approach to probing conformational space with molecular mechanics – random incremental pulse search. *Journal of the American Chemical Society*, **111**, 4371–78.

11. Chang, G., Guida, W.C., and Still, W.C. (1989) An internal coordinate monte-carlo method for searching conformational space. *Journal of the American Chemical Society*, **111**, 4379–86.

12. Saunders, M., Houk, K.N., Wu, Y.-D. *et al.* (1990) Conformations of cycloheptadecane – a comparison of methods for conformational searching. *Journal of the American Chemical Society*, **112**, 1419–27.

13. Ghose, A.K., Jaeger, E.P., Kowalczyk, P.J. *et al.* (1993) Conformational searching methods for small molecules. 1. Study of the sybyl search

14. Böhm, H.-J., Klebe, G., Lorenz, T. *et al.* (1990) Different approaches to conformational-analysis – a comparison of completeness, efficiency, and reliability based on the study of a 9-membered lactam. *Journal of Computational Chemistry*, **11**, 1021–28.

15. Taylor, R., Mullier, G.W., and Sexton, G.J. (1992) Automation of conformational-analysis and other molecular modeling calculations. *Journal of Molecular Graphics*, **10**, 152–60.

16. SYBYL Theory Manual, Tripos Associates, St. Louis, http://www.tripos.com.

17. Sippl, W. (1997) *Theoretische Untersuchungen zum Bindungsverhalten von Histamine H_2- und H_3-Rezeptor Liganden*, Ph. D. Thesis, Heinrich-Heine-University Duesseldorf, Germany.

18. Shenkin, P.S. and McDonald, D.Q. (1994) Cluster-analysis of molecular-conformations. *Journal of Computational Chemistry*, **15**, 899–916.

19. Höltje, H.-D. and Batzenschlager, A. (1990) Conformational-analyses on histamine h-2-receptor antagonists. *Journal of Computer-Aided Molecular Design*, **4**, 391–402.

20. Metropolis, N., Rosenbluth, A.W., Rosenbluth, M.N. *et al.* (1953) Equation of state calculations by fast computing machines. *Journal of Chemical Physics*, **21**, 1087–92.

21. Smellie, A., Kahn, S.D., and Teig, S.L. (1995) Analysis of conformational coverage. 2. Application of conformational models. *Journal of Chemical Information and Computer Sciences*, **35**, 295–304.

22. Catalyst, Accelrys Inc., San Diego, http://www.accelrys.com.

23. Verlet, L. (1967) Computer experiments on classical fluids. I. Thermodynamical properties of Lennard-Jones molecules. *Physical Review*, **159**, 98–103.

24. Beeman, D. (1976) Some multistep methods for use in

molecular-dynamics calculations. *Journal of Computational Physics*, **20**, 130–39.

25. Hockney, R.W. and Eastwood, J.W. (1981) *Computer Simulation Using Particels*, McGraw-Hill, New York.

26. van Gunsteren, W.F. and Berendsen, H.J.C. (1990) Moleküldynamik-Computersimulationen; Methodik, Anwendungen und Perspektiven in der Chemie. *Angewandte Chemie*, **102**, 1020–55.

27. Lybrand, T.P. (1990) Computer simulation of biomolecular systems using molecular dynamics and free energy perturbation methods, in *Reviews in Computational Chemistry* (eds K.B. Lipkowitz and D.B. Boyd), VCH Publishers, New York, Vol. 1, pp. 295–320.

28. Karplus, M. and Kuriyan, J. (2005) Molecular dynamics and protein function. *Proceedings of the National Academy of Sciences of the United States of America*, **102**, 6679–85.

29. Leach, R.A. (1991) A survey of methods for searching the conformational space of small and medium-sized molecules, in *Reviews in Computational Chemistry* (eds K.B. Lipkowitz and D.B. Boyd), VCH Publishers, New York, Vol. 2, pp. 1–47.

30. Friebolin, H., Schmid, H.G., Kabuß, S., and Faißt, W. (1969) Konformative Beweglichkeit Flexibler Ringsysteme-XI Untersuchungen mit Hilfe der Protonenresonanzspektroskopie Ringinversion bei Methyl- und Alkoxylcyclohexanen. *Organic Magnetic Resonance*, **1**, 147–62.

31. Kawai, T., Tomioka, N., Ichinose, T. *et al.* (1994) High-temperature simulation of dynamics of cyclohexane. *Chemical & Pharmaceutical Bulletin*, **42**, 1315–21.

32. Christensen, S.B. and Schaumburg, K. (1983) Stereochemistry and C-13 nuclear magnetic-resonance spectroscopy of the histamine-liberating sesquiterpene lactone thapsigargin – a modification

of horeau method. *The Journal of Organic Chemistry*, **48**, 396–99.

33. Auffinger, P. and Wipff, G. (1990) High-temperature annealed molecular-dynamics simulations as a tool for conformational sampling – application to the bicyclic-222 cryptand. *Journal of Computational Chemistry*, **11**, 19–31.

34. Kirkpatrick, S., Gelatt, C.D., and Vecchi, M.P. (1983) Optimization by simulated annealing. *Science*, **220**, 671–80.

35. Salvino, J.M., Seoane, P.R., and Dolle, R.E. (1993) Conformational-analysis of bradykinin by annealed molecular-dynamics and comparison to NMR-derived conformations. *Journal of Computational Chemistry*, **14**, 438–44.

36. Laughton, C.A. (1994) A study of simulated annealing protocols for use with molecular-dynamics in protein-structure prediction. *Protein Engineering*, **7**, 235–41.

37. Olga Kennard, F.R.S., *Cambridge Structural Database*, Cambridge Crystallographic Data Centre, http://www.ccdc.cam.ac.uk.

38. Sadowski, J., Gasteiger, J., and Klebe, G. (1994) Comparison of automatic three-dimensional model builders using 639 X-ray structures. *Journal of Chemical Information and Computer Sciences*, **34**, 1000–8.

39. Bernstein, F.C., Koetzle, T.F., Williams, G.J.B., *et al.* (1977) Protein data bank – computer-based archival file for macromolecular structures. *Journal of Molecular Biology*, **112**, 535–42.

40. Boström, J. (2001) Reproducing the conformations of protein-bound ligands: a critical evaluation of several popular conformational searching tools. *Journal of Computer-Aided Molecular Design*, **15**, 1137–52.

41. Boström, J., Greenwood, J.R., and Gottfries, J. (2003) Assessing the performance of OMEGA with respect to retrieving bioactive conformations. *Journal of Molecular Graphics & Modelling*, **21**, 449–62.

42. Vieth, M., Hirst, J.D., and Brooks, C.L. (1998) Do active site conformations of small ligands correspond to low free-energy solution structures? *Journal of Computer-Aided Molecular Design*, **12**, 563–72.

43. Boström, J., Norrby, P.-O., and Liljefors, T. (1998) Conformational energy penalties of protein-bound ligands. *Journal of Computer-Aided Molecular Design*, **12**, 383–96.

44. Kirchmair, J., Laggner, C., Wolber, G., and Langer, T. (2005) Comparative analysis of protein-bound ligand conformations with respect to catalyst's conformational space subsampling algorithms. *Journal of Chemical Information and Modeling*, **45**, 422–30.

45. Nicklaus, M.C., Wang, S.M., Driscoll, J.S., and Milne, G.W.A. (1995) Conformational-changes of small molecules binding to proteins. *Bioorganic & Medicinal Chemistry*, **3**, 411–28.

46. Perola, E. and Charifson, P.S. (2004) Conformational analysis of drug-like molecules bound to proteins: an extensive study of ligand reorganization upon binding. *Journal of Medicinal Chemistry*, **47**, 2499–510.

47. Sadowski, J. and Boström, J. (2006) MIMUMBA revisited: torsion angle rules for conformer generation derived from X-ray structures. *Journal of Chemical Information and Modeling*, **46**, 2305–9.

48. Mohamadi, F., Richards, N.G.J., Guida, W.C. *et al.* (1990) MacroModel – an integrated software system for modeling organic and bioorganic molecules using molecular mechanics. *Journal of Computational Chemistry*, **11**, 440–67.

49. Kirchmair, J., Wolber, G., Laggner, C., and Langer, T. (2006) Comparative performance assessment of the conformational model generators omega and catalyst: a large-scale survey on the retrieval of protein-bound ligand conformations. *Journal of Chemical Information and Modeling*, **46**, 1848–61.

50. Omega, Version 2.0, Openeyes Scientific Software, Santa Fe, http://www.eyesopen.com.

2.4
Determination of Molecular Interaction Potentials

The initial step in the formation of a complex, for example, a drug-receptor complex, is a recognition event. The receptor has to recognize whether an approaching molecule possesses the properties necessary for specific and tight binding. This recognition process occurs at rather large distances and precedes the formation of the final interaction complex. The 3D electrostatic field surrounding each molecule therefore plays a crucial role in recognition. Other molecular characteristics like polarizability or hydrophobicity come into play when the distance between the interacting surfaces gradually decreases. It is therefore easy to realize that molecular fields, which can be determined by systematic calculation and sampling of interaction energies between the molecules under study and different chemical probes, represent data sets of high value for the understanding of intermolecular interaction on any level of complexity of the molecular ensemble of interest.

In the subsequent sections, the methods for calculation and analysis of these molecular properties is described and evaluated.

2.4.1
Molecular Electrostatic Potentials (MEPs)

Knowledge about the molecular electrostatic potential (MEP) is critically important when molecular interactions and chemical reactions are to be studied. If molecules approach each other, the initial contact arises from long-range electrostatic forces. The long-range electrostatic forces can be separated into three components: the electrostatic part, the inductive part and the dispersive part. The first type of interaction appears between polar molecules, which carry a charge or possess a permanent dipole moment. The second type is found when one polar molecule interacts with a nonpolar molecule. Then the dipole of the polar molecule produces an electric field, which changes the distribution of the electrons in the nonpolar molecule, thereby inducing a dipole moment. Thirdly, even if both molecules are nonpolar and hydrophobic entities, the permanent fluctuations in the electron distribution of one molecule can induce a temporary molecular dipole moment in a neighboring molecule. This type of interaction is called *dispersion*. Dispersion forces are weak and drastically fall off with increasing distance between the interacting molecules (see Section 2.2.1). However, they constitute the main part of attraction between neutral nonpolar molecules. (The dispersion forces are also called *London* forces or *van der Waals* forces.)

The electrostatic interaction can be either attractive or repulsive; an electropositive portion of an approaching molecule will seek to dock with an electronegative region, while similarly charged portions will repel each other.

Molecular Modeling. Basic Principles and Applications. 3rd Edition
H.-D. Höltje, W. Sippl, D. Rognan, and G. Folkers
Copyright © 2008 Wiley-VCH Verlag GmbH & Co. KGaA, Weinheim
ISBN: 978-3-527-31568-0

Obviously, the noncovalent interaction is particularly large between charged regions of molecules. Because of the presence of charges and permanent dipole moments in a molecule, a 3D electrostatic field is generated in the surrounding environment. Therefore, at moderate distances from polar or even neutral molecules, a significant MEP exists. This electrostatic potential (ESP) can be represented as interaction energy between the molecular electron distribution and a positive point charge, which is located in a 3D grid at any point in space surrounding the molecule. For the determination of MEP, an accurate treatment of the electronic properties of the molecules is required. Therefore, we deal with these methods for the calculation of molecular charge densities in subsequent sections.

2.4.1.1 Methods for Calculating Atomic Point Charges

The electronic properties of molecules are defined by the electron distributions around the positively charged nuclei. Detailed information about the electron distribution can be obtained either through experimental results, for example, X-ray diffraction studies, or by calculations using quantum mechanical methods. However, with respect to the computational procedure, corresponding results only provide a probability distribution of the charge density throughout 3D space. For the purpose of interaction energy calculations, mostly point charges located at the center of the atom are needed. This, without doubt, produces a very simplified picture of the molecular electron distribution. To achieve the transformation, the electron density needs to be converted into so-called partial or point charges. This can be done by contracting the charge onto the atomic centers. Thus, the picture of a molecule consisting of atoms carrying the partial or point charges has emerged. The definition of these empirical partial charges bears some arbitrariness because the molecular electron distribution has to be assigned to individual atom centers; or to put it in a different way, a molecular characteristic is scaled down to an atomic property. Partial charges are not observable, so the method of assigning point charges is relevant and scientifically sound only when it can be used to correlate or predict physical or chemical properties of molecules. On the other hand, as stated before, the electrostatic part of the overall intermolecular interaction energy is very prominent, and therefore most of the commonly used molecular mechanics programs include a corresponding energy term that is dependent on atomic partial charges. The application of these methods allows the rapid computation of electrostatic energies even for macromolecules with more than some hundred atoms. For this reason, a variety of different techniques for the calculation of atomic partial charges have been developed (for review see Ref. [1]).

In principle, it must be distinguished between two methodologically different approaches:

  (1) topological procedures [2–6] like the Gasteiger–Hückel method [2]; and

(2) procedures that calculate atomic charges from the quantum chemical wavefunctions like the population analysis [7] or the potential-derived charge calculation methods [8–11].

Topological Charges

The topological methods are mainly based on the electronegativity of the different atom types. To allocate atomic charges to directly bonded atoms in a reasonable way, appropriate rules that combine the atomic electronegativities with experimental structural information on the bonds linking the atoms of interest are used. The topological methods do not need information about the molecular geometry or conformational status of a molecule. Only the connectivity matrix of the atoms is included in the calculation. The original method proposed by Del Re [3] exclusively for saturated molecules was extended to conjugated systems by Pullmann [4]. Both the methods are still implemented in some modeling programs. A newer approach, which gives more realistic results in comparison with experimental data, is the Gasteiger–Hückel method. It is a combination of the Gasteiger–Marsili method [2] for the calculation of the σ component of the atomic charge and the old Hückel theory [12]. The Hückel theory allows calculation of the π component of the atomic charge in a fast and fairly efficient way. Naturally, the total charge is the sum of σ and π elements. Formal charges on atoms included in π systems are assumed to be delocalized over the whole π system. For this reason, Hückel charges are calculated first, and the Gasteiger charge calculation is performed subsequently. The main advantage of the topological procedures is that they are computationally fast and, in many cases, do compare quite well with experimentally observable properties. However, the risk is that one cannot trust the results without validation for a particular group of molecules. Very often, the validation procedure is simply omitted. Of course, this renders the corresponding study useless.

Topological methods are often implemented into commercial software packages as standard tools for charge calculation.

Quantum Chemical Methods

All other methods for the calculation of atomic partial charges are based on the quantum mechanical computation of wavefunctions. Wavefunctions can be obtained using either semiempirical or ab initio methods, depending on the requested accuracy of the wavefunction and the available computational resources. Charge densities can be gained from wavefunctions using different procedures. The oldest and most widely used is the Mulliken population analysis [7], which is implemented as a standard method in various quantum mechanical programs [13–15]. The population analysis takes the electron density derived from the wavefunction and partitions it between the atoms on the basis of the occupancy of each atomic orbital. Although widely used, it has long been recognized in literature that the Mulliken method is strongly dependent on the basis set applied. Often it gives unrealistic results [16, 17] (see also Table 2.4.1). An improved technique that eliminates most of the problems associated with the Mulliken procedure is the natural population analysis [18], but it is effective on ab initio wavefunctions only.

A second, much more recently developed, technique yielding atomic charges from quantum mechanically calculated wavefunctions is the method of deriving charges by fitting the MEP (also called ESP fit method) [7–11]. The charge density is a well-defined function [19]. It contains important and detailed information about the molecule because all electrons contribute in some way to the distribution of the electronic charge in space. It is also experimentally accessible [20] from X-ray diffraction. However, this technique is extremely demanding as far as cost and time consumption are concerned and cannot be used as a standard procedure. A set of atomic charges able to reproduce the 3D electron density seems to be an excellent choice for generating a fairly correct picture of the electronic properties of any molecule. The mathematical technique underlying the ESP fit method involves least-square fitting of the atomic charges to reproduce, as closely as possible, the charge density, which is calculated quantum mechanically at a set of points in space surrounding the molecule. This yields much better results [9, 11] than the Mulliken population analysis.

Whether a charge distribution obtained with a particular method is reliable and able to realistically represent the electronic proportions of a molecule must be checked against experimental data. One rather easily accessible experimental property is the molecular dipole moment. On the basis of atomic point charges, a molecular dipole moment can be calculated in a simple and fast way, and can be compared with appropriate experimental values, which are listed for many compounds in literature (see e.g. [21]). Because the dipole moment decisively depends on the conformation of a molecule, only the values for rigid molecules should be taken into consideration for comparative purposes. To decide on the applicability of a particular method for the calculation of charges in a series of molecules, one often proceeds by investigating not the entire flexible molecule but only small and rigid fragments instead. Table 2.4.1 lists the calculated and experimental dipole moments for a representative set of small and rigid structures. The dipole moments have been calculated using various methods and basis sets as well as the different procedures discussed above. The dipole moment is a quantum mechanically defined property. Therefore, it can also be calculated directly from wavefunctions (marked as self consistent field (SCF) in Table 2.4.1). Corresponding results derived with a basis set, like 6-31G**, are in especially good agreement with the experimental values.

Table 2.4.1 Comparison of experimentally derived and theoretically calculated dipole moments. The theoretical dipole moments were calculated using several procedures: the Gasteiger–Hückel method was chosen as an example for simple topological methods; on quantum mechanical level, the dipole moments were calculated directly from the wavefunction (SCF) as well as using the Mulliken and potential-derived point charges (ESP).

	Experimental (gas phase)	Gasteiger–Hückel	AM1 SCF	AM1 Mulliken	AM1 ESP	PM3 SCF	PM3 Mulliken	PM3 ESP	STO-3G SCF	STO-3G Mulliken	STO-3G ESP	3-21G* SCF	3-21G* Mulliken	3-21G* ESP	6-31G** SCF	6-31G** Mulliken	6-31G** ESP
Imidazol	3.8 ± 0.4	3.118	3.508	2.129	3.575	3.861	2.412	3.869	3.535	2.213	3.494	4.025	2.855	3.962	3.855	2.822	3.810
Thiazol	1.61 ± 0.03	1.466	2.012	2.680	2.041	1.249	1.463	1.259	1.986	2.554	1.989	1.683	3.556	1.709	1.435	2.594	1.507
Furan	0.66	0.599	0.493	0.354	0.484	0.216	0.066	0.234	0.532	0.675	0.498	1.101	2.222	3.936	0.772	1.813	0.738
Methylsilan	0.735	–	0.374	0.276	0.331	0.432	0.175	0.402	–	–	–	0.702	1.572	0.238	0.672	0.027	0.658
NH$_3$	1.470	0.593	1.848	0.644	1.793	1.550	0.011	1.499	1.876	0.902	1.869	1.752	1.189	1.869	1.839	1.384	1.867
Dimethyl-ether	1.31	1.764	1.429	1.052	1.473	1.254	0.854	1.3194	1.333	1.181	1.384	1.847	3.109	1.901	1.475	2.512	1.531

The type of procedure to be used for the investigation of a particular molecular system depends on several factors. On the one hand, the size of the studied molecules plays an important role and on the other, the available computer power is the limiting factor for choosing a particular method.

Topological methods have the advantage over quantum chemical properties in that they are very fast and give reasonable estimates of physical properties associated with charge. These methods generally produce dipole moments that are in good agreement with experimental values – partly a consequence of their calibration against experimental results. However, the main disadvantage is the neglect of molecular geometries and conformations. Of course, topological methods must fail in case of molecules that contain atom types missing in the parameter list (see e.g. methylsilan in Table 2.4.1; silicon parameters are not included in the Gasteiger–Hückel method).

Calculation of atomic charges from the molecular charge densities is the best choice if the results are for use in empirical energy functions for the purpose of interaction energy calculations. As can be deduced from Table 2.4.1, it is not absolutely necessary to use large ab initio basis sets. Also, on the basis of smaller basis sets and even the semiempirical AM1 method, dipole moments are obtained, which compare quite well with experimental values. However, the quality of the resulting dipole moment does depend, very distinctly, on the procedure used for generating the atomic point charges. Results obtained directly from molecular charge distribution are more realistic than the results of the Mulliken population analysis, which for some basis sets yields absolutely crude and false dipole moments (see Table 2.4.1).

If a molecule of interest does contain more than 100 atoms, then a sufficiently accurate calculation of wavefunctions is not feasible for the entire molecule. This impediment can be overcome by dividing the large molecule into overlapping fragments. The fragment results are then transferred onto the large structure, hoping that the fragment properties correctly mirror the characteristics of the parent molecule.

But even if point charges of high quality have been determined for a series of molecules, these quantities are only weak arguments if the question of molecular similarity is the object of interest. Molecular similarity can be determined much more adequately on the basis of the 3D charge distribution. The most advantageous way to use this important and well-defined magnitude is through the MEPs.

2.4.1.2 Methods for Generating MEPs

MEPs are represented as interaction energies of a positively charged unit (a proton) with the charge density produced by the molecular set of nuclei and electrons at any point in space in the vicinity of the molecule. In general, a cut-off value is defined to limit the number of MEP points to be calculated. The MEP is a very useful tool in molecular modeling studies. It describes the electrostatic features of molecules and can be used for the analysis and prediction of molecular interactions. For the generation of MEPs, two different approaches can be followed. The most desirable way is to calculate the MEPs directly from the quantum mechanically derived wavefunction. This procedure is straightforward and more accurate, and hence time consuming. A simpler approach is the calculation of MEPs on the basis of the atomic partial charges representing the molecular charge distribution. The MEP is then calculated by applying the Coulomb equation for electrostatic interactions. Of course, the first procedure by far is superior and, by all means, should be used if sufficiently accurate wavefunctions are attainable for a particular molecule.

Many investigations of the basis set dependence of MEPs derived directly from wavefunctions are found in the literature [22–25]. It has also been shown that the ESP based on AM1 wavefunction correlates sufficiently well with ab initio results [22]. Therefore, AM1 can be used in all cases that cannot be handled owing to molecular size at the ab initio level.

For the display of the MEP, different techniques are in use. The major obstacle for a fast and easy utilization of MEPs for the comparison of different molecules is the large amount of data points associated with this property. One very widely used method to visualize MEPs is the display of the MEP in the form of a two-dimensional isocontour map in a particular plane of the molecule. The map may be displayed in color on a graphics screen. It can be manipulated in real time. A single contour line represents values with similar energy. Regions containing a high nuclear contribution produce positive fields corresponding to a repulsive interaction with a positive point charge, while those with a high electron density produce a negative potential corresponding to an attractive interaction with a positive point charge.

The next level of complexity is reached by switching from 2D to 3D display mode. In principle, nothing changes except that the molecule is completely wrapped by sets of isopotential shells. Each point on a particular shell experiences an ESP of the same sign and magnitude. With the help of this technique, the overall distribution of positively and negatively charged regions around a molecule could be visualized very distinctly. While two-dimensional charts, naturally, may not always reveal a complete picture of the MEP, the 3D isopotential surfaces effectively allow a qualitative interpretation and comparison between different compounds.

The third method for displaying the MEP is associated with the calculation and visualization of the molecular surfaces. We therefore dwell on the various definitions of the molecular surface. In the formal treatment of molecular surfaces, the atomic positions are treated as points, whereas the electron clouds are approximated by spheres centered on the atomic centers. If the electron spheres are represented by the van der Waals radii, then the surface generated by summing all spheres is called *the van der Waals surface*. Van der Waals surfaces approximately represent the 3D volume requirements of molecules. A different type of surface that is often used in molecular modeling studies is the solvent accessible surface, also called *Connolly surface* [26]. The Connolly surface is the surface encircled by the center of a solvent probe as the probe molecule rolls over the van der Waals surface.

The ESP can be color-coded either onto the van der Waals or onto the Connolly surface. Each color at a defined point on the surface indicates a distinct energy value of the ESP. This technique attempts to simultaneously display both the shape of the molecules as well as the electrostatic properties. However, when larger molecules are studied, then the images become very complex. A way out of this dilemma is sometimes found by using different techniques in a combined approach because areas hidden in one display mode may be perceptible in the other (see Figure 2.4.1).

The MEP is a much more reliable indicator of electrostatic reactivity than the concept of atomic point charges. MEPs and their 3D representation have proved to be effective tools for analyzing and predicting the interaction of ligands with their macromolecular receptors.

The ESPs of different molecules, which bind to the same receptor site in a similar way, must share common features. It has been shown that, in many cases, where an atom-by-atom fit of the corresponding molecules does not lead to a satisfactory result, the MEP-directed superimposition yields an acceptable solution of the problem (see Section 2.5.3).

As an example, it has been shown in a study of the ESP of histaminergic H_2-antagonists [28] that the imidazole ring of cimetidine and the guanidinothiazole ring of tiotidine can be superimposed on the basis of their ESP. This can be easily deduced from Figure 2.4.2.

2.4.2
Molecular Interaction Fields

Many biological processes are determined by noncovalent interactions between molecular structures. This is true for the docking of a ligand to a receptor, the interaction of a substrate with an enzyme or the folding of a protein. Also, in the world of crystals the noncovalent forces decisively determine the geometry and symmetry of the molecular arrangement. As a general rule, binding occurs only if the generated energy of interaction overcomes the repulsive van der Waals forces. One method to investigate the energetic conditions between molecules approaching each other is the generation of molecular interaction fields. These fields describe the variation of interaction energy between a target molecule and a chemical probe moved in a 3D grid, which has been set around the target. The probes reflect chemical characteristics of a binding partner or fragments of it. Using computer graphics, molecular interaction fields can be displayed as 3D isoenergy contours. Contours of large positive energies indicate regions from which the probe would be repelled, while those of large negative energies correspond to energetically favorable binding regions.

The calculation of molecular interaction fields can be carried out using a variety of programs like GRID [29], MOE [30], HINT [31] or ISOSTAR/SUPERSTAR [32–34].

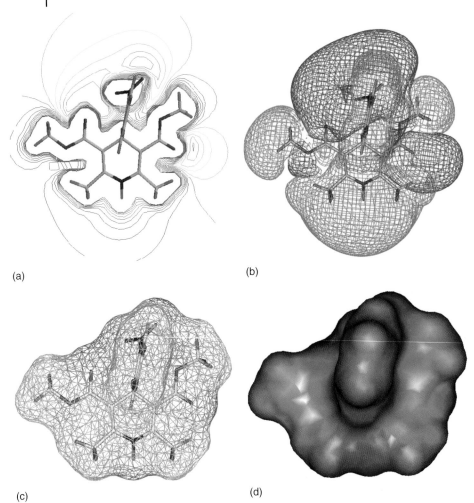

(a)

(b)

(c)

(d)

Figure 2.4.1 Visualization of the molecular electrostatic potential (MEP) of nifedipine, using different techniques. (a) The MEP is displayed as a two-dimensional isocontour map in the plane of the dihydropyridine ring system. The electrostatic potential has been calculated directly from the ab initio wavefunction (using a 6-31G** basis set) and is contoured from -50 kcal mol^{-1} (red) to 90 kcal mol^{-1} (blue). (b) The MEP is displayed in the form of isopotential surfaces. The electrostatic potential has been calculated by a point charge approach (ESP point charges have been derived from an ab initio calculation applying a 6-31G** basis set) and is displayed at the region of -5 kcal mol^{-1} (blue) as well as 5 kcal mol^{-1} (red). (The calculations have been performed using the quantum mechanical software package SPARTAN 3.0 [14].) Figures (c) and (d) show the electrostatic potential displayed on the Connolly surface of nifedipine. The values of the electrostatic potential have been calculated using ESP-derived point charges (the same as in case of b) and are displayed in the form of a simple dot surface (c) as well as in the more sophisticated form of a solid 'triangular' surface (d). Blue areas represent negative electrostatic potentials, whereas red areas represent positive values. (The calculations have been performed using the program MOLCAD [27].)

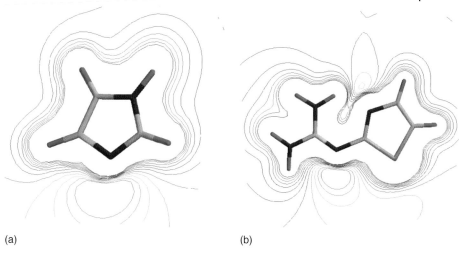

(a)   (b)

Figure 2.4.2 Electrostatic potential of imidazole (Figure 2.4.2a) and guanidinothiazole (Figure 2.4.2b). The electrostatic potentials have been calculated using the ab initio wavefunction (with a 6-31G** basis set) and are contoured from -50 kcal mol^{-1} (red) to 90 kcal mol^{-1} (blue). (The calculations are performed using the quantum mechanical software package SPARTAN 3.0 [14].)

GRID is one of the most widely used programs for investigating molecular interaction fields. It works for small molecules as well as for large protein molecules like enzymes. As input, only Cartesian coordinates are needed. The list of probes is very comprehensive. The interaction energy is calculated on a regular grid of points surrounding the target molecule. The grid can also be confined to a particular fragment of the target molecule if that is the only part of interest. The calculated energies are stored in a file and can be transferred for graphical display and analysis into most of the common molecular modeling programs [35–38]. 3D contour maps may then be generated at any selected energy level and studied together with the target molecule on a computer graphic system. The contouring is a quick process that allows the user to control, almost immediately, the graphical results.

In this chapter, we focus on the calculation of the interaction fields for small molecules, whereas the investigation of the fields for macromolecules will be discussed subsequently (see Section 4.6).

2.4.2.1 Calculation of GRID fields

The probes that can be used for the calculation are small molecules, chemical fragments or particular atoms, for example, a water molecule, a hydroxyl group or a calcium ion. These probes simulate the chemical characteristics of the corresponding binding partners, for example a potential receptor protein binding site or the neighbor molecule in a crystal. In the course of a GRID calculation, the probe is systematically moved through a regular 3D array of grid points around the target structure. At each point, the interaction energy between the probe and the target is calculated using the following empirical energy function:

$$E_{tot} = E_{vdw} + E_{el} + E_{hb} \tag{1}$$

where

E_{tot} = total interaction energy
E_{vdw} = van der Waals interaction energy
E_{el} = electrostatic energy
E_{hb} = interaction energy due to hydrogen bond formation.

The van der Waals interaction energy can be regarded as a combination of attractive and repulsive dispersion forces between nonbonded atoms. An atom of the probe is prevented from penetrating into an atom in the target molecule by atomic repulsion and electron overlap. Repulsion forces can be estimated by an empirical energy function that becomes large and positive when the interatomic distance between two atoms is less than the sum of their van der Waals radii. The attractive part of the dispersion interaction is due to the correlated motion of electrons around the nuclei, which results in induced dipole interactions. For nonpolar molecules, the balance between the attractive dispersion forces and the short-range repulsive forces can be described with the Lennard–Jones potential [38] (see e.g. Equation 5 in Section 2.2) which is implemented in the GRID program.

Electrostatic interactions are particularly important due to their long-range character for the attraction between ligand and macromolecular receptor.

The Coulomb equation (Equation 6 in Section 2.2) is widely used in molecular mechanics programs for the calculation of the electrostatic term because of its simple mathematical form. Its disadvantage, of course, is the fact that the heterogeneous media of molecular systems, which consist of molecules with different dielectric properties, are not sufficiently represented. The discontinuity between solute and solvent is taken into account by using an extended and more comprehensive form of Coulomb's law [29], which is used by GRID.

The directional properties of hydrogen bonds play a crucial role in determining the specificity of intermolecular interactions. It is therefore of utmost importance for a proper evaluation of interaction energies to describe this part of attractive forces between molecules in a correct form. A hydrogen bond can be regarded as intermediate-range interaction between a positively charged hydrogen atom and an electronegative acceptor atom [39]. The resulting distance between acceptor and donator atom is less than the sum of their van der Waals radii. In contrast to other noncovalent forces like dispersion and electrostatic point charge interactions, the hydrogen bonding interaction is directional – that is, depends on the propensity and orientation of the lone pairs of the acceptor heteroatom.

To comply with the requirements of these aspects, the GRID method uses an explicit energy term for hydrogen bonds [40]. The functional form of this term has been developed to fit experimental data. All parameters are founded on experimental crystallographic data – that is, direction, type and typical strength of these interactions are classified according to the real world of crystals.

The probes implemented in GRID are extensively defined by a variety of parameters, for example the hydrogen bonding possibility, the van der Waals radius or the atomic charge. The detailed description makes them very specific so that they can be regarded as realistic representatives of important functional groups found in the active site of macromolecules. As an example, properties and parameters for three important probe groups are shown in Table 2.4.2.

GRID also contains a table of parameters defining each type of atom possibly existent in target molecules. The respective parameters define the strength of the van der Waals, the electrostatic and the hydrogen bond interactions potentially formed by an atom. The careful parameterization and the large variety of implemented probes make the GRID program a precise and widely used method for the investigation of interaction fields for small molecules as well as for macromolecular structures.

The calculation of molecular interaction fields has been applied to a wide range of molecular modeling studies [41–46]. The strategy used depends on the available structural information for ligands and macromolecular targets. If the 3D structure of a macromolecule is known, then the interaction fields can be used to precisely locate favorable binding regions for the ligands. Subsequently, these regions can be taken as starting point for the design of new ligands for the particular receptor. A variety of successful applications of the GRID approach can be found in Ref. [47].

Table 2.4.2 Examples of the different parameters that are necessary to define GRID probe groups.

	Methyl probe	Hydroxyl probe	Carboxyl probe
Van der Waals radius (Å)	1.950	1.650	1.600
Effective number of electrons	8	7	6
Polarizabilty (Å^3)	2.170	1.200	2.140
Electrostatic charge	0.000	−0.100	−0.450
Optimal hydrogen bond energy (kcal mol^{-1})	0.000	−3.500	−3.500
Hydrogen bonding radius (Å)	–	1.400	1.400
Number of hydrogen bonds donated	0	1	0
Number of hydrogen bonds accepted	0	2	2
The hydrogen bonding type	0	4	8

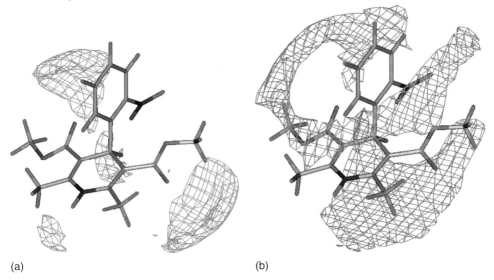

(a)   (b)

Figure 2.4.3 Visualization of the molecular interaction fields of nifedipine. (a) Shows favorable hydrogen bonding regions derived from GRID [29] calculations using a hydroxyl probe (contour level: -3.5 kcal mol^{-1}). The favorable hydrophobic interaction regions obtained with a methyl probe are displayed in (b) (contour level: -1.4 kcal mol^{-1}).

More often, situations are encountered where structural information about the macromolecular receptor does not exist but only the properties of the ligands are available. Under such circumstances, molecular interaction fields can help generate a more or less detailed picture of the potential receptor binding site. The prerequisite for this approach of course is that all investigated ligand molecules indeed bind to the same receptor site in an analogous mechanism. Only then they can be expected to exhibit a similar interaction pattern. Also, relative positions and size of the contours at any given energy level should be comparable. At which energy level the contours have to be compared is strongly dependent on the chosen probe type. In Figure 2.4.3, two different types of interaction fields for the Ca^{2+}-channel blocking agent, nifedipine, are shown as an example.

The interaction fields mark those parts of the corresponding binding region that possess particular chemical and physical properties. These properties can be translated into a model of the binding region of the macromolecule. If the macromolecular target is a receptor protein, the model is composed of single amino acid fragments that are located at the corresponding interaction regions. The amino acid fragments should mimic the different binding regions, which are common for all the active compounds. For example, the hydrophobic interaction fields possibly represent the location of hydrophobic amino acids like phenylalanine, tryptophan, valine, leucine or isoleucine. Of course, further investigations are necessary to specify the exact type of the amino acid in each case [48]. Information obtained by generating the interaction fields for a set of superimposed ligands (as an e.g., see Figure 2.4.3.11) can be used to refine the so-called pseudoreceptor models (see Section 2.6.3) or protein homology models (see Section 4.3).

If a large set of compounds has to be studied, it may become difficult to recognize all existing common interaction patterns. One way to solve such a problem is by calculation of common interaction regions for different target molecules, which are obtained in each case using the same probe. The common regions are mathematically detected in a gridpoint-by-gridpoint comparison of the fields. Only hits are saved in a file and used for the generation of a common interaction field [49].

A more profound technique for comparison and quantitative analysis of molecular interaction fields is the use of chemometric methods [50–53]. These techniques are discussed in detail in Section 2.6.

2.4.2.2 Hydrophobic Interactions

As already discussed, attraction and repulsion between molecules are controlled by different types of interactions. One type of interaction that has not been taken into account so far is the so-called hydrophobic interaction. Hydrophobic interaction between molecules is a complex process resulting primarily from entropic effects related to the change not only in the orientation of solvent molecules in the solvation shell wrapping the solute molecules but also in the bulk form of the solvent molecules. For an effective hydrophobic interaction, a close contact of the interacting hydrophobic surfaces is necessary [54, 55]. The following piece of fiction might lead to a better understanding of hydrophobic bonding: a nonpolar region of a deep binding cavity in a protein is not directly solvated. The nearby water molecules shield the cavity and are thought to form an iceberglike structure, which is stabilized by intermolecular hydrogen bonds between the water molecules. An interaction between the hydrophobic surfaces of the binding cavity and an entering substrate leads to a disruption of the ordered iceberglike structure. The disruption yields an increase in entropy, which results in a gain of free energy for the total system [55]. Of course, the desolvation of the substrate molecule also adds to the amount of newly formed bulk solvent molecules and must be taken into account. Till date the entropic effect is usually ignored in most of the modeling studies because a simple method of calculation is not available. On the other hand, it is generally accepted that the hydrophobic bonding or entropic effect indeed plays an important role in each drug–receptor interaction [56] as well as protein folding [57] event. As a natural follow-up, the hydrophobic interaction should, by all means, be included in the energy balance of these processes.

Hydrophobicity can also be regarded as an empirical property of molecules encoding specific thermodynamic information about a molecule's interaction with its environment. Hitherto several attempts were made for taking into account hydrophobic effects on the basis of experimental findings. The most important experimental measure of hydrophobicity is the solvent partition coefficient – expressed as log P – of a molecule between water and an organic phase. Since the log P can be determined experimentally, it is a very useful tool. It can also be used to control and improve empirically developed methods, which are reported by several authors [27, 58]. The prediction of log P can be achieved by transforming experimental solvent partition data for sets of variously substituted molecules into the so-called hydrophobic fragment constants. These fragment constants represent the relative lipophilicity of particular structural elements found in the original set of molecules. The total lipophilicity of a compound (given by log P) can then be calculated by summation of all fragment constants for a molecule under study. Today, fragment constants are available for a wide variety of organic species of biological importance.

It should be noted that log P is a simple 'one-dimensional' representation of hydrophobicity, and it only reflects an overall property. It is insufficient when a more detailed insight into molecular interactions between ligands and macromolecules is needed.

For this reason, several attempts were made to utilize solvent partition coefficients as foundation to create a 3D representation for the hydrophobicity. One route that can be followed to approach the problem is the generation of a hydrophobic field in analogy to the electrostatic field. This technique, for example, is implemented in the program HINT [31] or in the DRY probe in program GRID [29].

The HINT model of hydrophobic interactions is based on the fact that solubility data can be regarded as just another physical property capable of mirroring the molecular interactions between solute and solvent molecules. In the framework of the HINT theory, the fragment-level solvent partition data (the hydrophobic fragment constants) are reduced to hydrophobic atom constants [59]. These atomic descriptors are the key parameters and have to be assigned for each atom in a molecule under investigation. Since the hydrophobic atom constants are derived from experimental partition experiments and solubility data, the hydrophobic atom constants obtained model not only the hydrophobic interactions but also include other types of molecular interactions, like electrostatic and van der Waals terms. The generated field therefore incorporates hydrophobic as well as hydrophilic parameters. It is called a *hydropathic* field. The calculation is performed using an empirical function (for detailed description of the functional form, see Ref. [59]). The hydrophobic atom constants, the section of the solvent accessible surface created by each atom and a distance function are included in the algorithm. The distance function is necessary to adequately describe the distance dependence of the hydrophobic effect in the biological environment. HINT generates 3D molecular grid maps in a similar way as discussed for comparable programs.

The result of a HINT study is a combined contour map for the hydrophobic and hydrophilic field around a molecule. Grid points with a positive sign represent a hydrophobic region. The opposite is true for hydrophobic (polar) segments of space. Because of the empirical nature of the data, it is difficult to decide at which energy level the fields have to be contoured. It is self-evident that the selected energy level directly determines the size of the visualized part of the field. For a proportionally correct balance in the size of the displayed contours, it is usually advisable to contour the hydrophilic effect at a level two to five times higher than that of the hydrophobic effect [60].

The appearance of hydrophobic (contour level) and hydrophilic (contour level) fields is again shown for the well-known Ca^{2+}-channel blocker nifedipine in Figure 2.4.4(a).

The information obtained from the analysis of the hydropathic field can be exploited following different strategies. The qualitative information on the distribution of hydrophobic and polar properties in the vicinity of a series of molecules, for example, can be used to generate a 3D map of the unknown receptor macromolecule. If the investigated series is large and complex, an interface allows to read the produced data set directly into CoMFA for a more elaborate analysis [60].

If the structure of the macromolecular receptor is known, the generated hydropathic fields can also be used to optimize the structures of ligands for enhancement of the biological activity. For a review of other potential applications, see Ref. [61].

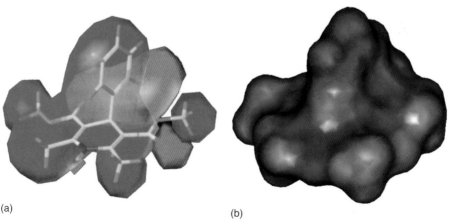

(a)

(b)

Figure 2.4.4 (a) Hydrophobic field map of nifedipine. The green colored surface represents the hydrophobic area, and the red colored surface represents the hydrophilic area of nifedipine. The map has been calculated by HINT 2.02 [31] and is contoured at -8 (red) and 4 (green).

(b) Molecular lipophilic potential of nifedipine displayed on the Connolly surface. Brown areas on the surface represent more lipophilic parts, and blue areas represent the hydrophilic parts of the molecule. (The calculation has been performed using the program MOLCAD [33].)

2.4.3
Display of Properties on a Molecular Surface

The display of hydrophobic and hydrophilic property distributions in the extra-molecular space can also be projected on a molecular surface. The program MOLCAD [62] employs, for example, the Connolly surface [26] of a molecule as a screen for mapping local molecular properties like lipophilicity by a color-coded representation. A distance-dependent function has to be defined to correctly reflect the influence of different atoms or fragments on the local lipophilicity at a certain point on the molecular surface. This can be realized, for example, by introducing a molecular lipophilicity potential [63], which can be regarded as a pendant to the MEP. As in the case of the MEP, the projection of any local properties onto a surface facilitates the perception and interpretation of the distribution of the visualized property descriptor. The main advantage of the surface-bound representation of hydrophobicity is the fact that the analysis of large molecular systems, like proteins, is much easier in comparison to the evaluation of hydropathic fields. Because the theoretical background of both methods is equivalent, the results obtained should be comparable qualitatively. For both methods, a test of reliability can be performed very well for all molecules with experimentally derived log P values available. However, the partition coefficient like the charge distribution is also drastically influenced by the conformation of a molecule. And what makes the situation even more complicated is the fact that conformation of a molecule can change when it migrates from the aqueous to the lipophilic environment. This fact unfortunately limits the amount of test molecules to a rather small collection of rigid or, at least, semirigid structures. An example for the appearance of MOLCAD hydrophobic surfaces is shown in Figure 2.4.4(b).

References

1. Williams, D.E. (1991) Net atomic charge and multipole models for the ab initio molecular electric potential, in *Reviews in Computational Chemistry* (eds K.B. Lipkowitz and D.B. Boyd), VCH Publishers, New York, Vol. 4, pp. 219–71.
2. Gasteiger, J. and Marsili, M. (1980) Iterative partial equalization of orbital electronegativity – a rapid access to atomic charges. *Tetrahedron*, **36**, 3219–28.
3. Del Re, G. (1958) A simple MO–LCAO method for the calculation of charge distributions in saturated organic molecules. *Journal of the Chemical Society*, 4031–40.
4. Berthod, H. and Pullman, A. (1965) Sur le calcul des caracteristiques du squelette sigma des molecules conjuguees. *Journal of Chemical Physics*, **62**, 942–46.
5. Abraham, R.J. and Hudson, B. (1985) Charge calculations in molecular mechanics. 3. Amino-acids and peptides. *Journal of Computational Chemistry*, **6**, 173–81.
6. Mullay, J. (1986) A simple method for calculating atomic charge in molecules. *Journal of the American Chemical Society*, **108**, 1770–75.
7. Mulliken, R.S. (1955) Electronic population analysis on lcao-mo molecular wave functions. 1.

Journal of Chemical Physics, **23,** 1833–40.

8. Momany, F.A. (1978) Determination of partial atomic charges from ab initio molecular electrostatic potentials – application to formamide, methanol, and formic-acid. *Journal of Physical Chemistry,* **82,** 592–601.

9. Cox, S.R. and Williams, D.E. (1981) Representation of the molecular electrostatic potential by a net atomic charge model. *Journal of Computational Chemistry,* **2,** 304–23.

10. Singh, U.C. and Kollman, P.A. (1984) An approach to computing electrostatic charges for molecules. *Journal of Computational Chemistry,* **5,** 129–45.

11. Chirlian, L.E. and Francl, M.M. (1987) Atomic charges derived from electrostatic potentials – a detailed study. *Journal of Computational Chemistry,* **8,** 894–905.

12. Purcell, W.P. and Singer, J.A. (1967) A brief review and table of semiempirical parameters used in the Hueckel molecular orbital method. *Journal of Chemical and Engineering Data,* **12,** 235–46.

13. Frisch, M., Binkley, J.S., Schlegel, H.B. *et al. GAUSSIAN,* Department of Chemistry, Carnegie Mellon University, Pittsburgh, http://www.gaussian.com.

14. JAGUAR, Schrödinger Inc., http://www.schrodinger.com.

15. Schmidt, M.W., Boatz, J.A., Baldrige, K.K. *et al.* GAMESS, Program No. 115, *Quantum Chemistry Program Exchange,* Indiana University, Bloomington, http://www.qcpe.chem.indiana.edu.

16. Williams, D.E. and Yan, J.M. (1987) Point-charge models for molecules derived from least-squares fitting of the electric-potential. *Advances in Atomic and Molecular Physics,* **23,** 87–130.

17. Wiberg, K.B. and Rablen, P.R. (1993) Comparison of atomic charges derived via different procedures. *Journal of Computational Chemistry,* **14,** 1504–18.

18. Reed, A.E., Weinstock, R.B., and Weinhold, F. (1985) Natural-population analysis. *Journal of Chemical Physics,* **83,** 735–46.

19. McWeeney, R. (1989) *Methods of Molecular Quantum Mechanics,* 2nd Edition, Academic Press, San Diego.

20. Destro, R., Bianchi, R., and Morosi, G. (1989) Electrostatic properties of l-alanine from x-ray-diffraction at 23-K and abinitio calculations. *Journal of Physical Chemistry,* **93,** 4447–57; Destro, R., Bianchi, R., Gatti, C., and Merati, F. (1991) Total electronic charge-density of l-alanine from x-ray-diffraction at 23-K. *Chemical Physics Letters,* **186,** 47–52.

21. McClellan, A.L. (1974) *Tables of Experimental Dipole Moments,* Rahara Enterprise, Vol. 2, El Cerrito.

22. Ferenczy, G.G., Reynolds, C.A., and Richards, W.G. (1990) Semiempirical AM1 electrostatic potentials and AM1 electrostatic potential derived charges – a comparison with abinitio values. *Journal of Computational Chemistry,* **11,** 159–69.

23. Rodriguez, J., Manaut, F., and Sanz, F. (1993) Quantitative comparison of molecular electrostatic potential distributions from several semiempirical and ab initio wave functions. *Journal of Computational Chemistry,* **14,** 922–27.

24. Ford, G.P. and Wang, B.Z. (1993) New approach to the rapid semiempirical calculation of molecular electrostatic potentials based on the AM1 wave-function – comparison with ab-initio hf/6-31g asterisk results. *Journal of Computational Chemistry,* **14,** 1101–11.

25. Price, S.L., Harrison, R.J., and Guest, M.F. (1989) An abinitio distributed multipole study of the electrostatic potential around an undecapeptide cyclosporine derivative and a comparison with point-charge electrostatic models. *Journal of Computational Chemistry,* **10,** 552–67.

26. Connolly, M.L. (1983) Solvent-accessible surfaces of proteins and nucleic-acids. *Science,* **221,** 709–13.

27. Rekker, R.F. and Mannhold, R. (1992) *Calculation of Drug Lipophilicity,* VCH Publishers, Weinheim.

28. Höltje, H.-D. and Batzenschlager, A. (1990) Conformational-analyses on histamine H-2-receptor antagonists. *Journal of Computer-Aided Molecular Design*, 4, 391–402.

29. Goodford, P.J. (1985) A computational-procedure for determining energetically favorable binding-sites on biologically important macromolecules. *Journal of Medicinal Chemistry*, 28, 849–57.

30. MOE, Chemical Computing Group, Montreal, http://www.chemcomp.com.

31. Kellogg, G.E., Semus, S.F., and Abraham, D.J. (1991) Hint – a new method of empirical hydrophobic field calculation for COMFA. *Journal of Computer-Aided Molecular Design*, 5, 545–52.

32. Bruno, I.J., Cole, J.C., Lommerse, J.P. *et al.* (1997) IsoStar: a library of information about nonbonded interactions. *Journal of Computer-Aided Molecular Design*, 11, 525–37.

33. Verdonk, M.L., Cole, J.C., and Taylor, R. (1999) SuperStar: a knowledge-based approach for identifying interaction sites in proteins. *Journal of Molecular Biology*, 289, 1093–108.

34. Boer, D.R., Kroon, J., Cole, J.C. *et al.* (2001) SuperStar: comparison of CSD and PDB-based interaction fields as a basis for the prediction of protein-ligand interactions. *Journal of Molecular Biology*, 312, 275–87.

35. INSIGHT/DISCOVER, Accelrys Inc., San Diego, http://www.accelrys.com.

36. SYBYL, Tripos Associates, St. Louis, http://www.tripos.com.

37. Mohamadi, F., Richards, N.G.J., Guida, W.C. *et al.* (1990) MACROMODEL – an integrated software system for modeling organic and bioorganic molecules using molecular mechanics. *Journal of Computational Chemistry*, 11, 440–67.

38. Jones, J.E. (1924) On the determination of molecular fields. II. From the equation of state of a gas. *Proceedings of the Royal Society of London*, 106A, 463–77.

39. Dean, P.M. (1986) *Molecular Foundations of Drug-Receptor Interaction*, Cambridge University Press, Cambrige.

40. Wade, R.C. (1993) Molecular interaction fields, in *3D QSAR in Drug Design – Theory Methods and Applications* (ed. H. Kubinyi), ESCOM Science Publishers B. V., Leiden, pp. 486–505.

41. Wade, R.C., Clark, K.J., and Goodford, P.J. (1993) Further development of hydrogen-bond functions for use in determining energetically favorable binding-sites on molecules of known structure. 1. Ligand probe groups with the ability to form 2 hydrogen-bonds. *Journal of Medicinal Chemistry*, 36, 140–47.

42. Sippl, W., Contreras, J.M., Parrot, I. *et al.* (2001) Structure-based 3D QSAR and design of novel acetylcholinesterase inhibitors. *Journal of Computer-Aided Molecular Design*, 15, 395–410.

43. Meng, E.C., Shoichet, B.K., and Kuntz, I.D. (1992) Automated docking with grid-based energy evaluation. *Journal of Computational Chemistry*, 13, 505–24.

44. Jendretzki, U.K., Elz, S., and Höltje, H.-D. (1994) Computer aided molecular analysis of 5-HT2A agonists. *Pharmaceutical and Pharmacological Letters*, 3, 260–63.

45. Wade, R.C. (1988) The use of molecular graphics in the design of anti-influenza agents. *British Journal of Pharmacology*, 95,(Suppl.), 588.

46. Cruciani, G. and Watson, K.A. (1994) Comparative molecular-field analysis using Grid force-field and Golpe variable selection methods in a study of inhibitors of glycogen-phosphorylase-B. *Journal of Medicinal Chemistry*, 37, 2589–601.

47. Cruciani, G. (2005) Molecular Interaction Fields, in *Methods and Principles in Medicinal Chemistry* (Series eds H. Kubinyi, G. Folkers and R. Mannhold), VCH Publishers, New York.

48. Höltje, H.-D. and Jendretzki, U.K. (1995) Construction of a detailed

serotoninergic 5-HT2A receptor model. *Archiv der Pharmazie*, **328**, 577–84.

49. Höltje, H.-D. and Anzali, S. (1992) Molecular modeling studies on the digitalis binding-site of the NA+/K+-ATPase. *Die Pharmazie*, **47**, 691–97.

50. Baroni, M., Costantino, G., Cruciani, G. *et al.* (1993) Generating optimal linear PLS estimations (GOLPE) – an advanced chemometric tool for handling 3D-QSAR problems. *Quantitative Structure-Activity Relationships*, **12**, 9–20.

51. Wold, S., Johansson, E., and Cocchi, M. (1993) PLS – partial least-squares projections to latent structures, in *3D QSAR in Drug Design – Theory Methods and Applications*, (ed. H. Kubinyi), ESCOM Science Publishers B. V., Leiden, pp. 523–50.

52. Klebe, G. and Abraham, U. (1993) On the prediction of binding-properties of drug molecules by comparative molecular-field analysis. *Journal of Medicinal Chemistry*, **36**, 70–80.

53. Folkers, G., Merz, A., and Rognan, D. (1993) CoMFA: scope and limitations, in *3D-QSAR in Drug Design – Theory Methods and Applications* (ed. H. Kubinyi), ESCOM Science Publishers B.V., Leiden, pp. 583–618.

54. Tanford, C. (1978) Hydrophobic effect and the organization of living matter. *Science*, **200**, 1012–18.

55. Tanford, C. (1980) *The Hydrophobic Effect.*, 2nd Edition, John Wiley & Sons, New York.

56. Suzuki, T. and Kudo, Y. (1990) Automatic log P-estimation based on combined additive modeling methods. *Journal of Computer-Aided Molecular Design*, **4**, 155–98.

57. Nicholls, A., Sharp, K.A., and Honig, B. (1991) Protein folding and association – insights from the interfacial and thermodynamic properties of hydrocarbons. *Proteins*, **11**, 281–96.

58. Hansch, C. and Fujita, T. (1964) Rho-sigma-pi analysis. Method for correlation of biological activity + chemical structure. *Journal of the American Chemical Society*, **86**, 1616–26.

59. Kellogg, G.E., Joshi, G.S., and Abraham, D.J. (1992) New tools for modeling and understanding hydrophobicity and hydrophobic interactions. *Medicinal Chemistry Research*, **1**, 444–53.

60. Kellogg, G.E. and Abraham, D.J. (1992) Key, lock, and locksmith – complementary hydropathic map predictions of drug structure from a known receptor receptor structure from known drugs. *Journal of Molecular Graphics*, **10**, 212–17.

61. Abraham, D.J. and Kellogg, G.E. (1993) Hydrophobic fields, in *3D QSAR in Drug Design – Theory Methods and Applications* (ed. H., Kubinyi), ESCOM Science Publishers B. V., Leiden, pp. 506–22.

62. Heiden, W., Moeckel, G., and Brickmann, J. (1993) A new approach to analysis and display of local lipophilicity hydrophilicity mapped on molecular-surfaces. *Journal of Computer-Aided Molecular Design*, **7**, 503–14.

63. Furet, P., Sele, A., and Cohen, N.C. (1988) 3D molecular lipophilicity potential profiles – a new tool in molecular modelling. *Journal of Molecular Graphics*, **6**, 182–89.

Further Reading

Ghose, A.K. and Crippen, G.M. (1986) Atomic physicochemical parameters for 3-dimensional structure-directed quantitative structure-activity-relationships. 1. Partition-coefficients as a measure of hydrophobicity. *Journal of Computational Chemistry*, **7**, 565–77.

2.5
Pharmacophore Identification

2.5.1
Molecules to be Matched

In the first few sections of this chapter, we have described how physicochemical characteristics of molecules can be calculated and visualized. Now we discuss how this knowledge can be used to understand or predict the pharmacological properties of a compound. In the majority of cases, the basis for a pharmacodynamic effect is the interaction of a certain substance with a protein of physiological importance. The macromolecule might be an enzyme or a receptor. In both cases, there must exist a highly specific three-dimensional cavity that would serve as binding site for the drug molecule. Compounds exerting qualitatively similar activities at the same enzyme or receptor must therefore possess closely related binding properties. That is, these molecules must present to the macromolecular binding partner structural elements of identical chemical functionality in sterically consistent locations. In short,, congeners of a defined pharmacological group do possess an identical pharmacophore, and one of the major tasks to be done using molecular modeling techniques is the determination of pharmacophores for congeneric groups of drug molecules. Because the three-dimensional structures of many receptors are still undiscovered, information on the corresponding hypothetical pharmacophore, as a matter of fact, is very important for understanding drug–receptor interactions at a molecular level.

When all physicochemical properties have been intensively studied, the question left is 'How do we superimpose the members of a series to find the pharmacophore?' To be able to answer this question, first we have to define the pharmacophoric elements. That is, we have to decide what functional groups or atoms have to be superimposed. Of course, this question cannot be answered completely in an automated procedure because one always has to decide in advance on the atom pairs that correspond between two molecules. This may produce a large amount of useless data if one does not include known structure-activity-relationship information. This facilitates the superpositioning procedure because it drastically limits the number of solutions. It should be noted that similarity between ligands must not comprise the whole molecule because most of the ligand molecules are not completely wrapped by receptor binding sites when they are bound to it. This also reduces the number of reasonable solutions.

If hydrogen bonds are considered to be of importance for the pharmacophore, then the direction and distance of lone pairs should be added to the atomic pattern of the molecules under study. This can be realized, for example, by locating dummy atoms in corresponding positions. These positions are then labeled by different flags as hydrogen-bond-acceptor or hydrogen-bond-donator sites (only hydrogens bound to heteroatoms), and can be used as a first test

Molecular Modeling. Basic Principles and Applications. 3rd Edition
H.-D. Höltje, W. Sippl, D. Rognan, and G. Folkers
Copyright © 2008 Wiley-VCH Verlag GmbH & Co. KGaA, Weinheim
ISBN: 978-3-527-31568-0

for a superposition mode (e.g. program AUTOFIT [1]). Furthermore, planar elements like aromatic ring systems can be treated as special structural units. In this case, for example, the center of the ring system can be defined as matching point instead of the ring system itself. Other planar groups can be handled analogously.

If a set of molecules does only contain very flexible congeners, then the search for a common pharmacophore is not only very difficult and tedious but might even yield none or an arbitrary, and therefore, useless result. This task can be extremely facilitated and earns much more significance if there exist rigid or, at least, semirigid compounds, which of course must be highly active, as otherwise they cannot be used as matrices for the flexible ligands. The consideration of highly active but conformationally restricted molecules by the same token relieves the need to prove that one indeed is dealing with bioactive conformations.

The selection of the molecules to be superimposed is very important in order to obtain significant results. The case that is easiest to perform but which is rather ineffective is the superimposition of structurally similar compounds. This does not provide much information. So it is much more effective to include the structures containing different skeletons in the series. As a natural follow-up, this leads to a situation that is highly desirable where a simple atom-to-atom superpositioning is not possible, but a matching of functionally equivalent elements or a matching of molecular fields has to be performed.

There is another point that must be addressed. Are inactive molecules or molecules with only low activity to be taken into consideration? It seems to be useful to first superpose highly active molecules alone. The derived pharmacophore can then be tested against and eventually be modified by inclusion of less active and inactive congeners. The same holds true for antagonists and agonists of one receptor type. Superpositions should be performed separately for both groups. However, both models can possibly be combined afterward because very often, competitive antagonists, at least partially, are bound in the agonist receptor binding site. But it should be noted that overlapping binding sites of agonists and antagonists indeed are common but do not necessarily exist.

One also has to keep in mind that indirect approaches incur serious limitations. First, the ligands must bind to the target protein at the same location and preferably adopt the same binding mode. If the former prerequisite is missing, the superimposition will be misleading (e.g. see [2, 3]). Also, pharmacophore models are usually restricted to low-energy conformations. Since the number of accessible conformations of a molecule increases dramatically with the energy tolerated, it is impossible to consider all conformations for the pharmacophore generation. Hence, strained conformations, such as the ones observed in the transition state of a chemical reaction, will usually not be covered by pharmacophore models. A detailed comparison between the energies of *in silico* generated and experimentally determined protein-bound ligand conformations has been published recently [4, 5].

Several different superpositioning procedures are available. They comprise manual or automatic fitting by rigid-body rotation or flexible-fitting procedures where both root-mean-square (rms) derivation between the fitted atom pairs and conformational energies are minimized. Other important superpositioning techniques perform alignments on the basis of equivalences detected in molecular surfaces or molecular field properties.

2.5.2
Atom-by-atom Superposition

The least-squares technique for superpositioning of corresponding atom positions is the most widely used method. Two molecules are superimposed by minimizing the rms of the distances between the corresponding atom pairs in the molecules. The rms value represents a measure for the goodness of the fit. This procedure is very powerful in discovering dissimilarity between molecules that seem to be similar on a first view. The weak point is that it has to be decided in advance regarding which are the atom pairs that match. It is obvious that different superpositions are obtained depending on the atoms used for the procedure. The method cannot be applied to those molecular systems in which atom correspondences are not detectable in advance. However, rigorous similarity in atomic structure is not a prerequisite for the interaction of different molecules with the same receptor. Therefore, for a large number of cases where pharmacological data and structure-activity studies advocate upon a common mechanism of action for a set of dissimilar molecules, the conventional least-squares superpositioning method is considered to be inadequate.

One may try to escape such a situation by performing a manual, interactive superposition if the test set is not too large. In principle, any number of molecules can be investigated directly on the graphics display and the fit may be judged visually. Certainly, this procedure is very creative and may stimulate new ideas about the underlying mechanism of experimentally detected structure-activity data. On the other hand, such a proceeding is obviously biased and often cannot be reproduced correctly because of nonapplicability of a computational optimization.

Active analog approach, an efficient and fast search technique that can be used very successfully for the generation of pharmacophore models, was developed by Marshall *et al.* [6, 7]. Active analog approach utilizes a systematic search algorithm for calculating a representative number of sterically and energetically allowed conformations for congeneric molecules. For each of these conformations, a set of distances between pharmacophoric groups believed to be important for the interaction with the receptor is generated. If each set of distances for one molecule is compared with all the sets of all other molecules with the intention of finding possible correspondences, the problem would not be solved except for small molecules. On the other hand, in the framework of pharmacophore identification the interest lies only in those subregions that are accessible to all active ligands and not in the complete conformer space of all compounds. As we have discussed above, it is of big advantage to include rigid or semirigid compounds in a conformational analysis for a series of flexible molecules. For this reason, the conformational search is started with the most rigid molecule. After determination of the respective distance map for this compound, these distances are used as constraints in the conformational search runs for the more flexible molecules. Following these lines, the results of a search on one active and rigid analog are taken as basis to explore the conformational space of all the other congeners of the series. As all the active compounds must fit the receptor model, the search is restricted to those regions of conformational space that correspond to the previously defined model. For example, if a pair of atoms, according to the model, must lie within a certain distance range to agree with the constraints, then only those torsions that permit this constraint to be satisfied have to be calculated. An example that has demonstrated the strength of the active analog approach dealt with 28 angiotensin-converting-enzyme (ACE) inhibitors in an effort to predict a model for the ACE active site [8]. Applying this searching technique, the search time was reduced by over 3 orders of magnitude in comparison with a previously performed conventional systematic search study on the same subject.

Another mapping procedure, which on the contrary does not explicitly use an atom-by-atom superposition approach, is SEAL [9]. This program allows a rapid pairwise comparison of dissimilar molecules. The similarity score, as indicator for the goodness of the fit, is calculated from a summation over all possible atom pairs between two molecules. Each atom pair is weighted by the relative distance between the contributing atoms. By doing so, the alignment function considers all theoretically possible atom pairs in the molecules in the comparison procedure and not only one atom pair like in an atom-by-atom fit approach. As a consequence, the resulting superposition reflects, to some extent, the properties associated with the global shape of the molecules. The program also offers the possibility to include physicochemical properties in the alignment procedure. Therefore, the terms used in the pairwise summation can be composed from any physicochemical quality considered to be important for the biological effect. In the original version, the authors use only van der Waals radii as an expression of the volume as well as point charges mimicking the electronic molecular properties to optimize the alignment.

An extended version of the original SEAL program has been developed by Klebe *et al.* [10]. Several structural alignment methods including rigid-body superposition based on an efficient overlap optimization were provided. Different molecular fields are described by sets of Gaussian functions. Consideration of the intramolecular conformational strain energy, as well as a flexible fitting routine, has been provided in addition [11].

There also exist mapping techniques that include the automatic and therefore unbiased identification of atomic centers or site points as correspondences used for superpositioning as one of the first steps in the computational protocol. Site points may include points of the molecular surface representing molecular features like hydrogen-bond-acceptor or hydrogen-bond-donator characteristics. Several commercial program packages like Catalyst [12] offer such functionalities. Others like DISCO [13], RECEPS [14] and AUTOFIT [1] have been intensively discussed in a recent review on common alignment programs [15]. As described earlier, the superposition is performed by matching the assigned corresponding atoms or site points in all possible combinations.

Most of the modern alignment programs handle the molecules to be superimposed flexibly, whereas one 'rigid' reference compound is needed. One popular program, namely, FlexS [16], uses a flexible superpositioning on the basis of combinatorial matching procedure. Pairs of molecules are aligned, one of which is considered rigid and the other one is flexibly fitted. The FlexS method originates from the related docking program FlexX [17]. FlexS tries to decompose the flexible structure into relatively rigid portions to start the placement of corresponding pharmacophoric features [18]. The remaining portions of the molecule are added in an iterative incremental procedure. The similarity between the superimposed molecules is calculated on the basis of energylike matching terms for paired intermolecular interactions and overlap terms of utilized Gaussian functions. The approach has been validated on experimental data obtained from X-ray structures [18]. Because of its computational efficiency, the method is extremely fast and can be used to screen large databases of compounds.

2.5.3
Superposition of Molecular Fields

Since molecules recognize each other by characteristic properties on or outside their van der Waals volume and not through their atomic skeleton, the determination of molecular similarity should take into account the molecular fields. As a natural follow-up, the superpositioning approaches should also concentrate on mapping and comparing these properties. For the purpose of matching, the molecules are located in a three-dimensional grid of equally spaced field points. Each grid point is loaded with a certain characteristic property measure like charge distribution, hydrophobic potential or simply information on the size of the volume. Similarity thresholds can be defined to guide the optimization procedure to a significant and unequivocal result. Single grid points or clusters of adherent grid points can be assigned different weights in order to pay close attention to structure-activity relationships. One molecule, preferentially with limited conformational freedom, is chosen as the template molecule. The grid loadings of the template serve as measure for the various properties, and all trial molecule grids are manipulated by rotation and translation to find the best fit of the grid values. The computational technique of orientational search that has to be used is extremely time-consuming. Different methods that not only mirror different levels of complexity but also utilize various field properties have been described. Sanz *et al.* reported an effective method that maximizes the similarity between molecular surfaces on the basis of the molecular electrostatic field [19]. Other authors like Clark [20] or Dean [21], in addition, use physicochemical field properties calculated using Lennard-Jones potentials, or replace the regular grid-based evaluation technique by an integration over Gaussian-type functions to approximate the electrostatic potential. Goodness-of-fit indices can be calculated – for example, as ratio of the number of commonly occupied grid points to the total number of grid points.

In summary, the tools for matching molecular surfaces do exist. Since the corresponding methods do not require any atom correspondences between molecules, they can be efficiently used for superposing dissimilar molecules. However, only if the complicated calculations can be made fast enough to deal with a large number of conformations for each molecule to be superimposed, it might become a routine technique. For a detailed comparison between atom-by-atom and field-based methods, the reader is referred to the literature [15, 22, 23]. An outstanding collection of articles and reviews dealing with the pharmacophore concept can be found in Ref. [24].

References

1. Kato, Y., Inoue, A., Yamada, M. *et al.* (1992) Automatic superposition of drug molecules based on their common receptor-site. *Journal of Computer-Aided Molecular Design*, **6**, 475–86.
2. Klebe, G. and Abraham, U. (1993) On the prediction of binding-properties of drug molecules by comparative molecular-field analysis. *Journal of Medicinal Chemistry*, **36**, 70–80.
3. Böhm, H.J., Klebe, G., and Kubinyi, H. (1996) *Wirkstoffdesign*, Spektrum Akademischer Verlag.
4. Boström, J. (2001) Reproducing the conformations of protein-bound ligands: a critical evaluation of several popular conformational searching tools. *Journal of Computer-Aided Molecular Design*, **15**, 1137–52.
5. Boström, J., Norrby, P.-O., and Liljefors, T. (1998) Conformational energy penalties of protein-bound

ligands. *Journal of Computer-Aided Molecular Design*, **12**, 383–96.
6. Marshall, G.R., Barry, C.D., Bosshard, H.E. *et al.* (1979) The conformational parameter in drug design: the active analog approach, in *Computer-Assisted Drug Design, ACS Monograph, Vol. 112* (eds E.C. Olsen and R.E. Christoffersen), American Chemical Society, Washington, DC, pp. 205–26.
7. Dammkoehler, R.A., Karasek, S.F., Shands, E.F.B., and Marshall, G.R. (1989) Constrained search of conformational hyperspace. *Journal of Computer-Aided Molecular Design*, **3**, 3–21.
8. Mayer, D., Naylor, C.B., Motoc, I., and Marshall, G.R. (1987) A unique geometry of the active site of angiotensin-converting enzyme consistent with structure-activity studies. *Journal of Computer-Aided Molecular Design*, **1**, 3–16.

9. Kearsley, S.K. and Smith, G.M. (1990) An alternative method for the alignment of molecular structures: maximizing electrostatic and steric overlap. *Tetrahedron Computer Methodology*, **3**, 615–33.

10. Klebe, G., Mietzner, T., and Weber, F. (1994) Different approaches toward an automatic structural alignment of drug molecules – applications to sterol mimics, thrombin and thermolysin inhibitors. *Journal of Computer-Aided Molecular Design*, **8**, 751–78.

11. Klebe, G., Mietzner, T., and Weber, F. (1999) Methodological developments and strategies for a fast flexible superposition of drug-size molecules. *Journal of Computer-Aided Molecular Design*, **13**, 35–49.

12. Catalyst Accelrys Inc., San Diego, http://www.accelrys.com.

13. Martin, Y.C., Bures, M.G., Danaher, E.A. *et al.* (1993) A fast new approach to pharmacophore mapping and its application to dopaminergic and benzodiazepine agonists. *Journal of Computer-Aided Molecular Design*, **7**, 83–102.

14. Kato, Y., Itai, A., and Iitaka, Y. (1987) A novel method for superimposing molecules and receptor mapping. *Tetrahedron*, **43**, 5229–36.

15. Lemmen, C. and Lengauer, T. (2000) Computational methods for the structural alignment of molecules. *Journal of Computer-Aided Molecular Design*, **14**, 215–32.

16. Lemmen, C. and Lengauer, T. (1997) Time-efficient flexible superposition of medium-sized molecules. *Journal of Computer-Aided Molecular Design*, **11**, 357–68.

17. Rarey, M., Kramer, B., Lengauer, T., and Klebe, G. (1996) A fast flexible docking method using an incremental construction algorithm. *Journal of Molecular Biology*, **261**, 470–89.

18. Lemmen, C., Lengauer, T., and Klebe, G. (1998) FLEXS: a method for fast flexible ligand superposition. *Journal of Medicinal Chemistry*, **41**, 4502–20.

19. Manaut, M., Sanz, F., Jose, J., and Milesi, M. (1991) Automatic search for maximum similarity between molecular electrostatic potential distributions. *Journal of Computer-Aided Molecular Design*, **5**, 371–80.

20. Clark, M., Cramer, R.D. III, Jones, D.M. *et al.* (1990) Comparative molecular field analysis (CoMFA). 2. Toward its use with 3D-structural databases. *Tetrahedron Computer Methodology*, **3**, 47–59.

21. Dean, P.M. (1990) Molecular recognition: the measurement and search for molecular similarity in ligand-receptor interaction, in *Concepts and Applications of Molecular Similarity* (eds M.A. Johnson and G.M. Maggiora), John Wiley & Sons, New York, pp. 211–38.

22. Mason, J.S., Good, A.C., and Martin, E.J. (2001) 3-D Pharmacophores in drug discovery. *Current Pharmaceutical Design*, **7**, 567–97.

23. Good, A.C. and Mason, J.S. (1995) Three-dimensional structure database search, in *Reviews in Computational Chemistry* (eds K.B. Lipkowitz and D.B. Boyd), VCH Publishers, New York, Vol. 7, pp. 73–95.

24. Langer, R. and Hoffmann, R. (2006) In *Pharmacophores and Pharmacophore Concepts, Methods and Principles in Medicinal Chemistry, Series* (eds H. Kubinyi, G. Folkers and R. Mannhold), VCH Publishers, New York.

2.6
3D QSAR Methods

The three-dimensional quantitative structure-activity relationship (3D QSAR) techniques are the most prominent computational means to support chemistry within indirect drug design projects. The primary aim of these techniques is to establish a correlation of biological activities of a series of structurally and biologically characterized compounds with the spatial fingerprints of numerous field properties of each molecule, such as steric demand, lipophilicity and electrostatic interactions. Typically, a 3D QSAR study allows identification of pharmacophoric arrangement of molecular fragments in space, and provides guidelines for the design of next-generation compounds with enhanced biological potencies.

The number of 3D QSAR studies has exponentially increased over the last decade, since a variety of methods are commercially available in user-friendly, graphically guided software [1–3]. Besides the commercial distribution, a major factor in the continuing enthusiasm for 3D QSAR comes from the proven ability of several of these methods to correctly forecast the biological activity of novel compounds [4]. However, the ease of application of 3D QSAR programs may, in particular, encourage beginners in modeling to apply the methods to all the available data sets. It is the aim of this chapter to not only provide an introduction of relevant 3D QSAR methods, but also to analyze the limitations of the various approaches.

2.6.1
The CoMFA Method

The comparative molecular field analysis (CoMFA) [1] method was developed as a tool to investigate 3D QSARs. 3D QSAR approaches use statistical methods (chemometrical methods) to correlate the variation in biological or chemical activity with information on the 3D structure for a series of compounds. A CoMFA analysis starts with traditional pharmacophore modeling to suggest a bioactive conformation of each molecule and ways to superimpose the molecules under study. This is not a trivial task as has been documented in the last section. The underlying idea of the CoMFA is that differences in a target property, for example, biological activity, are often closely related to equivalent changes in the shapes and strengths of the noncovalent interaction fields surrounding the molecules. In other words, the steric and electrostatic fields provide all the information necessary for understanding the biological properties for a set of compounds. As in the GRID approach, the molecules are located in a cubic grid and the interaction energies between the molecule and a defined probe are calculated for each grid point. Normally only two potentials, namely, the steric potential as a Lennard-Jones function and the electrostatic potential as a simple Coulomb function, are used within a CoMFA

Molecular Modeling. Basic Principles and Applications. 3rd Edition
H.-D. Höltje, W. Sippl, D. Rognan, and G. Folkers
Copyright © 2008 Wiley-VCH Verlag GmbH & Co. KGaA, Weinheim
ISBN: 978-3-527-31568-0

study. It is obvious that neither the description of molecular similarity nor the description of the process of interaction of ligands with corresponding biological targets is a trivial task. In the standard application of CoMFA, only enthalpic contributions of the free energy of binding are provided by the used potentials [5]. However, many binding effects are governed by hydrophobic and entropic contributions. Therefore, one has to characterize in advance the expected main contributions of forces and analyze whether under these conditions CoMFA will actually be able to find realistic results.

2.6.1.1 Biological Data Used for 3D QSAR Studies

As with any QSAR method, an important point is the question of whether the biological activities of all compounds studied are of comparable quality. Preferably, the biological data should have been obtained in the same laboratory under identical conditions. All compounds being tested in a system must have the same mechanism (binding mode), and all inactive compounds must be shown to be truly inactive. Only *in vitro* data should be considered, since only *in vitro* experiments are able to reach a real equilibrium; all other test systems undergo time-dependent changes by multiple coupling to parallel biochemical processes (for example membrane permeation). Another critical point is the existence of transport phenomena and diffusion gradients underlying all biological data. One has to bear in mind that CoMFA was developed to describe only one interaction step in the lifetime of ligands. In all cases where nonlinear phenomena result from drug transport and distribution, any 3D QSAR technique should be applied with caution.

The biological activities of the molecules used for a CoMFA study should ideally span a range of at least 3 orders of magnitude. For all molecules under study, the exact 3D structure has to be reported. If no information is given about the exact stereochemistry of the tested compounds (mixtures of enantiomers or diastereomers), then these compounds are not included in a CoMFA study.

2.6.1.2 Deriving the CoMFA Model

Once the biological activities have been checked and the molecules have been superimposed in their putative bioactive conformation, the CoMFA analysis continues by calculating the intermolecular interaction fields surrounding each molecule. For CoMFA this means that a lattice is constructed such that it surrounds all the molecules and the electrostatic and van der Waals energies with a chosen probe atom are calculated at the intersections of the lattice. Normally, the default lattice choice of 4 Å, beyond any dimension of any molecule in the dataset, is adequate for most of the CoMFA analyses [4]. The usual CoMFA calculation is performed using a lattice spacing of 2 Å. A controversial discussion about lattice spacing can be found in the literature [6]. Often, superior results are found using 2 Å as opposed to the more accurate 1 Å spacing. In addition, the CoMFA program provides a variety of other parameters (probe atoms, charges, energy scaling, energy cutoff, etc.) that can be adjusted by the user. This flexibility in parameter settings enables the user to fit the whole procedure as closely as possible to his problem. However, it enhances the possibility of chance correlations. Interestingly, nearly all the successful CoMFA analyses have been done with the default parameters. It is beyond the scope of this section to discuss all the results in modifying CoMFA parameters. For this purpose, the reader is referred to two articles dealing with an extensive analysis of parameter settings and their influence on the CoMFA model [6, 7].

2.6.1.3 Statistical Quality of CoMFA Models

The relationships between the biological activities and the generated interaction fields are evaluated by the special multivariate statistical technique of partial least squares (PLS). For a detailed description of the underlying mathematics of multivariate analysis, the reader is referred to the literature [8, 9]. The PLS method is able to build a statistical model even though there are more number of energy values than compounds because the various energy values are correlated with each other and many are unrelated to biological activity. These assumptions give PLS the power to extract a weak signal dispersed over many variables. Generally, no more than five or six linear combinations of the energy values are needed to build a realistic model. Since PLS operates on many variables (interaction energies), a realistic concern is that it would over-fit the data. For this reason, PLS models are validated by 'leave one out' cross-validation. This procedure involves calculating as many models as there are data points (molecules) in the data set. For each model, one of the compounds, in turn, is left out and its activity is predicted from the model without it. After each compound has been predicted once, a Q^2 value (square of the cross-validated correlation coefficient) and a standard deviation of error prediction (SDEP) value are calculated from the observed and predicted potencies of each compound (see Equations 1 and 2).

$$Q^2 = 1 - \frac{\sum \left(y_{obs} - y_{pred} \right)^2}{\sum \left(y_{obs} - y_{mean} \right)^2} \tag{1}$$

$$SDEP = \sqrt{\sum \frac{\left(y_{obs} - y_{pred} \right)^2}{N}} \tag{2}$$

where

Y_{obs} = experimental value,
Y_{pred} = predicted value,
Y_{mean} = average value, and
N = number of objects.

The SDEP value generally decreases for the first few latent variables, reaches a minimum, and then increases to indicate overfitting of the data. The number of latent variables to be used should be carefully decided upon. If adding an additional variable decreases the SDEP by less than 5%, the simpler model is preferred because it contains most of the signal in fewer variables. Using more variables results in adding more noise to the model. In CoMFA, a Q^2 value greater than 0.3 is usually considered statistically significant and acceptable [6]. However, several studies indicated that the statistical significance of CoMFA models should be carefully examined. To investigate the risk of chance correlation, normally a scrambling test is performed. In this test, the biological activities are randomly distributed among the training set molecules. On the basis of the randomly assigned activities, novel CoMFA models are generated and the Q^2 values are compared with the original model. If too many models resulting from this scrambling have a comparable quality (e.g. Q^2), then that is a strong indication of chance correlations for the original data set as well for the scrambled data set.

To investigate the risk of chance correlation, Krystek *et al.* used scrambled biological activities to test their CoMFA model for 36 endothelin subtype-A (ET_A) receptor ligands [10]. Scrambled biological activities yielded a model, using one latent variable, with a Q^2 value of 0.43 compared with a Q^2 value of 0.70 for the model with correct assigned activities. The R^2 values of randomly and correctly generated models were quite comparable, indicating that the R^2 value of a CoMFA model cannot be considered for validation. The results obtained also demonstrated that it is not possible to provide an exact Q^2-cutoff value for a chance correlation. From our experience, robust and predictive models should have Q^2 values of at least 0.5 [7, 11–13].

To overcome the problem of chance correlation, several other strategies have been suggested [6]. For example more robust cross-validation techniques are applied. In these methods, 10, 20 or 50% of the compounds were randomly selected, and a model is generated with the remaining 90, 80 or 50% of the compounds, which are then used to predict the randomly selected compounds. To get statistically reliable results, this procedure is repeated several times [12–14].

2.6.1.4 Interpreting the Results

One of the biggest advantages of 3D QSAR over classical QSAR is the graphical interpretability of the statistical results. Equation coefficients can be visualized in the region around the ligands. Upon visual inspection, regions of space contributing most to the activity can be easily recognized. The interpretation of the graphical results allows one to easily check the reliability of the models as well as design modified compounds with improved activity or selectivity. In this respect, 3D QSAR methods like CoMFA and GRID–GOLPE (graphic retrieval and information display–general optimal linear PLS estimation) have proved to be very useful.

The contours that represent 3D locations of fields with significant contribution to the model are displayed. Steric and electrostatic contributions are contoured separately and shown in different colors. The steric contours are relatively easy to interpret. Positive contoured maps show regions in space that, if occupied, increase potency; negative contours decrease potency. Interpretation of the electrostatic maps is more complicated because of the electroneutrality requirement, and also because of the fact that either positive or negative charges in electrostatics can increase potency. If a CoMFA analysis shows significant electrostatic effects, the user must examine the underlying electronic effects of corresponding functional groups to establish which is the true effect and which is an artificial correlation.

Normally, CoMFA contour maps are not considered to be equivalent to the corresponding attributes of the target protein, and such a comparison should be performed with extreme care. However, when the ligand alignment is based on the geometry of protein-bound conformations, the CoMFA steric and electrostatic coefficient contours may correspond to some extent to the steric and electrostatic environments of the binding site. For example, Oprea *et al.* [15] used inhibitor-bound enzyme X-ray structures not only to align the molecules, but also to evaluate the CoMFA results by comparing the CoMFA coefficient contour maps with the binding site structure. Several residues that are important to ligand binding were found to have corresponding steric and electrostatic CoMFA fields. However, the comparison also revealed some limitations of the model since not all key residues overlap with the CoMFA fields. Similar observations have been made in our CoMFA studies [11–13].

2.6.2
Other CoMFA-related Methods

2.6.2.1 CoMSIA
Because of the problems associated with the functional form of the Lennard-Jones potential used in most CoMFA methods [16], Klebe *et al.* [2] have developed a similar indices-based CoMFA method, which is named *CoMSIA* (comparative molecular similarity indices analysis). The method uses Gaussian-type functions instead of the traditional CoMFA potentials. Three different indices related to steric, electrostatic and hydrophobic potentials were used in their study of the classical Tripos steroid benchmark dataset. Models of comparable statistical quality with respect to internal cross-validation of the training set as well as predictabilities of the test set were derived using the CoMSIA method. The clear advantage of this method lies in the functions used to describe the studied molecules, as well as the resulting contour maps. The contour maps obtained with CoMSIA are easier to interpret compared with the ones obtained with the CoMFA approach. The CoMSIA procedure also avoids the cutoff values used in CoMFA to restrict the potential functions from assuming unacceptably large values. For a detailed description of the method as

well as its application, the reader is referred to the literature [17, 18]. Recently, the authors of CoMSIA have included a novel hydrogen-bond descriptor that could overcome the problem of underestimating hydrogen bonds in CoMFA studies [19].

2.6.2.2 GRID and GOLPE

The GRID program [20, 21] has been used by a number of authors [22, 23] as an alternative to the original CoMFA method for calculating the interaction fields. An advantage of the GRID approach, apart from the large number of chemical probes available, is the use of a 6-4 potential function, which is smoother than the 6-12 form of the Lennard-Jones type, for calculating the interaction energies at the grid lattice points. Good statistical results have been obtained, for example, in an analysis of glycogen phosphorylase b inhibitors by Cruciani *et al.* [24]. They used the GRID force field in combination with the GOLPE program [25], which performs the necessary chemometrical analysis. The particularly interesting aspect of this dataset is that the X-ray structures of all protein–ligand complexes have been solved. This allowed the authors to investigate the dataset using new and different methods to further develop 3D QSAR techniques.

A further refinement of the original CoMFA technique has been realized when introducing the concept of variable selection and reduction [3, 24]. As stated in Section 2.6.1.3, the large number of variables in the descriptor matrix (i.e. the interaction energies) represents a statistical problem of the CoMFA approach. These variables make it increasingly difficult for multivariate projection methods, such as PLS, to distinguish the useful information contained in the descriptor matrix from that of less quality or noise. Thus, approaches for selecting the useful variables from the less useful ones were needed. A statistical procedure named *GOLPE* was developed by Baroni *et al.* [3] to achieve the objective of improving the predictivity of QSAR models. In the GOLPE program, several variable selection methods, such as D-optimal design and fractional factorial design (FFD), are implemented. The predictivity of each variable is determined by generating a large number of 3D QSAR models and by calculating the SDEP. After completion of an FFD run, each variable is evaluated and classified into one of three categories: helpful for predictivity, detrimental for predictivity or uncertain. Only helpful variables are considered in the final PLS model. Applying variable selection, models with higher cross-validated Q^2-values were usually derived compared with the corresponding conventional CoMFA models [11–13, 24, 26]. For a detailed description of this method, see Refs [3, 24], and [26].

Although all CoMFA-related methods are of general use, a word of caution is necessary. There are a number of practical problems that emerge in the course of application. The results critically depend on the: chosen ligand conformation, reasonableness of the alignment, chemical parameters used to describe the interaction fields and, last but not the least, on the selected statistical evaluation method [14]. The reader should be aware of the fact that this program is a powerful tool in the hands of an experienced user, but may provide some difficulties to beginners.

2.6.2.3 Alignment-independent Methods

The most crucial and difficult step in a CoMFA-related analysis is that of aligning the studied molecules in a realistic manner. A recent development of the CoMFA method that is capable of circumventing the alignment problem has been recently described by several groups [27–29]. Silverman and Platt [27] used in their comparative molecular moment analysis (CoMMA) method descriptors that characterize shape and charge distribution such as the principal moments of inertia and properties derived from dipole and quadrupole moments, respectively. The authors investigated a number of datasets and obtained models with good consistency and predictivity. A comparable approach using the GRID force field for the generation of principal moments has been reported by Cruciani *et al.* [28]. They have integrated their methods in the commercially available programs, VOLSURF and ALMOND [30, 31]. For a detailed description of these relatively novel methods, the reader is referred to the literature [28, 29].

2.6.3
More 3D QSAR Methods

Several other 3D QSAR approaches have been developed during the last few years. Some of them are not based on properties calculated at a lattice, as performed in all CoMFA-related approaches. The GERM [32], COMPASS [33], receptor surface [34] and QUASAR [35] methods rely on properties calculated at discrete locations in the space at or near the union surface of the active ligands. The so-generated 'receptor surface' should simulate the macromolecular binding site. If all molecules of the data set bind in a manner that does not distort the residues at the binding site too much, then this can be a reasonable approach. The approach is supported by the fact that reasonable models have been obtained by all of them. However, there are two shortcomings that are associated with atomistic and receptor surface models based on averaged receptor entities: the adaptation of the shape of the binding site by means of induced fit and hydrogen-bond adaption. If the ligand–receptor interaction energy is determined with respect to an averaged receptor model, subtle effects associated with the adaptation of the receptor to the individual ligand molecules remain unaddressed. In addition, amino acid residues at a biological receptor bearing a conformationally flexible H-bond donor or acceptor can engage in differently directed H bonds with dissimilar ligand molecules. This effect also cannot be simulated with an averaged receptor.

Another way to derive a QSAR is the generation of the so-called binding site or pseudoreceptor models. The pseudoreceptor modeling approach attempts to generate a 3D model of the binding site of a structurally unknown target protein based on the superimposed structures of known ligand molecules in their bioactive conformation together with the experimentally determined binding affinities. The philosophy underpinning the 'pseudoreceptor concept' is to engage the bound species in sufficient, specific noncovalent binding so as to mimic the essential ligand–macromolecule interactions at the true biological receptor (see, for example, Refs [36–39]). In the first step of a pseudoreceptor generation, potential binding sites (anchor points) are identified for each of the molecules of the pharmacophore. Suitable binding partners (e.g. amino acids, metal ions, water molecules) are then selected and positioned in 3D space. After refinement, the ensemble of binding partners constitutes a pseudoreceptor for the ligand molecules that were used to generate it. In general, type and arrangement of the pseudoreceptor building blocks surrounding the pharmacophore model will bear no structural resemblance to the real biological target protein. Instead of reproducing the complex structure of the ligand-binding protein, the receptor surrogate should be regarded as a purely hypothetical model of the binding pocket, accommodating a series of structurally related ligands. The estimation of binding affinities relies on the evaluation of ligand–pseudoreceptor interaction energies, ligand desolvation energies and changes in ligand internal energy and entropy upon receptor binding. The pseudoreceptor concept, integrated in the PrGen program [40], has been validated by constructing receptor surrogates for the enzyme human carbonic anhydrase, the dopaminergic and the β_2-adrenergic receptor. Predicted differences in free energy of ligand binding toward the pseudoreceptor and experimental validation determined toward the biological receptor agree within 1.2 kcal mol^{-1} [41]. The advantages of the pseudoreceptor concept are the use of a directional force field that is capable of correctly treating hydrogen bonds and ligand–metal ion–protein interactions that are frequently found to be of prime importance for the binding of drug molecules and consideration of solvation and entropic processes, which are often missing in 3D QSAR approaches [42].

2.6.4
Receptor-based 3D QSAR

Structure-based methods, such as docking, are nowadays able to calculate the position and orientation of a potential ligand in a receptor-binding site with fair accuracy. Various docking approaches and programs are available (see Chapter 5), which have successfully applied for drug design (examples can be found in Refs [43–47]). The docking methods yield important information concerning the spatial orientation of the ligands in the binding site and also toward other ligands binding to the same target. The major problem of today's docking programs is the inability to evaluate binding free energies correctly in order to use the scores for the prediction of biological activities. The problem in predicting affinity has generated considerable interest in developing methods to calculate ligand affinity reliably for a widely diverse series of molecules binding to the same target protein of known structure [48–52]. For the calculation of free energies of binding, most approaches rely on molecular mechanics force fields that represent van der Waals and Coulombic interactions on the basis of empirical potentials. Other approaches use simpler scoring functions rather than calculating the affinity by molecular mechanics equations (for a detailed review, see Ref. [52]). These methods commonly use available experimental data to obtain parameters for some relatively simple functions that allow quick estimation of the binding energy. The estimated binding energies or scores are widely used to discriminate between active and inactive ligands, for example in virtual database screening, but are mostly not accurate enough to correctly predict biological activities. The main problem in the prediction of binding data is that the underlying molecular interactions are highly complex and various terms should be taken into account to quantify the free energy of the interaction process. Only rigorous methods, such as, free energy perturbation or thermodynamic integration, are able to predict the binding affinity correctly. While these methods clearly have the potential of providing accurate evaluation of relative binding free energies, they are very expensive in a computational sense (for more information, see Refs. [53] and [54]).

Considering the strengths of both approaches, the docking programs using protein information and the 3D QSAR methods to develop predictive models for related molecules, they have been combined in an automated unbiased procedure called as *receptor-based* 3D QSAR [11–13, 55–67]. In this context, the 3D structure of a target protein along with a docking protocol is used to guide alignment selection for CoMFA [11]. This approach has often led to predictive and meaningful models. Apart from the good predictive ability, the models derived are also able to point out the interaction sites in the binding pocket that might be responsible for the variance in biological activities. A collection of successful case studies that applied the receptor-based 3D QSAR can be found in Ref. [68].

Another interesting method that might overcome the problem of neglecting the protein information in a 3D QSAR analysis has been recently developed by Gohlke and Klebe [69]. In the approach named *AFMoC* (adaptation of fields for molecular comparison), potential fields derived from the scoring function DrugScore are generated in the binding pocket of a target protein. Methodologically, the program is related to CoMFA and CoMSIA, but it has the advantage of including the protein environment to the 3D QSAR analysis. Instead of using only the Coulomb or Lennard-Jones potential, AFMoC starts with a grid of preassigned values. The numbers at the individual grid points consider the DrugScore potential values. By using ligands with known binding mode and biological data, the ligand atoms that are actually placed introduce an activity-based weighting of the individual DrugScore potential values. The resulting interaction fields are then evaluated by classical PLS. It has been shown that AFMoC-derived QSAR models achieve much better correlation between experimentally derived and computed activities compared with the original scoring function DrugScore [70].

2.6.5
Reliability of 3D QSAR Models

The quality and reliability of any 3D QSAR model is strongly dependent on the careful examination of each step within a 3D QSAR analysis. As with any QSAR method, an important point is the question of whether the biological activities of all compounds studied are of comparable quality (as we already discussed in Section 2.6.1.1).

The search for the bioactive conformation and the molecular alignment constitutes a serious problem within all 3D QSAR studies. It is one of the most important sources of wrong conclusions and errors in all 3D QSAR analysis. The risk of deriving irrelevant geometries has been reduced by considering rigid analogs. Even then, the alignment poses problems because there are some cases of different binding modes of seemingly closely related compounds. Also, if the binding modes are comparable and only the conformations of the ligands are wrong, in the same manner, the results of a 3D QSAR analysis will nevertheless be reliable. The problems in generation of the conformations and the correct alignment could be avoided by deriving them from the 3D structures of ligand–protein complexes, which are known from X-ray crystallography, NMR or homology modeling (as described in Section 2.6.4).

The final stage of a 3D-QSAR analysis is the statistical validation, when the significance of the model, and hence its ability to predict biological activities of novel compounds, is established. In most of the 3D-QSAR case studies published in the literature, the leave-one-out (LOO) cross-validation procedure has been used for this purpose. The outputs of this procedure are the cross-validated q^2 and the SDEP, which are commonly regarded as an ultimate criterion of both robustness and predictive ability of a model. The simplest cross-validation method is LOO, where one object at time is removed and predicted. A more robust and reliable method is the leave-several-out cross-validation. For example, in the leave-20%-out cross-validation, five groups of approximately the same size are generated. Thus, 80% of the compounds were randomly selected for model generation, which were then used to predict the remaining compounds. This operation must be repeated a large number of times to obtain reliable statistical results. The leave-20%-out or also the more demanding leave-50%-out cross-validation results are much better indicators for the robustness and the predictive ability of a 3D QSAR model than the usually used LOO procedure [7, 12, 46]. LOO often yields too optimistic models, which fail when predicting real test set molecules.

Despite the known limitations of the LOO procedure, it is still uncommon to test 3D QSAR models for their ability to correctly predict the biological activities of compounds not included in the training set. Still, many authors claim that their models showing high LOO Q^2 values have high predictive ability in the absence of external validation (for a detailed discussion on this problem, see [14, 71–76]). In contrast to such expectations, it has been shown by several studies that a correlation between the LOO cross-validated Q^2 value for the training set and the correlation coefficient R^2 between the predicted and observed activities for the test set does not exist [72, 75]. Therefore, it is highly recommended to use demanding cross-validation procedures and external test sets to further validate an established 3D-QSAR model.

References

1. Cramer, R.D., Patterson, D.E., and Bunce, J.D. (1988) Comparative molecular-field analysis (COMFA). 1. Effect of shape on binding of steroids to carrier proteins. *Journal of the American Chemical Society*, **110**, 5959–67.

2. Klebe, G., Abraham, U., and Mietzner, T. (1994) Molecular similarity indexes in a comparative-analysis (comsia) of drug molecules to correlate and predict their biological-activity. *Journal of Medicinal Chemistry*, **37**, 4130–46.

3. Baroni, M., Costantino, G., Cruciani, G. *et al.* (1993) Generating optimal linear PLS estimations (GOLPE) – an advanced chemometric tool for handling 3D-QSAR problems. *Quantitative Structure-Activity Relationships*, **12**, 9–20.

4. Martin, Y.C. (1998) 3D QSAR: current state, scope, and limitations. *Perspectives in Drug Discovery and Design*, **12**, 3–23.

5. Klebe, G. and Abraham, U. (1993) On the prediction of binding-properties of drug molecules by comparative molecular-field analysis. *Journal of Medicinal Chemistry*, **36**, 70–80.

6. Kim, K.H., Greco, G., and Novellino, E. (1998) A critical review of recent CoMFA applications. *Perspectives in Drug Discovery and Design*, **12**, 257–315.

7. Folkers, G., Merz, A., and Rognan, D. (1993) CoMFA: scope and limitations, in *3D QSAR in Drug Design: Theory, Methods and Applications* (ed. H. Kubinyi), ESCOM, Leiden, pp. 583–618.

8. Wold, S. (1991) Validation of QSARs. *Quantitative Structure-Activity Relationships*, **10**, 191–93.

9. Wold, S., Johansson, E., and Cocchi, M. (1993) PLS – partial least squares projections to latent structures, in *3D QSAR in Drug Design: Theory, Methods and Applications* (ed. H. Kubinyi), ESCOM, Leiden, pp. 523–50.

10. Krystek, S.R., Hunt, J.T., Stein, P.D., and Stouch, T.R. (1995) 3-dimensional quantitative structure-activity-relationships of sulfonamide endothelin inhibitors. *Journal of Medicinal Chemistry*, **38**, 659–68.

11. Sippl, W. (2000) Receptor-based 3D QSAR analysis of estrogen receptor ligands – merging the accuracy of receptor-based alignments with the computational efficiency of ligand-based methods. *Journal of Computer-Aided Molecular Design*, **14**, 559–72.

12. Sippl, W., Contreras, J.M., Parrot, I. *et al.* (2001) Structure-based 3D QSAR and design of novel acetylcholinesterase inhibitors. *Journal*

of Computer-Aided Molecular Design,
15, 395–410.

13. Sippl, W. (2002) Binding affinity
prediction of novel estrogen receptor
ligands using receptor-based 3-D
QSAR methods. *Bioorganic &
Medicinal Chemistry*, **10**, 3741–55.

14. Oprea, T.I. and Garcia, A.E. (1996)
Three-dimensional quantitative
structure-activity relationships of
steroid aromatase inhibitors. *Journal of
Computer-Aided Molecular Design*, **10**,
186–200.

15. Oprea, T.I., Waller, C.L., and Marshall,
G.R. (1994) 3-dimensional quantitative
structure-activity relationship of
human-immunodeficiency-
virus-(i) protease inhibitors. 2.
Predictive power using limited
exploration of alternate binding
modes. *Journal of Medicinal Chemistry*,
37, 2206–15.

16. Norinder, U. (1998) Recent progress
in CoMFA methodology and related
techniques. *Perspectives in Drug
Discovery and Design*, **12**, 25–39.

17. Klebe, G. and Abraham, U. (1999)
Comparative molecular similarity
index analysis (CoMSIA) to study
hydrogen-bonding properties and to
score combinatorial libraries. *Journal
of Computer-Aided Molecular Design*,
13, 1–10.

18. Böhm, M., Stürzebecher, J., and
Klebe, G. (1999) Three-dimensional
quantitative structure-activity
relationship analyses using
comparative molecular field analysis
and comparative molecular similarity
indices analysis to elucidate selectivity
differences of inhibitors binding to
trypsin, thrombin, and factor Xa.
Journal of Medicinal Chemistry, **42**,
458–77.

19. Böhm, M. and Klebe, G. (2002)
Development of new hydrogen-bond
descriptors and their application to
comparative molecular field analyses.
Journal of Medicinal Chemistry, **45**,
1585–97.

20. Goodford, P.J. (1985) A
computational-procedure for
determining energetically favorable
binding-sites on biologically important

macromolecules. *Journal of Medicinal
Chemistry*, **28**, 849–57.

21. Wade, R.C., Clark, K.J., and Goodford,
P.J. (1993) Further development of
hydrogen-bond functions for use in
determining energetically favorable
binding-sites on molecules of known
structure. 1. Ligand probe groups with
the ability to form 2 hydrogen-bonds.
Journal of Medicinal Chemistry, **36**,
140–47.

22. Davis, A.M., Gensmantel, N.P.,
Johansson, E., and Marriott, D.P.
(1994) The use of the grid program in
the 3-D QSAR analysis of a series of
calcium-channel agonists. *Journal of
Medicinal Chemistry*, **37**, 963–72.

23. Kim, K.H., Greco, G., Novellino, E.
et al. (1993) Use of the hydrogen-bond
potential function in a comparative
molecular-field analysis (COMFA) on
a set of benzodiazepines. *Journal of
Computer-Aided Molecular Design*, **7**,
263–80.

24. Cruciani, G. and Watson, K.A. (1994)
Comparative molecular-field analysis
using grid force-field and golpe
variable selection methods in a study
of inhibitors of
glycogen-phosphorylase-B. *Journal of
Medicinal Chemistry*, **37**, 2589–601.

25. GOLPE, Multivariate Infometric
Analysis, Perugia.
http://www.miasrl.com.

26. Cruciani, G., Clementi, S., and
Pastor, M. (1998) GOLPE-guided
region selection. *Perspectives in Drug
Discovery and Design*, **12**, 71–86.

27. Silverman, B.D. and Platt, D.E. (1996)
Comparative molecular moment
analysis (CoMMA): 3D-QSAR without
molecular superposition. *Journal of
Medicinal Chemistry*, **39**, 2129–40.

28. Cruciani, G., Crivori, P., Carupt, P.A.,
and Testa, B. (2000) Molecular fields
in quantitative structure-permeation
relationships: the VolSurf approach.
Theochem, **503**, 17–30.

29. Pastor, M., Cruciani, G., McLay, I.
et al. (2000) GRid-INdependent
descriptors (GRIND): a novel class of
alignment-independent
three-dimensional molecular

descriptors. *Journal of Medicinal Chemistry*, **43**, 3233–43.

30. VOLSURF, Molecular Discover Ltd., Oxford, http://www.moldiscovery.com.

31. ALMOND, Multivariate Infometric Analysis, Perugia. http://www.miasrl.com.

32. Walters, D.E. (1998) Genetically evolved receptor models (GERM) as a 3D QSAR tool. *Perspectives in Drug Discovery and Design*, **12**, 159–66.

33. Jain, A.N., Koile, K., and Chapman, D. (1994) Compass – predicting biological-activities from molecular-surface properties – performance comparisons on a steroid benchmark. *Journal of Medicinal Chemistry*, **37**, 2315–27.

34. Hahn, M. and Rogers, D. (1998) Receptor surface models. *Perspectives in Drug Discovery and Design*, **12**, 117–33.

35. Vedani, A. and Zbinden, P. (1998) Quasi-atomistic receptor modeling: a bridge between 3D QSAR and receptor fitting. *Pharmaceutica Acta Helvetiae*, **73**, 11–18.

36. Sippl, W., Stark, H., and Höltje, H.-D. (1998) Development of a binding site model for histamine H-3-receptor agonists. *Pharmazie*, **53**, 433–37.

37. Höltje, H.-D. and Jendretzki, U.K. (1995) Construction of a detailed serotoninergic 5-HT2A receptor model. *Archiv der Pharmazie*, **328**, 577–84.

38. Greenidge, P.A., Merz, A., and Folkers, G. (1995) A pseudoreceptor modelling study of the varicella-zoster virus and human thymidine kinase binding sites. *Journal of Computer-Aided Molecular Design*, **9**, 473–78.

39. Schmetzer, S., Greenidge, P.A., Kovar, K.A. *et al.* (1997) Structure-activity relationships of cannabinoids: a joint CoMFA and pseudoreceptor modelling study. *Journal of Computer-Aided Molecular Design*, **11**, 278–92.

40. PrGen, Biographics Laboratory, Basel. http://www.biograf.ch.

41. Vedani, A., Zbinden, P., Snyder, J.P., and Greenidge, P.A. (1995) Pseudoreceptor modeling – the construction of 3-dimensional receptor surrogates. *Journal of the American Chemical Society*, **117**, 4987–94.

42. Schleifer, K.J. (2006) In *Concepts and Applications of Pseudoreceptors, Series: Methods and Principles in Medicinal Chemistry – Pharmacophores and Pharmacophore Concepts* (eds T. Langer, R. Hoffmann, H. Kubinyi *et al.*), VCH Publishers, New York, pp. 117–30.

43. Kramer, B., Rarey, M., and Lengauer, T. (1997) CASP2 experiences with docking flexible ligands using FLEXX. *Proteins*, **28** (Suppl. 1), 221–25.

44. Böhm, H.J. (1994) The development of a simple empirical scoring function to estimate the binding constant for a protein-ligand complex of known three-dimensional structure. *Journal of Computer-Aided Molecular Design*, **8**, 243–56.

45. Verdonk, M.L., Cole, J.C., Hartshorn, M.J. *et al.* (2003) Improved protein-ligand docking using GOLD. *Proteins*, **52**, 609–23.

46. Meng, E.C., Shoichet, B.K., and Kuntz., I.D. (1992) Automated docking with grid-based energy evaluation. *Journal of Computational Chemistry*, **13**, 505–24.

47. Kontoyianni, M., McClellan, L.M., and Sokol, G.S. (2004) Evaluation of docking performance: comparative data on docking algorithms. *Journal of Medicinal Chemistry*, **47**, 558–65.

48. Tame, J.R.H. (1999) Scoring functions: a view from the bench. *Journal of Computer-Aided Molecular Design*, **13**, 99–108.

49. Böhm, H.J. (1998) Prediction of binding constants of protein ligands: a fast method for the prioritization of hits obtained from de novo design or 3D database search programs. *Journal of Computer-Aided Molecular Design*, **12**, 309–23.

50. Wang, R., Lu, Y., Fang, X., and Wang, S. (2004) An extensive test of 14 scoring functions using the PDBbind

refined set of 800 Protein-Ligand complexes. *Journal of Chemical Information and Computer Sciences*, **44**, 2114–25.

51. Perola, E., Walters, W.P., and Charifson, P.S. (2004) A detailed comparison of current docking and scoring methods on systems of pharmaceutical relevance. *Proteins*, **56**, 235–49.

52. Gohlke, H. and Klebe, G. (2002) Approaches to the description and prediction of the binding affinity of small molecule ligands to macromolecular receptors. *Angewandte Chemie International Edition*, **41**, 2644–76.

53. Masukawa, K.M., Kollman, P.A., and Kuntz, I.D. (2003) Investigation of neuraminidase-substrate recognition using molecular dynamics and free energy calculations. *Journal of Medicinal Chemistry*, **46**, 5628–37.

54. Huang, D. and Caflisch, A. (2004) Efficient evaluation of binding free energy using continuum electrostatics solvation. *Journal of Medicinal Chemistry*, **47**, 5791–97.

55. Sippl, W., Contreras, J.M., Rival, Y., and Wermuth, C.G. (2000) In *Molecular Modelling and Predicting of Bioactivity* (eds K. Gundertofte and F.S. Jorgensen), Plenum Press, New York, pp. 53–58.

56. Sippl, W. (2002) Development of biologically active compounds by combining 3D QSAR and structure-based design methods. *Journal of Computer-Aided Molecular Design*, **16**, 825–30.

57. Cinone, N., Höltje, H.-D., and Carotti, A. (2000) Development of a unique 3D interaction model of endogenous and synthetic peripheral benzodiazepine receptor ligands. *Journal of Computer-Aided Molecular Design*, **14**, 753–68.

58. Hammer, S., Spika, I., Sippl, W. *et al.* (2003) Glucocorticoid receptor interactions with glucocorticoids: evaluation by molecular modeling and functional analysis of glucocorticoid receptor mutants. *Steroids*, **68**, 329–39.

59. Pastor, M., Cruciani, G., and Watson, K.A. (1997) A strategy for the incorporation of water molecules present in a ligand binding site into a three-dimensional quantitative structure-activity relationship analysis. *Journal of Medicinal Chemistry*, **40**, 4089–102.

60. Tervo, A.J., Nyrönen, T.H., Rönkkö, T., and Poso, A. (2003) A structure-activity relationship study of catechol-O-methyltransferase inhibitors combining molecular docking and 3D QSAR methods. *Journal of Computer-Aided Molecular Design*, **17**, 797–810.

61. Pandey, G. and Saxena, K.A. (2006) 3D QSAR studies on protein phosphatase 1B inhibitors: Comparison of the quality and predictivity among 3D QSAR models obtained from different conformer-based alignments. *Journal of Chemical Information and Modeling*, **46**, 2579–90.

62. Waller, C.L., Oprea, T.I., Giolitti, A., and Marshall, G.R. (1993) 3-dimensional QSAR of human-immunodeficiency-virus-(I) protease inhibitors. 1. A COMFA study employing experimentally-determined alignment rules. *Journal of Medicinal Chemistry*, **36**, 4152–60.

63. De Priest, S.A., Mayer, D., Naylor, C.B., and Marshall, G.R. (1993) 3D-QSAR of angiotensin-converting enzyme and thermolysin inhibitors – a comparison of COMFA models based on deduced and experimentally determined active-site geometries. *Journal of the American Chemical Society*, **115**, 5372–84.

64. Cho, S.J., Garsia, M.L.S., Bier, J., and Tropsha, A. (1996) Structure-based alignment and comparative molecular field analysis of acetylcholinesterase inhibitors. *Journal of Medicinal Chemistry*, **39**, 5064–71.

65. Vaz, R.J., McLean, L.R., and Pelton, J.T. (1998) Evaluation of proposed modes of binding of (2S)-2-4-(3S)-1-acetimidoyl-3-pyrrolidinyloxyphenyl-3-(7-amidino-2-naphthyl)propanoic acid hydrochloride and some analogs to Factor Xa using a comparative

molecular field analysis. *Journal of Computer-Aided Molecular Design*, **12**, 99–110.

66. Ortiz, A.R., Pisabarro, M.T., Gago, F., and Wade, R.C. (1995) Prediction of drug-binding affinities by comparative binding-energy analysis. *Journal of Medicinal Chemistry*, **38**, 2681–91.

67. Lozano, J.J., Pastor, M., Cruciani, G. *et al.* (2000) 3D-QSAR methods on the basis of ligand-receptor complexes. Application of COMBINE and GRID/GOLPE methodologies to a series of CYP1A2 ligands. *Journal of Computer-Aided Molecular Design*, **14**, 341–53.

68. Sippl, W. (2006) In *Application of Structure-based Alignment Methods for 3D-QSAR, Series: Methods and Principles in Medicinal Chemistry – Pharmacophores and Pharmacophore Concepts* (eds T. Langer, R. Hoffmann, H. Kubinyi *et al.*), VCH Publishers, New York, pp. 223–49.

69. Gohlke, H. and Klebe, G. (2002) DrugScore meets CoMFA: adaptation of fields for molecular comparison (AFMoC) or how to tailor knowledge-based pair-potentials to a particular protein. *Journal of Medicinal Chemistry*, **45**, 4153–70.

70. Silber, K., Kurz, T., Heidler, P., and Klebe, G. (2005) AFMoC enhances predictivity of 3D QSAR: a case study

with DOXP-reductoisomerase. *Journal of Medicinal Chemistry*, **48**, 3547–63.

71. Golbraikh, A. and Tropsha, A. (2002) Beware of q(2)! *Journal of Molecular Graphics & Modelling*, **20**, 269–76.

72. Kubinyi, H., Hamprecht, F.A., and Mietzner, T. (1998) Three-dimensional quantitative similarity-activity relationships (3D QSiAR) from SEAL similarity matrices. *Journal of Medicinal Chemistry*, **41**, 2553–64.

73. Golbraikh, A., Shen, M., Xiao, Z.Y. *et al.* (2003) Rational selection of training and test sets for the development of validated QSAR models. *Journal of Computer-Aided Molecular Design*, **17**, 241–53.

74. Norinder, U. (1996) Single and domain mode variable selection in 3D QSAR applications. *Journal of Chemometrics*, **10**, 95–105.

75. Doweyko, A.M. (2004) 3D-QSAR illusions. *Journal of Computer-Aided Molecular Design*, **18**, 587–96.

Further Reading

Vedani, A., Zbinden, P., and Snyder, J.P. (1993) Pseudo-receptor modeling – a new concept for the 3-dimensional construction of receptor-binding sites. *Journal of Receptor Research*, **13**, 163–77.

3

A Case Study for Small Molecule Modeling: Dopamine D$_3$ Receptor Antagonists

In this chapter we describe the determination of a pharmacophore model and a subsequent three-dimensional quantitative structure-activity relationship (3D QSAR) analysis (Section 2.6.2.2) for dopamine D$_3$ receptor antagonists. We used the steric and electrostatic information coming from partially rigid, high affine ligands to build up the pharmacophore model. After defining the pharmacophoric features, the model was validated by closer examination of molecular fields that were produced by the ligands superimposed onto each other in their pharmacophoric conformations. In the last step, the molecular interaction fields were used to establish a 3D QSAR. For this purpose, the fields that were produced by the program GRID were correlated with the binding affinities by means of a Partial Least Square (PLS) model. This PLS model was validated using different cross-validation techniques and the predictive power was tested by an external test set of ligands.

For building the pharmacophore model, sterically restricted dopamine D$_3$ receptor antagonists from our project [1] and antagonists from literature were taken [2–4]. Forty ligands were used for the subsequent GRID–GOLPE analysis (Tables 3.1–3.3). These ligands exclusively stemmed from our coworker's laboratory (Prof H. Stark, University of Frankfurt/Main, Germany) to ensure consistency in the binding data.

3.1
A Pharmacophore Model for Dopamine D$_3$ Receptor Antagonists

Five ligands that bind as antagonists to the dopamine D$_3$ receptor have been examined in detail to define their bioactive conformation (Table 3.4). The molecular structure of the antagonists that are included in this analysis can be regarded as a composition of three fragments. They consist of an aromatic–basic element, an aromatic–amidic portion and an aromatic or alkyl spacer. Figure 3.1 shows the decomposition of substance BP897 [2] into the mentioned fragments. Variations of this structure are shown together with their binding affinities in Table 3.1.

Molecular Modeling. Basic Principles and Applications. 3rd Edition
H.-D. Höltje, W. Sippl, D. Rognan, and G. Folkers
Copyright © 2008 Wiley-VCH Verlag GmbH & Co. KGaA, Weinheim
ISBN: 978-3-527-31568-0

Table 3.1 Dopamine D3 antagonists with variations in their amidic portions and the length of the spacer.

$R1-(CH_2)_n$

Compound	R1	R2	n	D3 pKi	Compound	R1	R2	n	D3 pKi
ST-63		H	4	7.59	ST-64		H	4	7.37
ST-65		H	4	8.00	ST-66		OCH3	3	6.67
ST-67		OCH3	3	6.52	ST-68		OCH3	3	6.40
ST-69		OCH3	4	8.11	ST-70		OCH3	4	8.04

Compound	Substituent		pK_i
ST-71	OCH$_3$	4	8.41
ST-82	OCH$_3$	4	8.55
ST-84	OCH$_3$	4	7.42
ST-85	OCH$_3$	4	7.63
ST-86	OCH$_3$	4	7.54
ST-88	OCH$_3$	3	6.25
ST-92	OCH$_3$	4	7.49
ST-93	OCH$_3$	3	6.97
ST-95	OCH$_3$	4	8.60
ST-96	OCH$_3$	4	9.00

(continued overleaf)

Table 3.1 (continued).

Compound	R1	R2	n	D$_3$ pK$_i$	Compound	R1	R2	n	D$_3$ pK$_i$
ST-98		OCH$_3$	4	8.18	ST-99		OCH$_3$	4	8.83
ST-100		OCH$_3$	4	8.02	ST-101		OCH$_3$	4	8.53
ST-144		OCH$_3$	4	8.74	ST-150		OCH$_3$	4	7.38
ST-152		OCH$_3$	4	8.20	ST-167		OCH$_3$	4	9.30
ST-168		OCH$_3$	4	9.21	ST-188		OCH$_3$	4	9.16
ST-189		OCH$_3$	4	8.67	ST-317		OCH$_3$	4	7.99

Table 3.2 Dopamine D$_3$ antagonists with variations in their spacer.

Compound	X	D$_3$ pK$_i$	Compound	X	D$_3$ pK$_i$
ST-81		7.40	ST-176		7.00
ST-177		6.12	ST-205		7.43

Table 3.3 Dopamine D$_3$ antagonists with aminotetralines in their aromatic-basic portion.

Compound	R	D$_3$ pK$_i$	Compound	R	D$_3$ pK$_i$
ST-124		6.63	ST-125		7.32
ST-126		7.55	ST-127		7.22
ST-185		7.42			

Most of the ligands described as dopamine D$_3$ antagonists fit into this scheme. However, there are some ligands like compound 1 (Table 3.4) that have their amidic portions replaced by other moieties that are able to accept

Table 3.4 Dopamine D_3 antagonists used for the definition of the pharmacophore.

Compound	Structure	D_3 pK_i
1 [3]		7.59
2 [4]		9.00
ST-205		7.43
ST-84		7.42
ST-85		7.63

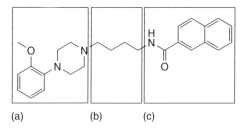

(a)   (b)   (c)

Figure 3.1 BP 897 decomposed in three fragments. (a) The aromatic–basic fragment, (b) the spacer and (c) the aromatic–amidic fragment.

hydrogen bonds. Therefore, all ligands can roughly be described as represented in Figure 3.2. As all ligands examined contain the same pharmacophoric descriptors (e.g. basic nitrogen, aromatic moieties, H-bond acceptors), we assume that they bind in a very similar mode to the same binding pocket of

Figure 3.2 Blueprint for dopamine D$_3$ receptor antagonists.

the dopamine D$_3$ receptor. To detect the bioactive conformation they adopt in the binding pocket, we concentrated on the analysis of the conformational space of those ligands that show rigidization in parts on their structures.

Since none of the ligands examined is completely rigidized, but some are partially rigidized, those molecules were, in the first step, decomposed into three fragments and the conformational restricted fragments were then examined individually. The fragments chosen were overlapping to some extent in order to determine the bioactive conformation of their linking parts. After the putative bioactive conformations for the fragments were defined, the structures were recomposed to yield the putative binding conformations.

3.1.1
The Aromatic–Basic Fragment

Most of the structures in our dataset (Tables 3.1–3.4) exhibit an *N*-(4-(2-methoxyphenyl)piperazine-1-yl) moiety, which is quite flexible, since various energetically favorable conformations can be adopted by the ring system. In contrast, ligand compound 2 is rigidized, instead, in that part of the structure because it contains a ring system without any rotatable bonds instead.

Therefore, in the first step, the conformational space of the octahydrobenzo-quinoline ring system of compound 2 was examined in detail. For this purpose, a simulated annealing (see Section 2.3.3) was accomplished. The 4-methyl-1,2,3,4,4a,5,6,10b-octahydrobenzo[f]quinoline-7-ol fragment was heated up 10 times to 2000 K and was subsequently annealed to 0 K. The low temperature conformations were collected and inspected visually. Two clusters of very similar low energy conformations that exhibit the possible conformations of this ring system were found. Representatives of these clusters are shown in Figure 3.3.

The 1-(2-methoxyphenyl)-4-methylpiperazine ring system was fitted onto both structures shown in Figure 3.4 using the FlexS [5] program (see Section 2.5.2). FlexS not only includes the steric and electronic demands of the fragments in the superposition procedure, but also proposes virtual interacting points that may serve as counterions or hydrogen bond partners. In Figure 3.4 the superimposition of the phenylpiperazine ring system onto both conformations of the tricyclic system of compound 2 is depicted. The

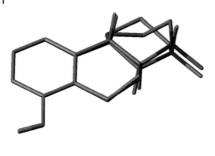

Figure 3.3 Representative structures of the octahydrobenzoquinoline ring system.

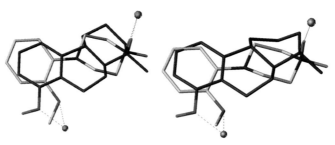

Figure 3.4 Superimposition of 1-(2-methoxyphenyl)-4-methylpiperazine (white carbon atoms) onto both low energy conformers of 4-methyl-1,2,3,4,4a,5,6,10b-octahydrobenzo[f]quinoline-7-ol (gray carbon atoms) by FLEXS. Virtual interacting points are shown as orange balls.

conformation of the phenylpiperazine is the same in both cases. As it can be precisely seen, there is a possibility for the different fragments to interact with the same putative hydrogen bond donors and salt-bridge partners.

3.1.2
The Spacer

The fragments representing the spacer are the most flexible parts of the ligands in the dataset. Therefore it is difficult to decide which conformation of this part is the one that binds to the receptor. Fortunately, the corresponding spacer fragments of compound 1 and ST-205 (Figure 3.5) are, at least, rigidized to some extent and were therefore examined in detail.

In the first step, the conformational space of the ST-205 spacer was determined. The bicyclic ring system was subjected to a simulated annealing, which led to three different conformations that are displayed in Figure 3.6.

The ring system was then extended to include parts of the neighboring fragments, and the rotatable bonds of the extended fragment were examined by a systematic search (Figure 3.7).

The simulated annealing procedure of the bicyclic ring system in combination with the systematic search led to 992 possible conformations of

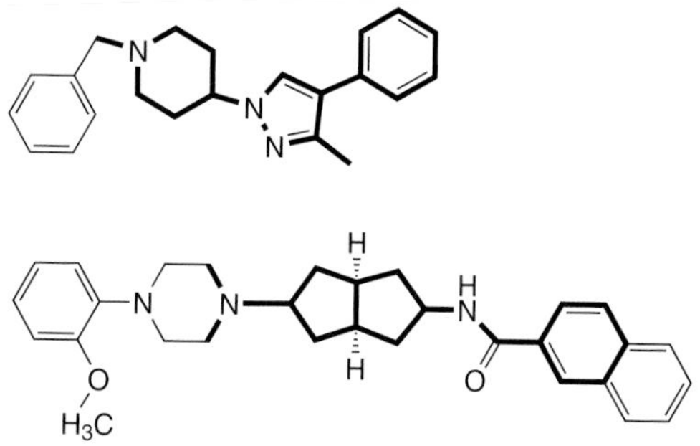

Figure 3.5 Compounds 1 and ST-205. Fragments that were examined in detail to define the conformation of the spacer are displayed in bold.

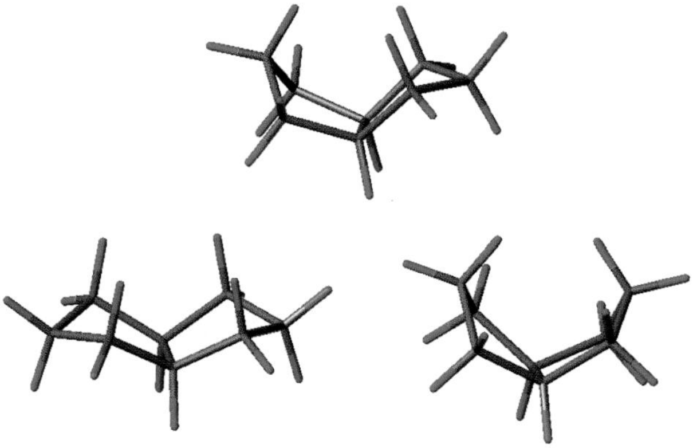

Figure 3.6 Fragment of ST-205. Three conformations are possible for the ring system.

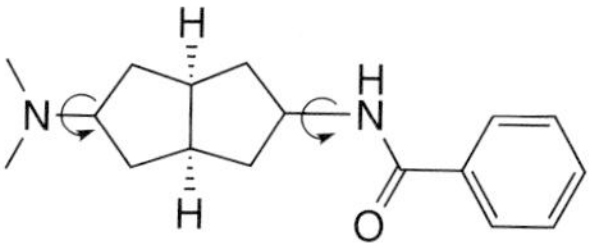

Figure 3.7 Extended spacer of ST-205. The indicated bonds are rotated in steps of 10°.

the extended spacer fragment of ST-205. The analogous fragment of compound 1 was then flexibly fitted onto each of the 992 possible conformations. Each superimposition was rated by a scoring function implemented in FlexS. The superimposition that was rated the best was assumed to be the binding conformation of these fragments. This conformation is shown in Figure 3.8.

3.1.3
The Aromatic–Amidic Residue

In several of the compounds that bind with high affinity to the dopamine D$_3$ receptor, the so-called 'aromatic–amidic fragment' is represented by a rigid phthalimide moiety. Therefore it is easy to determine the pharmacophoric

Figure 3.8 Fragments of compound 1 and ST-205. The superimposition that was rated best by FLEXS is shown.

conformation for this part of the ligands. The planar phthalimide systems present in ST-84 and ST-85 were taken to determine the pharmacophoric conformation of the 'aromatic–amidic' part of all antagonists.

3.1.4
Resulting Pharmacophore

After determination of the preferred conformation of all fragments, the ligands were reassembled. The resulting pharmacophore is presented in Figure 3.9 together with the relevant distances between the pharmacophoric features. Figure 3.10 shows four ligands superimposed onto each other in their putative bioactive conformation. The ligands adopt a stretched conformation. The basic nitrogen atoms, the H-bond acceptors and the lipophilic residues occupy distinct areas. The pharmacophoric points that can form directional interactions with the receptor, the H-bond acceptors and the basic nitrogen atoms are about 6.5 Å apart from each other. The aromatic region at the aromatic–amidic part of the ligand can be considerably elongated.

3.1.5
Molecular Interaction Fields

As mentioned in Section 2.5.3, molecules recognize each other by characteristic properties that can increase their van der Waals volume. Therefore pharmacophoric superimpositions should not only be estimated by the consideration of superimposed atomic skeletons, but the interaction fields generated by them should also be taken into account.

Figure 3.11 shows four high affinity ligands in their pharmacophoric conformations superimposed onto each other. Molecular interaction fields for each ligand are calculated by program GRID [6] using different probes. According to their situation, the basic nitrogen atoms of the ligands are protonated for this analysis under physiological conditions. As can be seen in Figure 3.11b–d, the protonated nitrogen atoms of all ligands interact in a similar way favorably with the 'ionized alkyl carboxyl group' probe (Figure 3.11c), and they also interact

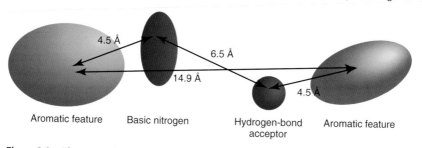

4.5 Å   6.5 Å   14.9 Å   4.5 Å

Aromatic feature   Basic nitrogen   Hydrogen-bond   Aromatic feature
acceptor

Figure 3.9   Pharmacophore model for dopamine D₃ receptor antagonists.

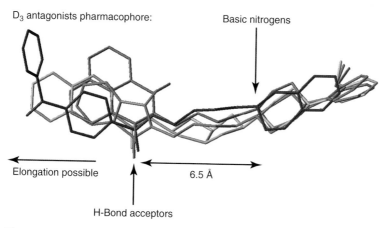

D₃ antagonists pharmacophore: Basic nitrogens

Elongation possible   6.5 Å

H-Bond acceptors

Figure 3.10   Superimposition of four ligands in their pharmacophoric conformation. ST-127 (blue), ST-205 (yellow), ST-84 (cyan) and ST-86 (red).

with the 'sp² NH with lone pair' probe (Figure 3.11b) and show favorable lipophilic interactions with the 'sp² CH' probe (Figure 3.11d). In all cases, the corresponding fields occupy similar regions. The described interactions seem to be crucial for the binding with the receptor. Some other favorable ligand–probe interactions are only formed by some of the members of the series. This behavior is demonstrated in Figure 3.11a, which shows the inter- action of ligands with the hydrogen-bond-accepting 'carbonyl oxygen' probe. Only ligands ST-127 and ST-205 are able to donate hydrogen bonds in their amidic portion and therefore produce favorable interaction energies in this area. However, since all ligands displayed in Figure 3.11 bind with high affinity to the receptor, this ability seems to be of less importance. The information obtained from the GRID fields can be used to develop ideas for the binding site of the receptor. In this case under study, an amino acid that can donate a hydrogen bond, another that forms a salt-bridge and some additional lipophilic amino acids are most likely the interacting partners of D₃ receptor antagonists. The relative spatial arrangement of these amino acids is probably defined by the corresponding GRID fields.

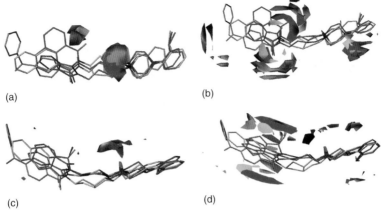

(a)

(b)

(c)

(d)

Figure 3.11 Four ligands ST-205 (yellow carbons), ST-84 (violet carbons), ST-127 (orange carbons) and ST-86 (green carbons) are displayed with their GRID contours in corresponding colors. (a) GRID contours derived from the 'carbonyl oxygen'-probe. Energy contoured at -2.5 kcal mol^{-1}. (b) GRID contours derived from the 'sp^2 NH with lone pair' probe. Energy contoured at -4 kcal mol^{-1}. (c) GRID contours derived from the 'ionized alkyl carboxyl group' probe contoured at -3.5 kcal mol^{-1}. (d) GRID contours derived from the aromatic 'sp^2 CH' probe contoured at -1.2 kcal mol^{-1}.

3.2
3D QSAR Analysis

Forty D$_3$ antagonists (Tables 3.1–3.3) were superimposed onto each other in their pharmacophoric conformations. The superimposition was conducted using FlexS and Sybyl's Multifit routine for the refinement of the superimpositions (for a detailed description of the methodology used, see Ref. [1]). The resulting superimposition of all 40 ligands is shown in Figure 3.12. GRID interaction fields are calculated subsequently using several GRID probes placed on each node of a grid box surrounding the ligands. The size of the grid box was defined so that it extends about 4 Å from the structures of the ligands. GRID calculations are carried out using 1 Å grid spacing, thus giving 14 580 probe–ligand interactions for each compound.

3.2.1
Variable Reduction and PLS Model

Using principle component analysis (PCA), we first analyzed within GOLPE which GRID probe is able to best explain the D$_3$ antagonist data set. It was found that the PCA based on the fields obtained with the hydroxyl (OH) probe showed the highest percent explanation of the data set. Therefore, further analysis was based on the hydroxyl-probe-based model.

Figure 3.12 All dopamine D$_3$ antagonists superimposed in the pharmacophoric conformation.

As stated in Section 2.6.2.2, the large number of variables (i.e. interaction energies) in the descriptor matrix represents a statistical problem for multivariate projection methods. Only some of the calculated interaction energies contain useful information and others only introduce noise into the statistical analysis. Therefore, the program GOLPE [7] has been used to conduct a variable selection and set up a PLS model.

Starting with 14 580 variables, the data was pretreated in such a way that those variables that only take two or three values and those with absolute values below 10^{-7} kcal mol^{-1} were removed from the model. This first step led to a reduction to 13 665 variables that were considered further on. The variables were then classified using PCA and a PLS model was first generated and cross-validated using the 'leave one out' (LOO) method. We continued to reduce the variables several times using the D-optimal preselection method implemented in the program and obtained a PLS model built up with 1682 variables. This model showed no decrease in the quality compared to the first one, when it was checked using the LOO cross-validation method. At that step, the fractional factorial design (FFD) method combined with the smart region definition (SRD) [8] was chosen to reduce the number of variables for the last time and generate the final model. As mentioned in Section 2.6.2.2, this variable selection very carefully assorts those variables helpful for predictivity of the model. The final model was validated using the LOO and the 'leave 20% out' method. The 'leave groups out' (e.g. leave 20% out) cross-validation has been shown to yield better indices for the robustness of a model than the normal LOO procedure.

$$R^2 = 1 - \frac{\sum (y_{obs} - y_{calc})^2}{\sum (y_{obs} - y_{mean})^2} \tag{1}$$

$$SDEC = \sqrt{\sum \frac{(y_{obs} - y_{calc})^2}{N}} \tag{2}$$

$$Q^2 = 1 - \frac{\sum (y_{obs} - y_{pred})^2}{\sum (y_{obs} - y_{mean})^2} \tag{3}$$

$$SDEC = \sqrt{\sum \frac{(y_{obs} - y_{pred})^2}{N}} \tag{4}$$

where

y_{obs} = experimental value,
y_{calc} = calculated value,
y_{pred} = predicted value,
y_{mean} = average value,
N = number of objects.

The results of the 3D QSAR analysis are shown in Table 3.5. The LOO as well as the 'leave 20% out' cross-validation yielded high Q^2 values indicating the robustness and internal predictivity of the derived QSAR model. Three principal components have been used to generate the models. The characteristics used to describe the quality of the models are the number of variables, the R^2 values (correlation coefficient) (Equation 1), the SDEC (standard error of correction) (Equation 2), the Q^2 values (cross-validated correlation coefficient) (Equation 3) and the SDEP (standard error

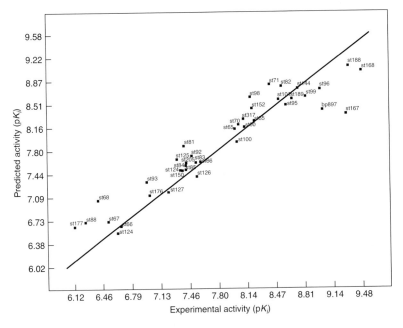

Figure 3.13 Cross-validated PLS model (leave one out).

Table 3.5 PLS models generated with variables from three principal components.

Number of variables	Selection method	R^2	SDEC	Q^2	SDEP	Validation method
13 665	–	0.9545	0.1863	0.7041	0.4753	LOO
6728	D-optimal	0.9545	0.1863	0.7041	0.4753	LOO
3364	D-optimal	0.9545	0.1863	0.7041	0.4753	LOO
1682	D-optimal	0.9545	0.1683	0.7044	0.4751	LOO
799	SRD–FFD	0.9673	0.1580	0.8743	0.3098	LOO
799	SRD–FFD	0.9673	0.1580	0.8549	0.3328	L 20% O

of prediction) (Equation 4). Figure 3.13 displays the result of the LOO' cross-validation of the final model. The correlation of the predicted and the experimental determined pK_i values is shown.

3.2.2
Validation of the Model

In the next step, the 3D QSAR method itself has been validated. As mentioned in Section 2.6, the treatment of a vast amount of data by application of statistical methods can result in chance correlation. The underlying dataset has been tested using a scramble test to check whether the obtained correlation has been generated by chance or can be considered as reasonable. The binding affinities of the ligands were scrambled and have been assigned randomly to the ligands. A PLS model was generated and the variables were reduced as mentioned above. The final PLS model was checked using the LOO cross-validation method. Ten models with scrambled affinities were set up. The characteristics of the resulting models are shown in Table 3.6.

Table 3.6 Models that are obtained with scrambled binding affinities.

Model	R^2	SDEC	Q^2	SDEP
1	0.7449	0.4413	−0.4612	1.0562
2	0.7871	0.4032	0.1757	0.7923
3	0.7874	0.4028	0.2367	0.7634
4	0.8521	0.3360	0.3356	0.7122
5	0.8066	0.3843	0.2233	0.7700
6	0.8719	0.3129	0.3979	0.6780
7	0.7481	0.4385	−0.4564	1.0545
8	0.8176	0.3732	−0.2714	0.9852
9	0.9128	0.2581	−0.1091	0.9202
10	0.8010	0.3898	−0.0810	0.9085

Interestingly, rather high correlation coefficient values (R^2) are obtained by each PLS model, but none of the models survives the cross-validation, which is indicated by the Q^2 values that span a range from -0.4564 to 0.3979. The SDEC and SDEP values are also in all cases very high compared to the model with correctly assigned pK_i values. These results show clearly that the method is able to generate good models only with correctly assigned binding affinity values.

3.2.3
Prediction of External Ligands

In the last step, the predictivity of the model has been tested on a set of ligands not included in the model (Table 3.7). The binding affinities of 12 ligands that were synthesized and tested in the same laboratories as the training set were predicted and compared to their actual pK_i values [9–11]. Since these

Table 3.7 Structures of the external test set ligands.

Compound	Structure	Predicted pK_i	Observed pK_i
ST-73		7.33	6.62
ST-75		7.10	6.66
ST-76		7.32	6.71
ST-78		7.02	6.39

Table 3.7 (continued).

Compound	Structure	Predicted pK_i	Observed pK_i
ST-87		7.91	7.62
ST-104		8.04	8.69
ST-106		7.57	7.12
ST-109		7.70	6.97
ST-111		7.32	6.75
ST-115		7.59	7.13
ST-128		7.50	7.66
ST-129		8.10	8.55

compounds were synthesized and tested later in the project and were not known when we established the QSAR model, they represent a real external test set. The chemical structures of these molecules are shown in Table 3.7.

The SDEP value for the external prediction is 0.57. In most cases, the prediction of the pK_i values appears to be within the range of ± 0.5 compared to the values determined experimentally, and this can be rated as a reasonable result. As often observed in external predictions with 3D QSAR models, the deviation between actual and predicted values is larger for the most active and the least active compounds. The most active antagonists were predicted too little active whereas the least active compounds were predicted too active.

References

1. Hackling, A., Ghosh, R., Perachon, S. *et al.* (2003) N-(omega-(4-(2-methoxyphenyl)piperazin-1-yl)alkyl)carboxamides as dopamine D2 and D3 receptor ligands. *Journal of Medicinal Chemistry*, **46**, 3883–99.

2. Pilla, M., Perachon, S., Sautel, F. *et al.* (1999) Selective inhibition of cocaine-seeking behaviour by a partial dopamine D3 receptor agonist. *Nature*, **400**, 371–75.

3. Moore, K.W., Bonner, K., Jones, E.A. *et al.* (1999) 4-N-linked-heterocyclic piperidine derivatives with high affinity and selectivity for human dopamine D4 receptors. *Bioorganic & Medicinal Chemistry Letters*, **9**, 1285–90.

4. Avenell, K.Y., Boyfield, I., Coldwell, M.C. *et al.* (1998) Fused aminotetralins: novel antagonists with high selectivity for the dopamine D3 receptor. *Bioorganic & Medicinal Chemistry Letters*, **8**, 2859–64.

5. Lemmen, C., Lengauer, T., and Klebe, G. (1998) FLEXS: a method for fast flexible ligand superposition.

Journal of Medicinal Chemistry, **41**, 4502–20.

6. GRID, Molecular Discovery Ltd., London, http://www.moldiscovery.com.

7. GOLPE, Multivariate Infometric Analysis, Perugia, http://www.miasrl.com/golpe.htm.

8. Cruciani, G., Clementi, S., and Pastor, M. (1998) GOLPE-guided region selection. *Perspectives in Drug Discovery and Design*, **12**, 71–86.

9. Mach, U., Hackling, A.E., Perachon, S. *et al.* (2004) Development of novel 1,2,3,A-tetrahydroisoquinoline derivatives and closely related compounds as potent and selective dopamine D3 receptor ligands. *Chembiochem*, **5**, 508–18.

10. Hackling, A.E. and Stark, H. (2002) Dopamine D3 receptor ligands with antagonist properties. *Chembiochem*, **3**, 946–61.

11. Ghosh, R. (2002) *Molecular Modelling Untersuchungen am Dopamin D3 Rezeptor und Seinen Liganden*, Ph. D. Thesis, Heinrich-Heine-University Düsseldorf, Germany.

4
Introduction to Comparative Protein Modeling

4.1
Where and How to Get Information on Proteins

After dealing with small molecules up to this chapter in the book, we focus on biopolymers in the next part. Since most of the known receptors and target molecules are polypeptides, we predominantly discuss modeling of protein structures.

Each modeling study strongly depends on the quality of the available experimental data, which always serve as the basis of a hypothetical model. Therefore, the first step should always be a careful search in the literature and databases to get a clear picture about the level of knowledge on the biopolymer structure of interest. A valuable information, for example, would be the complete 3D structure of the receptor or enzyme ideally derived from crystal data or nuclear magnetic resonance (NMR) measurements. After refinement, such structures can be used directly to calculate different properties of the protein or to investigate possible ligand–protein interactions. Despite the rapid growth in the number of solved 3D structures of proteins, the rate of increase of sequence data continues to be greater, resulting in an even larger number of sequences that have no known 3D structure.

Since the beginning of the 1990s, many laboratories are analyzing the full genome of several species such as bacteria, yeasts, mice and humans. During these collaborative efforts, enormous amounts of data are collected and stored in databases, most of which are accessible by public. Besides gathering all these data, it is all the more necessary to compare the nucleotide or amino acid sequences to find similarities and differences. Since the number of published sequences and structural information is increasing rapidly, an efficient search can only be done by using computer software suitable for this purpose. Several algorithms have been developed and implemented providing graphical user interfaces to existing databases. Thus, comparison of a newly found sequence with those already stored in a database can now be done in matter of minutes. Nevertheless, it is still necessary to carefully analyze the results and to fine-tune a database search if needed. Thereby, it is possible

Molecular Modeling. Basic Principles and Applications. 3rd Edition
H.-D. Höltje, W. Sippl, D. Rognan, and G. Folkers
Copyright © 2008 Wiley-VCH Verlag GmbH & Co. KGaA, Weinheim
ISBN: 978-3-527-31568-0

to quickly determine the differences among species, and the differences between a healthy versus a diseased individual. One of these well-known programs is the GCG program [1] developed by the Genetic Computer Group, Wisconsin, and now implemented in the Accelrys software [2]. This software package allows working with several databases that can be used for the search of an individual protein or DNA structure. The search can be accelerated and specified by using keywords like author names, journals or families of proteins.

A variety of nucleotide and protein sequence databases maintained by the scientific community are available via the World Wide Web. The EMBL (European molecular biology laboratory) Nucleotide Sequence Database [3] (also known as EMBL-Bank) constitutes Europe's primary nucleotide sequence resource. The database is produced in an international collaboration with GenBank (USA) and the DNA Database of Japan (DDBJ). Each of the three groups collects a portion of the total sequence data reported worldwide, and all new and updated database entries are exchanged between the groups on a daily basis. The main sources for DNA and RNA sequences are direct submissions from individual researchers, genome sequencing projects and patent applications.

The central repository for protein sequences and functions is the Universal Protein Resource (UniProt) Database (http://www.ebi.uniprot.org) [4]. It contains accurately annotated protein sequences, with extensive cross-references and querying interfaces freely accessible to the scientific community. The Uniprot database is maintained by a consortium comprising the European Bioinformatics Institute (EBI), the Swiss Institute of Bioinformatics (SIB) and the Protein Information Resource (PIR). Each consortium member is involved since years in protein database maintenance and annotation. In the past, EBI and SIB together produced the Swiss-Protein (Swiss-Prot) and the Translated European Molecular Biology Labaratory (TrEMBL) database, while PIR produced the Protein Sequence Database (PIR–PSD) [5, 6]. These databases coexisted with differing protein sequence coverage and annotation priorities. Swiss-Prot was recognized as the gold standard of protein annotation, with extensive cross-references, literature citations and computational analyses provided by experts. Recognizing that sequence data were being generated at a pace exceeding Swiss-Prot's capacity, TrEMBL (based on the EMBL-Bank) was created to provide automated annotations for proteins that were not in Swiss-Prot. Meanwhile, PIR maintained the PIR–PSD and related databases, including a database of protein sequences and curated families. To create one central repository system for protein sequences, it was decided to pool the overlapping and their complementary resources from the different consortium members.

The most important and standard database for all 3D structural information on macromolecules is the Protein Data Bank (PDB) [7], which is available via the World Wide Web (http://www.rcsb.org). In the PDB, atomic coordinates of protein or DNA structures are collected. Because of the continuously growing number of experimentally resolved structures, the database is regularly updated (~41 000 entries at the end of 2006). Information can be searched for in the PDB by specifying particular keywords. The author's name, a journal name or a part of a sequence, for example, can serve as the search subject.

On the basis of the PDB, some smaller structural databases have been created. Examples are the HSSP (Homology-Derived Secondary Structure of Protein) [8] and the SCOP (Structural Classification Of Proteins) [9] databases. HSSP contains homology-derived structures of proteins. It combines information from the PDB and sequences of proteins derived from a sequence database like Swiss-Prot. The SCOP database is a comprehensive ordering of all proteins of known structure according to their evolutionary and structural relationships. Protein domains in SCOP are grouped into species and hierarchically classified into families, superfamilies, folds and classes.

In general, the format, the organization and the information contained in different structural data files are very similar. As PDB is widely used, the standard format of a protein data file will be described in detail in the following. The header of the data file comprises some general information about the protein. It includes the official name, references, resolution of the crystal structure and some useful remarks about the secondary structure composition of the protein. Adjacent to the header, the atomic coordinates are listed. Atoms belonging to standard amino acid residues are labeled as ATOM. To distinguish between individual peptide chains, the ATOMS are separated by an additional line starting with the abbreviation TER. Between ATOMS, a bond is generally built when the file is read into the modeling program. This is important as the atoms that do not belong to standard amino acid residues are labeled as HETATM. Between HETATMS no connectivity is established. Therefore, an additional connectivity table is included at the end of the data file. It is advisable to be careful at this point because whether HETATMS are displayed properly and connectivities are correctly assigned would be program dependent.

HETATMS can either belong to nonstandard amino acids, or in the case of complexes, to the ligand molecule involved in the ligand–protein interaction. When reading a PDB file into a standard modeling program, the atom type assignment is often incorrect. A variety of programs offer the possibility to automatically assign atom types for ligand molecules. However, the user should be aware of the fact that this procedure often results in erroneous molecular structures. Therefore, it is absolutely necessary to check carefully all atom types to avoid mistakes resulting in wrong geometries of the ligands (this has already been discussed in Section 2.1.2).

Usually, all structures from the PDB do not include hydrogen atoms. For some types of investigations, hydrogen atoms can be neglected, but for the study of ligand–protein interactions, it is inevitable to add the hydrogen atoms. The ligand molecules, in particular, have to be checked explicitly to confirm that the correct degree of hybridization (atom type) and protonation (formal charges) has been assigned in case of acidic or basic substances.

Also, hydrogen atoms are never allocated to all water molecules. As a consequence, they are only displayed as single points representing the oxygen positions. Water molecules can present as either simple crystal water distributed near the surface of the protein or they can be located in the active site. In the latter case, it is absolutely necessary to include their complete coordinates into further investigations because they can crucially influence the conformation of the active-site structure. This is also true for cations implemented in the crystal structure as they can play an important role for ligand binding or enzyme activity if they are located in the active site.

Most of the modeling programs are able to read the PDB data files without any problem and are also able to transform the structural information into a 3D picture of the protein. However, some points of caution should be kept in mind when using experimentally derived information.

In principle, the resolution of a crystal structure should be at least between 2.5 and 1.5 Å or better. Otherwise the structural information is not very valuable. The purification process of proteins is a difficult and time-consuming task, and it may happen that as a result of proteolytic activity some information is lost before the crystallization process has finished. Therefore, sometimes, amino residues might be missing, leading to incomplete information in the data file.

Some enzymes and proteins do fulfill their biochemical function only in the dimeric or trimeric form. The modeler should be aware of this fact because it makes no sense to investigate the functionality of the active site of an enzyme that consists of a dimer when only the monomer structure is present in the PDB file. For this reason, the recently updated PDB web site also contains information about the biological unit and the crystal form of each entry.

Recently, the NMR technique is being increasingly used for obtaining structural information on proteins. NMR has a special bearing on cases where a protein has withstood all efforts to grow sufficiently large crystals. An additional advantage of NMR-derived data is that the conformation of the protein is not influenced by packing forces of the crystal environment. As the NMR measurements are performed in solution, the results are highly dependent on the solvent. Experiments in nonpolar solvents, for example, lead to an overestimation of hydrogen bonding phenomena. Measurements in aqueous environment should yield a more realistic picture of the protein structure.

The pool of information on proteins is already immense and continuously growing. But still, most of the available databases merely contain information on primary structures. In order to obtain a 3D protein model from these data, the application of alignment techniques, knowledge-based and comparative modeling approaches, is necessary. A detailed discussion on these subjects is given in Section 4.3.

References

1. Devereux, J., Haeberli, P. and Smithies, O. (1984) A comprehensive set of sequence analysis programs for the VAX. *Nucleic Acids Research*, **12**, 387–95.

2. Genetics Computer Group (GCG), Accelrys Inc., San Diego. http://www.accelrys.com.

3. Emmert, D.B., Stoehr, P.J., Stoesser, G. and Cameron, G.N. (1994) The European Bioinformatics Institute (EBI) databases. *Nucleic Acids Research*, **22**, 3445–49.

4. Bairoch, A., Apweiler, R., Wu, C.H. *et al.* (2005) The universal protein resource (UniProt). *Nucleic Acids Research*, **33**, D154–59.

5. Bairoch, A. and Boeckmann, B. (1994) The SWISS-PROT protein sequence data bank: current status. *Nucleic Acids Research*, **22**, 3578–80.

6. George, D.G., Barker, W.C., Mewes, H.-W. *et al.* (1994) The PIR-international protein sequence database. *Nucleic Acids Research*, **22**, 3569–73.

7. Berman, H.M., Westbrook, J., Feng, Z. *et al.* (2000) The protein data bank. *Nucleic Acids Research*, **28**, 235–42.

8. Sander, C. and Schneider, R. (1994) The HSSP database of protein structure-sequence alignments. *Nucleic Acids Research*, **22**, 3597–99.

9. Lo Conte, L., Brenner, S.E., Hubbard, T.J.P. *et al.* (2002) SCOP database in 2002: refinements accommodate structural genomics. *Nucleic Acids Research*, **30**, 264–67.

4.2
Terminology and Principles of Protein Structure

The complex 3D structure of proteins is characterized by four general levels of structural organization: the primary, secondary, tertiary and quaternary structures.

(1) The primary structure represents the linear arrangement of the individual amino acids in the protein sequence.

(2) The secondary structure describes the local architecture of linear segments of the polypeptide chain (i.e. α-helix, β-sheet), without regarding the conformation of the side chains. Another level of structural organization, which was introduced very recently, is the so called *supersecondary structure*. It describes the association of secondary structural elements through side chain interactions. Another term for the same matter is *motif*.

(3) The tertiary structure portrays the overall topology of the folded polypeptide chain.

(4) The quaternary structure describes the arrangement of separate subunits or monomers in the functional protein.

Owing to the remarkable capability of polypeptide chains not only *in vivo*, but also *in vitro* to fold into functional proteins, it is currently accepted that most aspects of protein architecture and stabilization directly derive from the properties of the particular sequence of amino acids that make up the polypeptide chain (i.e. the primary structure). These properties include the individual characteristics of the side chains of every residue and the influence of the polypeptide backbone on the protein conformation. The 3D structure of a protein can be understood only on the basis of this information. It is not in the scope of this introduction to provide a detailed description of all the properties which determine the conformation of a protein, but to explain the main features necessary to understand the contents of the chapters that follow. For a comprehensive description of the principles of protein conformation, the reader is referred to the literature [1–4].

4.2.1
Conformational Properties of Proteins

Generally, only twenty different amino acids are found in naturally occurring proteins. The physicochemical properties of their side chains like size, shape, hydrophobicity, charge and hydrogen bonding span a considerable range. They avoid, however, the extremes of high chemical reactivity and also, except for proline, strongly restricted degrees of freedom. The question most relevant in view of the 3D shape of proteins is, how the individual side chains interact

Molecular Modeling. Basic Principles and Applications. 3rd Edition
H.-D. Höltje, W. Sippl, D. Rognan, and G. Folkers
Copyright © 2008 Wiley-VCH Verlag GmbH & Co. KGaA, Weinheim
ISBN: 978-3-527-31568-0

with the backbone as well as with one another, and what roles they play within particular types of secondary and tertiary structures. The predominant influences of the sequential order on protein conformation are (aside from the linear connectivity and the steric volume) the hydrogen bonding capabilities and the chirality of all (except glycine) amino acid residues. All 19 chiral amino acids possess the L-configuration or according to the Cahn–Ingold–Prelog scheme the S-configuration with the exception of L-cystein, which due to a change in ligand priority possesses the R-configuration.

An important convention needed for understanding much of the information available for a particular protein, is the designation of the individual atoms and structural elements of a protein. All atoms, angles and torsion angles that describe the 3D structure of a protein are named using letters of the Greek alphabet. The central carbon atom in amino acid residues is termed α, and the side chain atoms are commonly designated β, γ, δ, ε and ζ in alphabetical order starting from the α carbon atom. The backbone of a protein consists of a repeated sequence of three atoms, belonging to one amino acid residue–the amino N, the C^{α} and the carbonyl C; these atoms are generally represented as N_i, C_i^{α} and C_i' respectively, where i is the number of the residue, starting from the amino end of the chain. As an example, a portion of the backbone of a polypeptide chain is shown in Figure 4.2.1. It illustrates the conventions used in describing the protein conformation.

The main chain torsion angles in proteins are named ϕ (phi), ψ (psi), and ω (omega). Rotation about the N–C^{α} bond is described by the torsion angle ϕ, rotation about the C^{α}–C' bond by ψ and rotation about the peptide bond C'–N by ω. Torsion angles of the side chains are designated by χ_j (chi$_1$, chi$_2$, etc.) where j is the number of the bond counting outward from the C^{α}-atom.

The peptide bond is usually planar, because of its partial double bond character and nearly always has the trans configuration ($\omega = 180°$) which is energetically more favorable than cis ($\omega = 0°$). The cis-configuration sometimes is found to occur with proline residues (about 10%). Small

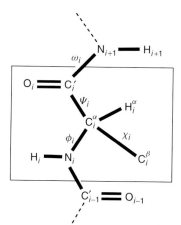

Figure 4.2.1 Designation of atoms and torsion angles in a protein.

deviations from planarity of the cis or trans form with $\Delta\omega = -20° - 10°$, seem to be energetically acceptable.

The variations in ϕ and ψ are constrained geometrically due to steric clashes between neighboring but nonbonded atoms. The permitted values of ϕ and ψ were first analyzed and determined by Ramachandran *et al.* [5] In their work computer models of small peptides were used to vary systematically ϕ and ψ with the purpose of detecting stable conformations. Each conformation, represented by a particular ϕ, ψ combination, was examined for close contacts between atoms. In this rough model the atoms were treated as hard spheres with fixed geometries for the bonds. Only values of ϕ and ψ, for which no close contacts between atoms have been discovered, are permitted and usually are presented in a two-dimensional map, the so called *Ramachandran plot*. Since ϕ and ψ constitute a virtually complete description of the backbone conformation, the two-dimensional Ramachandran plot is an important and easy way to analyze test for the validity of 3-D protein structure.

In Figure 4.2.2 the Ramachandran plot of poly-alanine is shown as an example. The area outside the solid lines corresponds to conformations where atoms in the polypeptide chain are located at distances closer than the sum of their van der Waals radii. These regions are sterically disallowed for all amino acids, except glycine. Glycine, which lacks a side chain, is evenly distributed over the complete plain of the Ramachandran plot. The shaded regions correspond to conformations where no steric clashes are found, that is, these are the allowed regions (or favored regions). The area directly outside the boundaries of this region includes conformations which are permitted if

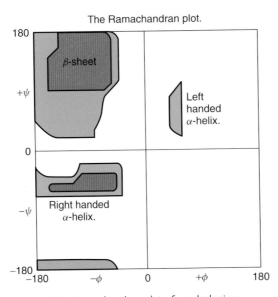

Figure 4.2.2 Ramachandran plot of a polyalanine.

slight alterations of bond angles are accepted. Ramachandran plots for other amino acids look comparable in the shape of the various regions.

Subregions of ϕ,ψ–space are generally named after the secondary structure elements which result when the corresponding ϕ,ψ–angles occur repeatedly. The right-handed α-helix for example resides in the lower left near $-60°, -40°$, the broad region of extended β-sheets in the upper left around $-120°, 140°$ and the slightly unfavored left-handed α-helical region in the upper right near $60°, 40°$. Conformational properties and other relevant parameters of these secondary structures are described in the following sections.

4.2.2
Types of Secondary Structural Elements

4.2.2.1 The α-Helix

The right-handed α-helix is the best known and most easily recognized secondary structural element in proteins [6, 7]. Approximately 32–38% of the residues in known globular proteins are involved in α-helices [8]. α-helices are characterized by a repetitive secondary structure; that is, all C_α-atoms of α-helical amino acids are in identical relative positions. Thus, the ϕ, ψ torsion angle pairs are the same for each residue in the helix. The structure of an α-helix repeats itself every 5.4 Å along the helix axis; this means that the α-helices have a pitch of $p = 5.4$ Å. α-Helices have 3.6 amino acid residues per turn, that is, a helix of 36 amino acids would form 10 turns.

The α-helical structure is mainly formed and stabilized by repeated hydrogen bonds between the carbonyl function of residue n and the NH of residue $n + 4$ (see Figure 4.2.3). This results in a very regular and energetically favored state.

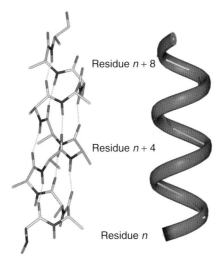

Residue $n + 8$

Residue $n + 4$

Residue n

Figure 4.2.3 General architecture of an α-helix.

α-Helices observed in protein structures are always right-handed. L-Amino acids cannot form extended regions of a left-handed α-helix because the C^β-atoms would collide with the following turn. Only individual residues which possess the ϕ, ψ torsion angles of a left-handed α-helix are found. Therefore, when referring to an α-helix, we usually mean the right-handed α-helix.

The exact geometry of the α-helix is found to vary somewhat in natural proteins, depending on its environment. The ideal α-helix ($\phi = -57°$ and $\psi = -47°$) is only one version of a family of similar structures [6]. More usually, a slightly different α-helix geometry ($\phi = -62°$ and $\psi = -41°$) can be observed in proteins [7]. This conformation is more favorable than the ideal α-helix because it permits each carbonyl oxygen to make hydrogen bonds with both the NH of residue $n + 4$ and the aqueous solvent (or other hydrogen-bond donors).

The side chains of an α-helix point outward into the surrounding space. Several restrictions exist for side chain conformations, especially for side chains with branched C^β-atoms (Val, Ile, Thr). Proline residues normally are incompatible with the α-helical structure because due to the cyclic structure the amide nitrogen lacks the hydrogen substituent necessary for hydrogen bonding. Nevertheless, if single proline residues appear in long α-helices (e.g. in some of the transmembrane α-helices of bacteriorhodopsin), this appearance yields a local distortion of the α-helical geometry.

Variations of the classical α-helix in which the protein backbone is either more tightly or more loosely coiled (with hydrogen bonds to residues $n + 3$ and $n + 5$), are named 3_{10}-helix and π-helix, respectively. In general, these helix types only play a minor role in the architecture of proteins. However, 3_{10}-helices frequently form the last turn of a classical α-helix.

4.2.2.2 The β-Sheet

Besides the α-helix, the second most regular and recognizable secondary structural motif is the β-sheet [9, 10]. Like the α-helix it is a periodic element. β-sheets are formed from β-strands. A β-strand develops when a linear extended conformation of a polypeptide backbone ($\phi = -120°$, $\psi = 140°$) turns up [9]. Since interactions between residues of the same strand are not possible, a β-strand is only stable as part of a more complex system, the β-sheet. Like in α-helices all hydrogen bond donor or acceptor groups of the peptide backbone are engaged in the formation of hydrogen bonds; however, because these bonds appear not intra- but intermolecularly, β-strands are energetically less favored. In contrast to α-helices which consist of a singular stretch of directly bound residues, β-sheets possess a much more pronounced structure-modulating effect because they are composed of several β-strands which can be distributed over a large part of the sequence.

Adjacent β-strands can be arranged either in a parallel or in an antiparallel fashion. In parallel sheets the strands all run in the same direction (see Figure 4.2.4a), whereas in antiparallel sheets they run in opposite directions (see Figure 4.2.4b).

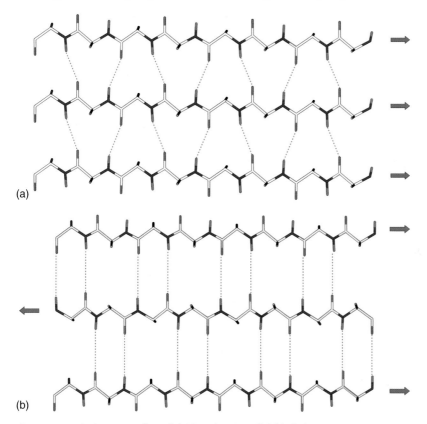

Figure 4.2.4 Architecture of parallel (a) and antiparallel (b) β-sheets.

The side chains of β-strands are located nearly perpendicular to the plane of the hydrogen bonds (between the single strands). Along the strand they alternate from one side to the other. For antiparallel β-sheets typically one side is buried in the interior of the protein and the other side is exposed to the solvent. Therefore, the physicochemical character of the amino acids tends to alternate from hydrophobic to hydrophilic. Parallel β-sheets, on the other hand are usually buried on both sides, so that the central residues tend to be hydrophobic, and hydrophilic amino acids are abundantly found toward the ends. For both types of β-sheets, edge strands can be much more hydrophilic than central strands.

β-sheets are very common in globular proteins (20–28%) [8]. They can consist exclusively of parallel or antiparallel strands or are formed from a mixture of them. Purely parallel sheets are less frequent; while purely antiparallel sheets are very common. Antiparallel sheets often consist of as few as just two or three strands, whereas parallel sheets always have at least four. Mixed sheets usually contain 3–15 strands.

4.2.2.3 **Turns**

Approximately one-third of all residues of globular proteins are involved in turn regions. The general function of turns is to reverse the direction of the polypeptide chain. Often turns are located on the protein surface and therefore, contain predominantly charged and polar amino acids.

Various different types of reverse turns have been observed in proteins. Their specific features depend, for example, on the type of secondary structural motifs which are linked by them. For a detailed description of all observed turn types the reader is referred to the literature [1–4, 11, 12].

Turns often connect antiparallel β-strands. In this case they are named β-turns or hairpin bends [12]. Seventy percent of the hairpin turns are shorter than seven residues in length; most often they include only two residues. Larger loops have less well defined conformations, which are often influenced by interactions with the rest of the protein. In all reverse turns the peptide groups are not paired by regular hydrogen bonds, but are accessible to the solvent. For this reason reverse turns often appear on the protein surface.

In general, the periodic secondary structural elements in proteins (α-helices and β-sheets) are rather short. The length of an α-helix is usually 10–15 residues (12–22 Å). A single β-strand is found to include 3–10 residues (7–30 Å). Most of the described ideal geometries of helices and sheets are only rarely observed in nature. Often, the geometries of secondary structures are more or less distorted. For example, solvent-exposed α-helices very often show a curved helix axis. Most β-sheets in folded proteins are rather twisted than planar with a twist of $0°–30°$ between the single strands.

The common properties of proteins described in this paragraph provide only some general rules of protein architecture. Each naturally occurring protein, on the other hand, is unique and attains its functional and structural character by means of specific noncovalent interactions. It is, therefore, necessary to compare each computer-generated structure with 'real' 3D structures of proteins, and to include as much as possible information about protein structures in the process of protein modeling.

The exclusive presentation of secondary structural motifs of a complex protein in a schematic form is a very helpful tool for comprehending the overall structure. Usually, in this kind of representation the side chains are omitted to yield a clearer picture of the whole protein including the various secondary structural elements. Helices are often described by cylinders or coiled ribbons and extended strands of β-sheets by broad arrows indicating the amino-to-carboxy direction of the backbone. The 3D structure of triose phosphate isomerase is presented in such simplified form in Figure 4.2.5.

4.2.3
Homologous Proteins

It has since long been recognized that the evolutionary mechanism of gene duplication, which is associated with mutations, leads to divergence and

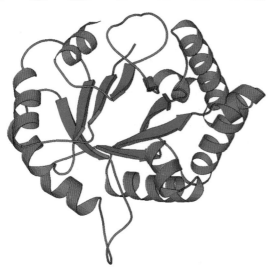

Figure 4.2.5 The 3D structure of triose phosphate isomerase presented in simplified form using MOLSCRIPT [13].

thereby to the foundation of families of related proteins with similar amino acid sequences and similar 3D structures. The proteins that have evolved from a common ancestor are said to be homologous. Two homologous sequences can be nearly identical, similar to varying degrees or dissimilar because of extensive mutations. As a matter of fact, the sequence similarity in homologous proteins is less preserved than the structural similarity. In other words,, 3D structures of homologous proteins have been remarkably conserved during their evolution because the common structure is crucial for the specific function of the proteins. The conservation of protein structure has been detected in many protein families. As one example, the 3D structures of α-chymotrypsin and trypsin, belonging to the family of serine proteases can be cited. They are remarkably similar, although only 44% of the amino acid residues shared by them are identical. This topological similarity can easily be observed in Figure 4.2.6. Other members of the family of serine proteases have changed more drastically during evolution. Bacterial serine proteases for example show only 20% sequence identity when compared to the mammalian enzymes like thrombin, trypsin or chymotrypsin. However, if the 3D structural similarity is considered the main features are still present.

The question which immediately comes to ones mind in this respect is how such large dissimilarities in the primary sequences are compatible with the observed structural similarity. The answer was found empirically and can be summarized as follows. The most pronounced dissimilarities generally appear in regions close to the protein surface, the so called *loop regions*. In these regions even the physicochemical properties of the side chains have often changed. Residues located in the interior of proteins however, vary less frequently and

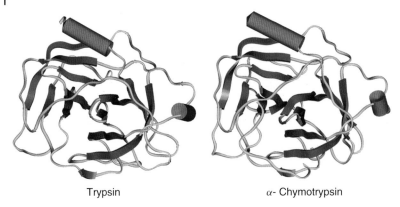

Trypsin   α- Chymotrypsin

Figure 4.2.6   3D structure of two homologue enzymes. (Color code: red = α-helix, blue = β-strand, yellow = peptide backbone.)

less distinctly. This leads to the situation that generally a common core of residues comprising the center of the protein and the main elements of the secondary structure remain highly conserved within a family of homologous proteins.

Within homologous proteins, the elements of secondary structure can move relative to each other, can change in length or can even disappear completely. But usually an α-helix is not replaced by a β-sheet or vice versa. In general, neither the order nor the orientation (parallel or antiparallel) of β-strands has ever been recognized to differ between proteins of the same family.

To summarize, the overall conformations of homologous proteins appear to have been highly conserved during evolution. This fact forms the basis for the development of the comparative protein modeling approach, which will be described in the next chapter.

References

1. Creighton, T.E. (1993) *Proteins: Structures and Molecular Properties*, 2nd edn, W. H. Freeman and Company, New York.
2. Branden, C. and Tooze, J. (1991) *Introduction to Protein Structure*, Garland Publishing, New York.
3. Schulz, G.E. and Schirmer, R.H. (1979) *Principles of Protein Structure*, Springer-Verlag, New York.
4. Fasman, G.D. (1989) *Prediction of Protein Structure and the Principles of Protein Conformation*, Plenum Press, New York.

5. Ramachandran, G.N. and Sasisekharan, V. (1968) Conformation of polypeptides and proteins. *Advances in Protein Chemistry*, **23**, 283–437.
6. Pauling, L., Corey, R.B., and Branson, H.R. (1951) The structure of proteins; two hydrogen-bonded helical configurations of the polypeptide chain. *Proceedings of the National Academy of Sciences of the United States of America*, **37**, 205–11.
7. Barlow, D.J. and Thornton, J.M. (1988) Helix geometry in proteins. *Journal of Molecular Biology*, **201**, 601–19.

8. Kabsch, W. and Sander, C. (1983) Dictionary of protein secondary structure: pattern recognition of hydrogen bonded and geometrical features. *Biopolymers*, **22**, 2577–637.

9. Chou, K.C., Pottle, M., Nemethy, G. *et al.* (1982) Structure of beta-sheets. Origin of the right-handed twist and of the increased stability of antiparallel over parallel sheets. *Journal of Molecular Biology*, **162**, 89–112.

10. Pauling, L. and Corey, R.B. (1951) Configurations of polypeptide chains with favored orientations around single bonds: two new pleated sheets.

Proceedings of the National Academy of Sciences of the United States of America, **37**, 729–40.

11. Rose, G.D., Gierasch, L.M., and Smith, J.A. (1985) Turns in peptides and proteins. *Advances in Protein Chemistry*, **37**, 1–109.

12. Sibanda, B.L. and Thornton, J.M. (1985) Beta-hairpin families in globular proteins. *Nature*, **316**, 170–74.

13. Kraulis, P.J. (1991) MOLSCRIPT: a program to produce both detailed and schematic plots of protein structures. *Journal of Applied Crystallography*, **24**, 946–50.

4.3
Comparative Protein Modeling

As discussed in Section 4.1, extensive information on primary and secondary structure of proteins are stored in various data bases. Protein sequence determination is now routine work in molecular biology laboratories. As a result of the Human Genome project, the rate of publication of primary sequences has increased dramatically in the last few years. Sequences of more than three million proteins are now available in the UniProt database. The translation of sequences into 3D structure on the basis of X-ray crystallography or NMR investigations, however, takes much more time. The 3D structures of more than 40 000 proteins are available in the PDB (as at the end of 2006). In certain circumstances it can take, depending on the kind of proteins, more than a year to perform a complete structure determination [1]. This is the reason that the number of known protein sequences is much larger than the number of complete 3D structures that have been determined. Because of the technical problems related to experimental elucidation of 3D structures, theoretical procedures for predicting protein 3D structure on the basis of the respective amino acid sequences are urgently needed. Since a general rule for the folding of a protein has not yet been developed, it is necessary to base structural predictions on the conformations of available homologous reference proteins [2–4] (see Section 4.2 for the underlying principles).

When a sequence is found homologous to another one, for which the 3D structure is available, the comparative modeling approach (which is also called *homology modeling approach*) is the method of choice for predicting the structure of the unknown protein. The underlying idea of comparative modeling is to make use of the collected body of knowledge about already resolved proteins. The first step is to compare the sequence of a new protein to all sequences of structurally known proteins stored in a data base. Proteins in the data base which are identified to be homologous to the unknown, are retrieved and used as templates for the structural prediction of the unknown protein. This approach was developed by several authors and is described in detail in the following paragraphs [5–8].

Successful comparative protein modeling, however, strongly depends upon how closely the structure that one is attempting to model fits the chosen template [9]. Our understanding of protein folding patterns still being rather limited at present, the only criterion that can be applied for structure prediction is the examination of the extent of sequence identity between the known and the unknown protein. Although many studies in the past have concluded that structural homology persists even when sequence identity is hardly detectable, for the purpose of knowledge-based modeling, the reverse is important. The prediction of structural similarity between different proteins can only be based on the detection of sequence identities in their sequences. Thus, the comparison of sequences using alignment methods is a

Molecular Modeling. Basic Principles and Applications. 3rd Edition
H.-D. Höltje, W. Sippl, D. Rognan, and G. Folkers
Copyright © 2008 Wiley-VCH Verlag GmbH & Co. KGaA, Weinheim
ISBN: 978-3-527-31568-0

central technique in comparative modeling and will be described in detail in Section 4.3.1.

The process of traditional comparative modeling involves the following steps:

(1) determination of proteins which are related to the protein being studied;
(2) identification of structurally conserved regions (SCRs) and structurally variable regions (SVRs);
(3) alignment of the sequence of the unknown protein with those of the reference protein(s) within the SCRs;
(4) construction of SCRs of the target protein using coordinates from the template structure(s);
(5) construction of SVRs;
(6) side-chain modeling;
(7) structural refinement using energy minimization methods and molecular dynamics.

4.3.1
Procedures for Sequence Alignments

The first step in comparative modeling is the assignment of the unknown protein structure to a protein family. In many cases this information is already known because the sequence to be modeled belongs to a well-known protein family. It may be, however, that this is not true. Then it is necessary to compare the new sequence with thousands of sequences already stored in protein data bases and to identify, if possible, homologous ones.

In the past, identifying new proteins through data base searches was difficult and time-consuming. Computer programs required several hours or had to make far reaching compromises in sensitivity or selectivity of the search. However, in the last decade heuristic methods were developed to tackle this problem [10–13]. They do not guarantee always that the globally optimal solution will be found, but in practice they rarely miss a particularly significant match. The two major methods are FASTA [12] and BLAST [13]. Both methods are implemented in many commercially available software packages (for example, HOMOLOGY [14], MODELLER [15], COMPOSER [16], WHAT IF [17], and GCG [18]). In addition, they are often integrated as search tools on biologically related databases and web sites (see, for example, the www.expasy.org site).

The central technique used for amino acid sequence comparison is called *sequence alignment*. In the framework of comparative modeling the sequence alignment procedure is of importance for several reasons. Firstly it is used to search data bases to find related sequences and to identify which regions of the detected proteins are conserved, thus suggesting where the unknown protein may be structurally conserved as well. This, for example, can be performed employing the above mentioned FASTA, BLAST or PSI-BLAST programs. Secondly, sequence alignment is used for detection of correspondences between amino acids of the structurally known reference protein and those of the protein to be modeled. These correspondences are the basis for transferring the coordinates of the reference protein(s) to the model protein. For this task the more sensitive and selective alignment procedures described below are needed.

A very natural procedure for aligning sequences would be to simply write them in a tabular form for visual inspection. Of course, this would be not only unsystematic, but also quite time-consuming, especially if more than two sequences are to be compared. For this reason, many programs have been developed which are able to perform alignments automatically [18–21]. Because the alignment of amino acid sequences is such a crucial step in homology modeling of proteins, many different methods and programs have been published and are still being developed. It is beyond the scope of this book to discuss all of them, but the reader is referred to the literature [12, 13, 18, 19].

One of the earliest attempts to clarify whether the structural similarity existing between proteins is due to homology or by chance, was carried out by Needleman and Wunsch [20]. Variants of the algorithm used by these authors have been further developed independently by others and applied in many fields. These programs are more sensitive in detecting similarities than the data base search programs, but on the other hand, are slower in finding an optimal alignment. However, the great advantage of the Needleman and Wunsch algorithm is that final detection of the best alignment for two sequences is guaranteed. As a consequence, computer programs based on this method (ALIGN, BESTFIT and GAP which are included in the GCG program package [18]) have been widely used for biological sequence comparison. Whereas the original Needleman and Wunsch algorithm is only able to align two sequences, many up to date programs handle the alignment of more than two sequences. The so called *multiple alignment methods* are significantly more difficult than the pairwise alignment techniques. This is because the number of possible alignments increases exponentially with the number of the sequences to be compared. Several programs have been derived to provide approximate solutions to this problem (for example, ClustalW, ClustalX [18] or MAXHOM [21]).

In contrast to the procedures described above, which search for the global optimal similarity of sequences, other approaches seek to identify the best local similarities between two sequences. These are called *optimal local alignment methods*, and are based on a modified Needleman and Wunsch algorithm. They represent an important tool for comparing sequences. This is especially true for the location of highly conserved regions dispersed over long sequences [22–24]. The basic idea of these methods is to consider only relatively conserved subsequences of homologous proteins; dissimilar regions do not contribute to the comparison (Figure 4.3.1).

```
ENTCL : PVSEKQLAEVVANTITPLMKAQSVPGMAVAVIY--QGKPHYYTFGKADIAANKPVTPQTLFELGSISKIFTGVLGGGAIA-
CITFR : AKTEQQIADIVNRTITPLMQEPAIPGMAVAIIY--EGKPYYFTWGKADIANNHPVTQQTLFELGSVSKTFNGVLGGRIA-
MEN1  : QTADVQQKLAELERQSG-GRLGVALINTADNSQILYR------------ADERFAMCSTSKVMAAVAVLKSE-
STAU  :            KELNDLEKKYN-AHIGVYALDTKSGKEVKFN-------------SDKRFAYASTSKAINSAILLEQVP-
BALI  :            DDFAKLEEQFD-AKLGIFALDTGTNRTVAYR-------------PDERFAFASTIKALTVGVLLQQKS-


ENTCL : -RGEISLDDAVTRYWPQLTGKQWQ---------GIRMLDLATYTAGGLPLQVPDEVTDNASLLRFYQNWQPQWKPGTTRLYANASIGLFGALAVKPSGMPYE
CITFR : -RGEIKLSDPVTKYWPELTGKQWR---------GISLLHLATYTAGGLPLQIPGDVTDKAELLRFYQNWQPQWTPGAKRLYANSSIGLFGALAVKSSGMSYE
MEN1  : -SEPNLLNQRVEIKKSDLVNYNPIAEKHVDGTMSLAELSAAALQ-----------------YSDNVAMNKLISHVGGP--ASVT
STAU  : ---YNKLNKKVHINKDDIVAYSPILEKYVGKDITLKALIEASMT-----------------YSDNTANNKIIKEIGGI--KKVK
BALI  : ---IEDLNQRITYTRDDLVNYNPITEKHVDTGMTLKELADASLR-----------------YSDNAAQNLILKQIGGP--ESLK


ENTCL : QAMTRVLKPLKLDHTWINVPKAEEAHYAWGYRDGKAVRVSPGMLDAQAYGVKTNVQDMANWVMANMAPENVADASLKQGIALAQSRYWRIGSMYQGLGW
CITFR : EAMTRRVLQPLKLAHTWITVPQSEQKNYAWGYLEGKPVHVSPGQLDAEAYGVKSSVIDMARWVQANMDASHVQEKTLQOGIELAQSRYWRIGDMYQGLGW
MEN1  : AFARQLG------DETFRLDRVEPTLNTAIPGDPRD-----------------TTSPRAMAQTLRNITLGKALG----DSQRAQLVTWMKGNTGAASIQA
STAU  : QRLKELG------DKVINPVRYEIELNYYSPKSKKD-----------------TSTPAAFGKTINKLIANGKLS----RENKKFLIDLMLNNKSGDTLIKD
BALI  : KELRKIG------DEVINPERFEPELNEVNPGETQD-----------------TSTARALVTSLRAFALEDKLP----SEKRELLIDWMKRNTTGDALIRA


ENTCL : EMLNWPVEANTVVEGSDSKVALAPLPVAEVNPPAPPVKASWVHKTGSTG--GFGSYVAFIPEK----QIGIVMLANTSY-----PNPARVEAAYHILEAL
CITFR : EMLNWPLKADSIINGSDSKVALAALPAVEVNPPAPAVKASWVHKTGSTG--GFGSYVAFVPEK---NLGIVMLANKSY-----PNPARVEAAWRILEKL
MEN1  : GLPAS------------------------WVVGDKTGSGD-YGTTNDIAVIWPKD-RAPLILVYFTQPQPKAESRRDVLASAAAKIVTNGL
STAU  : GVPKD------------------------YKVADKSQAITYASRNDVAFVYPKGQSEPIVLVIFTNKDNKSDKPNDKLISETAKSVMKEF
BALI  : GVPDG------------------------WEVADKTGAAS-YGTRNDIAIIWPPK-GDPVLLAVLSSRDKDAKYDDKLIAEATKVVMKAL
```

Figure 4.3.1 Multiple sequence alignment of cephalosporinases from *Enterobacter cloacae* (ENTCL) and *Citrobacter freundii* (CITFR), with penicillinases from *Escherichia coli* (MEN1), *Bacillus licheniformis* (BALI) and *Staphylococcus aureus* (STAU). Red letters indicate the determined SRCs.

In the course of comparison of two sequences, the alignment procedures at least, in effect, seek to duplicate the evolutionary process involved in converting one sequence into another. For this operation, a scoring scheme that dictates the weight for aligning a particular type of amino acid with another is required. This kind of scoring scheme is provided by matrices. The matrices make use of the most probable amino acid substitutions according to physical, chemical or statistical properties. High numerical values in the matrix imply that a substitution is probable, whereas low values indicate that a substitution is unlikely to occur. From the various kind of matrices which are in use [25–29] the following are the ones applied most often:

- Identity matrix. This is the simplest matrix that gives a score of 1 to identical pairs and 0 to all others.
- Codon substitution matrix. The scoring values for this matrix are derived from the DNA base triplets coding for the amino acid pairs. For each pair, all possible nucleotide triplets are examined and the number of point mutations required to change one amino acid into the other are evaluated. Identical amino acids get a score of 9, one required mutation gives a score of 3 and two mutations yield a score of 1.
- Mutation matrix. This matrix is also known as the Dayhoff or PAM250 matrix [25]. The matrix was obtained by counting the number of substitutions from one particular amino acid by others observed in related proteins, across different species. Large scores are given to identities and substitutions which are found frequently, and low scores are assigned to mutations that are not observed. Due to this procedure, larger scores are used for certain nonidentical pairs than for some identical ones. The Dayhoff matrix, which is shown in Figure 4.3.2, is the most widely used scoring scheme. It is often applied for finding an initial alignment for two unknown sequences. An advanced form of the Dayhoff matrix was suggested by Gribskov [26]. The Gribskov matrix assigns the highest score always to identical amino acid pairs.
- Physical property matrices. The scoring values of corresponding matrices are based on similarity indices for certain physical properties of amino acids, such as hydrophobicity, polarizability or helical tendency [28].

Differences in sequence lengths or variations in the locations of conserved regions complicate the alignment procedure. If one or both of the mentioned problems are found, gaps are introduced into the sequence to allow the simultaneous alignment of all conserved regions. To limit the total number

	Ala	Arg	Asn	Asp	Cys	Gln	Glu	Gly	His	Ile	
Ala	2	-2	0	0	-2	0	0	1	-1	-1	
Arg	-2	6	0	-1	-4	1	-1	-3	2	-2	
Asn	0	0	2	2	-4	1	1	0	2	-2	
Asp	0	-1	2	4	-5	2	3	1	1	-2	
Cys	-2	-4	-4	-5	12	-5	-5	-3	-3	-2	
Gln	0	1	1	2	-5	4	2	-1	3	-2	
Glu	0	-1	1	3	-5	2	4	0	1	-2	
Gly	1	-3	0	1	-3	-1	0	5	-2	-3	
His	-1	2	2	1	-3	3	1	-2	6	-2	
Ile	-1	-2	-2	-2	-2	-2	-2	-3	-2	5	
.											

Figure 4.3.2 Dayhoff evolutionary mutation matrix.

of inserted gaps (a large number would render the alignment increasingly unrealistic), an additional factor is implemented into the alignment algorithms; this is known as the *gap penalty function*. The overall balance between the number of aligned amino acids and the smallest number of required gaps leads to an optimal alignment.

The combination of an alignment algorithm, a scoring matrix and a gap weighting function constitutes a system which can optimally align two or more sequences. The quality of a particular alignment is described by the alignment score. It is important to know that a derived alignment for related sequences is optimal only for the chosen parameters; changing the values can lead to a different alignment and a different score. Thus, it should be borne in mind that automatic sequence alignment methods are far from being perfect. The resulting alignment should always be verified for reasonableness. All known information on all levels on protein organization (primary, secondary and tertiary structure) have to be incorporated in the examination. The derived alignment can be used as basis for the generation of a protein model only when it agrees with all known structural data.

Another fundamental problem of all sequence alignments is the fact that recognizable sequence similarity is lost more rapidly during evolution than the underlying structural similarity. Thus, it is difficult to give simple rules for the degree of similarity necessary to demonstrate unambiguously that two protein sequences are homologous. This depends strongly on the lengths of the sequences and their amino acid compositions. Several investigations have been performed in the last decade to quantify the relation between sequence and structural similarity [30–32].

Doolittle *et al.* have defined some rules of thumb which can ease the decision [30]. If the sequences are longer than 100 residues and are found to be more than 25% identical (with appropriate gaps), then they are very likely to be related. If the identity is in the range of 15–25%, then the sequences may still be related. If the sequences are less than 15% identical, they are probably not related.

In order to be able to take a decision in the undecided range between 15 and 25% sequence identity, it has to be proved that the alignment is statistically meaningful. One way to evaluate this point is by comparing the actual alignment score, which reflects the amount of similarity between two sequences, with the average alignment score of randomly permuted sequences (which were generated by randomly exchanging the amino acid residues in the original sequences). This procedure preserves the exact length and amino acid composition of the proteins, and the statistical variation of the random comparison provides a measure of the significance of the observed similarity. A number of, say n, randomizations will be generated for both sequences. Each derivative of the first sequence is then aligned against each derivative of the second, resulting in a total of n^2 alignments. Both the mean and the standard deviations of the alignments are normally reported and can be compared with the original score. As an approximate guide: if the alignment score is more than six times the standard deviation for the random alignments, most of the residues in secondary structures will be correctly aligned [31].

In the mid-1980s Chothia and Lesk performed an investigation on a test set of proteins to quantify the relation between sequence identity and 3D similarity in core regions of entirely globular proteins [32]. They have found that the success to be expected in modeling the structure of a protein from its sequence (using the 3D structure of a homologous protein as template) depends, to a high degree, upon the extent of sequence identity. They concluded that a protein structure provides a close general model for other proteins, if the sequence identity is above 50%. If the sequence identity drops to 20%, large structural differences can occur (see Figure 4.5.3). However, they found that the active site of distantly related proteins can have very similar geometries. Thus, in cases where the sequence identities are low, the structure of the active site in a protein may provide a reliable model for those in related proteins.

Aligning multiple sequences is a highly nontrivial task (in both a biological and computational sense). The accuracy of the alignment, in practice, depends largely on the choice of input sequences, the objective function and the heuristics employed. Therefore, the application of an alignment refinement algorithm to an existing or automatically generated alignment can be helpful for detecting alignment problems. Alignment refinement as a postprocessing operation is particularly worthwhile, considering the increasing importance of high quality alignments for comparative modeling. Alignment refinement has mainly relied on iterative approaches [33]. It was reported that the performance of alignment algorithms can be improved by including iteration steps during the progressive alignment. Another refinement program, RASCAL, implemented by Thompson *et al.* [34], uses a knowledge-based strategy to improve alignments; an alignment is decomposed into reliable and unreliable regions and only unreliable alignment regions are modified. For more information on this topic the reader is referred to literature [35].

4.3.2
Determination and Generation of Structurally Conserved Regions (SCRs)

Building a protein model using the homology approach is based on the fact that there are regions in all proteins belonging to the same protein family that are nearly identical in their 3D structures. These regions tend to be located at the inner cores of proteins where differences in peptide chain topology would have significant effects on the tertiary structure of the protein [36]. Accordingly, it has been observed that the secondary structural units of strongly related proteins, above all α-helices and β-strands, occupy the same relative orientations throughout the whole protein family. As a natural follow up, these regions lend themselves to being used as the basic framework for the assignment of atomic coordinates for one of the other proteins belonging to the same family. These segments are called *SCRs*.

The accurate assignment of SCRs within a family of homologous proteins is affected by several factors. How to proceed depends on the number of available crystal structures of related proteins. It is fortunate when more than one crystal structure at atomic resolution is available. In this situation, one can examine all structures in order to discover where the proteins are conserved structurally even with regard to the 3D structure. To recognize the conserved parts of the proteins, they have to be superimposed relative to each other. This is normally done using least-square fitting methods. The main problem, in this context, is the selection of the corresponding fitting atoms; this means one does not know a priori, which part of the protein should be aligned to receive the best 3D overlap. As a first approximation, the structures can be superimposed by least-square fitting of the C^α atoms [3]. The initial superposition then can be optimized using only matching points located in secondary structural elements that are found to be conserved. Several approaches, which try to solve the fitting problem automatically, have been developed [37–43].

Rossmann *et al.* [43] have suggested a method which uses the least-square fitting procedure. In the first step, two protein structures, which have to be aligned, are least-square fitted using an initial set of equivalent residues. The equivalences are then updated according to the distances between potentially equivalent residues and local directions of the main chain. The superposition and updating is repeated until no increase in the number of equivalences can be obtained.

In general, the resulting superimposed 3D structures show that large parts of the two proteins are very similar in structure and hence appear to be the SCRs, while other sections differ considerably. It should be noted, that the applied algorithms do not take the secondary structure solely into account. Because according to the definition, SCRs must be terminated at the end of a secondary structural unit, so that, for example, each single strand of a β sheet is comprising a separate SCR, secondary structural elements of the proteins have to be assigned before SCRs are determined. According to the definition, SCRs must be terminated at the end of a secondary structural unit, for example, each single strand of a β sheet comprises of a separate SCR; therefore, secondary structural elements of the proteins have to be assigned before SCRs are determined. The easiest way to obtain information about the secondary structure of any known protein is from crystal data files (for example, from the PDB files), which include the secondary structural elements detected by the crystallographers. However, as the assignment of secondary structures in crystal structure files is often subjective and sometimes incomplete, it is more convenient to use objective methods which are able to assign the secondary structural elements correctly. Programs like DSSP [44] or STRIDE [45] detect secondary structural elements on the basis of geometrical features, that is the hydrogen bonding pattern or the main-chain dihedral angle. Using these programs, which are accessible via the EMBL web site in Heidelberg, one can rapidly assign the secondary structures to all proteins if atomic coordinates exist.

The situation is more complicated when only a single homologous protein is known and can be used as reference structure for the target sequence. This is because with only one known template protein, a basis for a structural comparison does not exist. Under these circumstances, one has to detect the SCRs manually, using both sequence and structural information of the proteins. As was described before, conserved regions frequently are detected in stable secondary structure elements. Therefore, it is reasonable to carefully study as many of those elements as possible in the reference protein with the aim to discover potential clues for the existence of SCRs. Residues in the hydrophobic core tend to be more conserved with regard to sequence and 3D structure than residues at the protein surface. Amino acids involved in salt bridges, hydrogen bonds, disulfide bridges are most likely to be conserved within a protein family. The same holds true for amino acids located in the active site. Also, information derived from multiple sequence alignments can be used, with benefit, to locate the SCRs more accurately.

In many investigations on homologous proteins, it was found that SCRs show strong sequence similarities, while the variable regions show little or no sequence identity and are the sites of addition and deletion of residues. For that reason, the determined SCRs should have identical or closely related sequences. Due to the structural similarity of these regions no gaps are allowed in conserved areas.

In cases where the SCRs of the reference proteins are already known, one only has to locate the regions of the model protein that correspond to these SCRs. This is accomplished by aligning the target sequence with the sequences of the SCRs in the homolog structures. The alignment procedure which has to be applied for this purpose is slightly different from the one already described. Because SCRs cannot contain, by definition, insertions or deletions, an algorithm which disallows the introduction of gaps within SCRs, is needed. Unfortunately, the standard Needleman and Wunsch method does not have the measure for treating SCRs in a special manner. It places a gap at any location if this results in an optimized amino acid matching. For this reason, procedures [3, 22, 46] which can handle each SCR independently, have been developed. Corresponding programs generate alignments without gaps appearing within any conserved region. Once the correspondence between the reference and the target sequences has been established, the coordinates for the SCRs can be assigned. The coordinates of the reference proteins are used as basis for this assignment. In segments with identical side chains detected in reference and target proteins, all coordinates of the amino acids are transferred. In diverse regions only the backbone coordinates are transferred. The corresponding side chains are added after complete backbone (SCRs and SVRs) generation (see Section 4.3.4).

4.3.3
Construction of Structurally Variable Regions (SVRs)

Since significant differences in protein structures occur predominantly in loop regions, the construction of these SVRs is a more challenging task. Insertions and deletions due to differences in the number of amino acids additionally complicate the modeling procedure. A variety of methods for generating loops have been developed and described comprehensively in the literature [5–7, 47–49]. A good guide for modeling the missing region can be the structure of a segment of equivalent length in a homologous protein. Extensive investigations of variable regions in homologous proteins have shown that in cases where particular loops possess the same length and amino acid character, their conformation will be the same. The coordinates then can be directly transferred to the model protein in the same way as described for the SCRs. If no comparable loop exists in the protein family, two other strategies can be applied for modeling the SVRs. The coordinates for the SVRs can be either retrieved from peptide segments which are found in other proteins and that fit properly into the model's spatial environment [5–7], or by generating a loop segment *de novo* [44–46]. The former approach, which is known as the *loop search method*, looks for peptide segments in proteins which meet the specified geometrical criterion. Usually, the loop search programs scan the PDB for possible peptide segments. The specified geometry input for the data base search is given by distances and coordinates, including the residues of the regions embracing the loop segment in the model. The output of a particular search is a collection of loops satisfying the specific geometrical constraints. Usually, the 10–20 best loop fragments are retained for further examination. The loops are ranked according to goodness of fit to the desired structure. However, additional criteria not explicitly used during the loop search, can be a guide to ascertain the preference of one loop candidate over another. The retrieved fragments can be analyzed on the basis of quality of fit to the residues confining the loop region, by determining sequence identity between the original loop sequence and the sequence of the retrieved fragment, or via evaluation of steric interactions and energy criteria.

The loop search method offers the advantage that all loops found are guaranteed to possess reasonable geometry and resemble known protein conformations. It is not certain that the chosen segment fits properly into the already existing framework of the model, so that severe steric overlaps may be detected. If this happens, as an alternative, the *de novo* generation technique would be the method of choice.

Using this approach, a peptide backbone chain is built between two conserved segments using randomly generated numerical values for all the backbone dihedral angles. Several algorithms have been developed to optimize the search strategy and to reduce the computing time. Due to the complexity of this type of search method, the approach can only be used for loops smaller than seven residues.

All loops generated by data base or random search methods are usually far off from optimal geometry. For this reason, all loop regions (including confining residues) have to be subsequently refined by energy minimization techniques in order to remove steric hindrance and relax the loop conformations (see Section 4.4.3).

4.3.4
Side-Chain Modeling

Once the peptide backbone has been constructed, the next step is to add the side chains. The prediction of the numerous side-chain conformations is by far a more complex problem than the prediction of the backbone conformation of a homologous protein. Many of the side chains possess one or more degrees of freedom and therefore, can adopt a variety of energetically allowed conformations.

Several strategies have been developed in the past to find a solution for this multiple minima problem [50–57]. It has been generally assumed that identical residues in homologous proteins adopt similar conformations. Also, when the substituted side chain belongs to an amino acid pair that shows high similarity (for example, Ile and Val, or Gln and Glu), it is assumed that the side chains adopt the same orientation in the protein [50]. The situation will become more complicated if the amino acids to be substituted are not related. When the side chain to be considered is longer than its counterpart in the homologous protein or is structurally dissimilar, the side chains have to be positioned at random but in a conformation that avoids unfavorable contacts with other side chains [51]. An alternative way to obtain a suitable side-chain conformation is to select the calculated minimum conformations of the appropriate dipeptide potential energy surface [52].

A more reliable procedure was developed by analyzing the relation between the side-chain positions in homologous structures of globular proteins. It has been found that the side chains usually adopt only a small number of the many possible conformations [53, 54]. Side chains with for example two χ (chi) angles have been observed to exist in four to six common conformations. All observed rotamers are combinations of the familiar gauche and antiforms. On the basis of such statistical evaluations rotamer libraries have been developed [53, 56]. An often applied side-chain library is the one created by Ponder and Richards [53] which contains 67 rotamers for 17 amino acids. Several homology modeling programs make use of this library for generating the side chains of homologous proteins. Selecting the most probable conformation out of a rotamer library for side-chain modeling might be problematical, because this process disregards the information that is available from the equivalent side chain of the reference structure. Apart from that, the correct conformation of a side chain essentially depends on the local environment met by the amino acid in the real protein. This has been shown by several authors who have investigated well resolved protein structures [57, 58]. In the interior of a protein, hydrophobic interactions are predominant and result in tight packing of amino acid residues. Factors such as the secondary structure and tertiary contacts with other residues can influence the side-chain conformation. For this reason, methods have been developed which take into account information about the local environment and other constraints which may determine the positions of side chains. Blundell *et al.*, for example, have developed rules for mutual substitution of all twenty naturally occurring side chains in α-helical, β-sheet and loop regions—a total of $20 \times 20 \times 3 = 1200$ rules [57]. In order to determine which atom positions are preserved when substituting one amino acid for another at a topologically equivalent position, the study was performed on several sets of homologous proteins. All residues corresponding to a particular topologically equivalent position were aligned on their backbone atoms and were inspected to determine which atoms are correlated in spatial position.

A further refinement of the Ponder and Richards approach has been developed by Dunbrack *et al.* [59]. Their SCWRL program recognizes that side-chain conformations depend upon the conformation of the main chain. However, all available side-chain prediction methods invariably keep the backbone fixed.

As we have discussed, various methods for the modeling of side chains do exist. All of them can greatly assist the modeler by providing appropriate side-chain conformations. On the other hand one has to refine side-chain positions manually in several situations. Modifications must be applied, for example, when amino acids are involved in specific interactions like ion-pair formations, disulfide bridges, buried charge interactions or internal hydrogen bonds. Variations also occur when the residues are located on the protein surface and are fully accessible. Such exceptions must be treated on a case-by-case basis.

Once the final model has been built, a refinement of the structure is typically desirable. Regions where SCRs and SVRs are connected usually suffer from high steric strain and have to be minimized. Several side chains may also adopt positions which result in bad van der Waals contacts. A stepwise approach for the structure refinement is likely to produce the best result. Overall simultaneous optimization of all side chains possibly would destroy important internal hydrogen bonds and may cause a conformational change within conserved regions. In order to remove steric overlaps, conformational searches are applied for residues which show bad van der Waals contacts. Energy minimizing and/or molecular dynamics of the model are useful routes to explore the local region of conformational space and may produce a more refined structure. The details about energy minimization and molecular dynamics used for structure refinement will be described comprehensively in Section 4.4.3.

4.3.5
Distance Geometry Approach

While several reference structures are often used in the traditional homology modeling process, only one set of coordinates can be used for the construction of a particular SCR (see Section 4.3.2). The distance geometry approach in comparative modeling [38, 60, 61] offers the possibility of examining all the reference proteins at once to impose structural constraints that in turn can be used to generate conformations consistent with the data set. The first step in this procedure is the same as in the traditional homology modeling approach. The SCRs are identified and the sequence of the target protein is aligned with the sequences of the known proteins. The distance geometry method applies rules by which a multiple sequence alignment can be translated into distance and chirality constraints, which are then used as input for the calculation. By this means, one yields an ensemble of conformations for the unknown structure, where each member of the ensemble contains regions (which were constrained during the calculation) showing similar conformations and regions (which were free during the calculation) with varying arrangements. The structures of the ensemble then are energy minimized in order to eliminate structural irregularities that sometimes come forward during distance geometry calculations. The differences among the derived conformations provide an indication of the reliability of the structure prediction. A detailed description of this technique is given in a study reporting the application of the method to predict the structure of flavodoxin from *Escherichia coli* [62].

4.3.6
Secondary Structure Prediction

The best method for the generation of a structure proposal for a protein with unknown 3D structure is to base it on a homologous protein whose 3D structure is available, that is, by means of the knowledge-based approach as described in the preceding chapters. However, in cases where a homologous protein does not exist, several other methods have been developed that have concentrated on the prediction of the secondary structure. The underlying idea evolves from the fact that 90% of the residues in most proteins are engaged either in α-helices, β-strands or reverse turns. As a consequence, if the secondary structural elements are predicted accurately, it seems to be possible, to combine the predicted segments in an effort to generate the complete protein structure. Obviously, the reliability of this approach is much lower than homology modeling. Thus, it should be applied with extreme caution. However, the prediction of the secondary structure from the amino acid sequence has been widely practiced (for review see [63–69]).

Basically, three different kinds of methods can be employed for this task: statistical, stereochemical and homology/neural network-based methods. All different prediction methods rely, more or less, on information derived from known 3D structures stored in the PDB. The correct assignment of secondary structural regions in the crystal structure (see Section 4.3.2) is therefore necessary for a reliable validation of all prediction methods.

Statistical based methods were among the first which have been developed. The underlying idea takes advantage of the observation that many of the 20 amino acids show statistically significant preferences for particular secondary structures. Ala, Arg, Gln, Glu, Met, Leu and Lys for example are preferentially found in α-helices, whereas Cys, Ile, Phe, Thr, Trp, Tyr and Val more frequently occur in β-sheets. The most simple statistical method for secondary structure prediction is the one proposed by Chou and Fasman [64]. The prediction is done by calculating the probability that an amino acid belongs to a particular type of secondary structure, such as α-helix, β-sheet or turn, simply based on its frequency of occurrence as part of the respective secondary structure elements as found in the PDB. Another commonly used statistical method is that of Garnier, Osguthorpe and Robson (GOR) [65]. The success of algorithms of this type is difficult to verify because some of them merely produce tendencies toward a particular secondary structure instead of an absolute prediction. Therefore, the methods are open to divergent interpretations resulting in the situation that different authors obtain different results. The scope and limitations of the statistical methods have been demonstrated by Kabsch and Sander [66] in an analysis of three commonly used prediction methods, showing that these methods are below 56% accuracy in predicting helix, sheet and loop.

Another kind of secondary structure prediction method is based on the interpretation of the hydrophobic, hydrophilic and electrostatic properties of side chains in terms of the formulation of rules for the folding of proteins [67–69]. The method of Lim, for example, takes into account the interactions between side chains separated by up to three residues in the sequence [67] in view of their packing behavior in either the α-helical or β-sheet conformations. A sequence with alternating hydrophobic and hydrophilic side chains, for example, is likely to be found in a β-sheet strand, with hydrophilic residues exposed to the solvent and hydrophobic residues buried in the interior of the protein. Correspondingly, the stereochemical based methods have been applied successfully for the prediction of amphiphilic helices [68] or membrane spanning segments [69].

Rost and Sander have reported an algorithm which uses evolutionary information contained in multiple sequence alignments as input to neural networks [70, 71]. Neural networks potentially have a methodical advantage compared to other prediction methods because they can be trained. This means that rules determining the behavior of the studied systems are not needed in advance, but are formed by the network itself on the basis of known facts. The developed neural network method (called PHD) showed more than 70% accuracy in the prediction of three classes of secondary structure (helix, sheet, loop) on the basis of only one known homologous sequence [71, 72]. Other neural network-based prediction methods have also been reported to reach up to 80% accuracy [73, 74]. The neural network methods PHD [69] and PSIPRED [73] are at the moment the methods of choice for the prediction of unknown sequences and are integrated on several bioinformatics-related web sites. An evaluation of secondary prediction methods can be found on the EVA web-server (http://cubic.bioc.columbia.edu/eva).

Information derived from secondary structure prediction of homologous proteins are often used in addition to the results obtained in a primary sequence alignment in order to improve the location of the SCRs in a class of homologous proteins. Even when only the structure of one homologous protein is known (which can be used as template for the comparative protein modeling approach), as against several homologous sequences, it is helpful to include the predicted secondary structural elements for the homologous sequences to assign the SCRs. All available prediction methods should be applied in order to find the most probable assignment for the secondary structural elements. Of course, different methods do not yield absolutely the same result. This is demonstrated in Figure 4.3.3 using five methods (CHOU, GOR, ALB, JAMSEK, PHD) for the prediction of the known secondary structure of a cephalosporinase from *Enterobacter cloacae*. The prediction is also compared to the result of the DSSP program, which assigns the secondary structure on the basis of the known atomic coordinates.

Most of the described prediction methods are implemented in commercially available protein modeling programs or are integrated on molecular biology-related web sites. To get further information on secondary structure prediction methods see the molecular biology server at the Swiss Institute of Bioinformatics (http://www.expasy.org) or at the EBML (http://www.embl.org).

4.3.7
Threading Methods

The most welcome situation in protein modeling occurs when, for a query protein, it is possible to find another protein that is highly homologous (30% or larger sequence similarity) and for which the structure has been already solved experimentally. In such cases, the above described classical comparative modeling approach allows for the construction of a protein model with sufficient accuracy. Another typical situation is when the sequence methods, or threading procedures can detect only weak similarity [75–79]. Consequently, the similarity of the unknown 3D structure of the query sequence to the template structure cannot be quantified a priori. The two proteins may have identical topology; however they may differ in their structurally not conserved regions. Also, particular secondary structure elements may be of different size, and there may be different packing between secondary structure elements. Frequently, the actual structural similarity may be limited to only part of the structure having a common structural motif, while the remainder of the protein is completely different [79]. In these cases traditional comparative modeling methods fail and the so called *fold recognition or threading methods* have to be applied [75–81]. The earliest fold recognition methods were designed specifically to recognise folds in the absence of sequence similarity, and indeed, the sequence of the template protein was usually not taken into account at all. However, comparative modeling and threading approaches are often used simultaneously nowadays. Threading methods are closely related to the ab initio methods to protein structure prediction [82], but, whereas the ab initio methods have to explore all possible conformations, threading methods limit the search space to the conformations of known structures. Thus, threading methods fail for any protein which adopts a completely new fold.

The general threading approach involves taking a sequence and testing it on each member of a library of known protein structures. On each template, one must find the optimal sequence to structure alignment according to some score or force field. These alignments are then ranked on the calculated scores and the best ones are taken as reliable candidates [75, 83]. A collection of often applied threading methods can be found in literature [75–79].

A variety of different scoring functions have been developed for threading, [75, 76, 83, 84] most of them sharing some common properties. The applied functions have to be simple since threading calculations can require a large

```
          1                                                  50
Sequence  MMRKSLCCALLLGISCSALATPVSEKQLAEVVANTITPLMKAQSVPGMAV

CHOU          EEEEEEEEE      HHHHHHHHHHHHH  HHHHHHHHHHTTEEE
GOR       H  HH     E              HHHHHHHHH HH  HH        EEE
ALB          HHHHHHHHHHTTTTT TTT   HHHHHHHHHHHHHHH    TTEEE
JAMSEK    HHHHHHHH      TTT      HHHHHHHTTT                  E
PHD       TTTT HHHHHHHHHHHH   T  HHHHHHHHHHHHHHHH   TTTTT EE
DSSP                  HHHHHHHHHHHHHHHHHHH    EEEEEEE

          51                                                 100
Sequence  AVIYQGKPHYYTFGKADIAANKPVTPQTLFELGSISKTFTGVLGGDAIAR

CHOU      EEEEE    EEEEE HHHHH    EEEEEEE       EEEEEETTTHHHH
GOR       EEEE     EEE HHHHH      EHEEHE H  EEEEEE      EH
ALB       EEEE TTTTEEEE        TT TT   HHHHHHHHHHHHHTTHHHHH
JAMSEK    EEEE TTT EEEETTT   TTT  TTT        TTTEEEE  HHHHHH
PHD       EEEE TT EEEE     TTTTTTTTT                     H
DSSP      ETTEEEEEEEEEEEETTTT   EE TTTTEEEEE    HHHHHHHHHHHH

          101                                                150
Sequence  GEISLDDAVTRYWPQLTGKQWQGIRMLDLATYTAGGLPLQVPDEVTDNAS

CHOU      HHHHHHHEEEEEE     TT      HHHHHHHH  TT EEEETT  TTTTT
GOR        HE  HHHEE E HH      HHHEEHHHH  H            HHH HH
ALB        THHHHHHHHH                   TTTEEEEETTT    TTHH
JAMSEK    HH HHHHHHHHH TTT TTTHHHHHHHHHHEEE      TTT  TTHH
PHD         TTT   TT H      T      HHHHH TTTTT TT   H HHH
DSSP        TTTT     TTTTTT    TTT HHHHHH TTTTTTTTTTTT  HHH

          151                                                200
Sequence  LLRFYQNWQPQWKPGTTRLYANASIGLFGALAVKPSGMPYEQAMTTRVLK

CHOU      EEEEEEEEETT  TTEEEEEEEEEEEEEEHHHHHHTTT    HHHHHHHHHH
GOR       HHEEHH          EEEHH  HHHHHH          HHHHHHHHH
ALB       HHHHH    TTT TTT       HHHHHHHH  TTT HHHHHHHHHHH
JAMSEK    HHHHHHTTTTTTTTT   EEETTT EEEEE    TTT        HHHHHH
PHD       HHHHHHH TTTTTTTT EE  TT   HHHHHH    TTT HHHHHHHHHHHH
DSSP      HHHHHHH       TTTTEE  HHHHHHHHHHH    HHHHHHHH

          201                                                250
Sequence  PLKLDHTWINVPKAEEAHYAWGYRDGKAVRVSPGMLDAQAYGVKTNVQDM

CHOU      HHHHHHEEEEEHHHHHHH        TTTHHHHHHTTHHHHHH    EEEEEEHH
GOR        H H HHHHH  HHHHHHHHH H    EEE   HHHHHHH   E HHHH
ALB           EEEEETTT          TTT EEE HHHHHHHHHHHHHHHHH
JAMSEK    HHHTTTEEEEE           TTT  EEE         EEEEE
PHD       H TTTT   TTHHHHHHHHH  TTTT EE TTT T    TTT  HHHHH
DSSP       TTTEETTT        EETTTTEE   TTTHHHHHHEEEEHHHH
```

Figure 4.3.3 Comparison of secondary structure predictions using different methods for a crystallographically resolved cephalosporinase from *Enterobacter cloacae*.

```
            251                                               300
Sequence    ANWVMANMAPENVADASLKQGIALAQSRYWRIGSMYQGLGWEMLNWPVEA

CHOU        HHHHHHHHHHHHHHHHHHHH EEEEEEEEEEE   EEEEE HHHHHHHHH
GOR         HHHHHH  H   H HHHHHHHHHHHHHHH  EEEEEEE       HHHE
ALB         HHHHHHHH HHHHHHHHHHHHHHHHHHHH    HHHHHHHHHHHH
JAMSEK        EEE      HHHHHHHHHHHHHHEEEETTTEEEEEETTT            T
PHD         HHHHHH  T   T  HHHHHHHHHHH              T       TTTTT
DSSP        HHHHHHHH      HHHHHHHHHHH EEEEETTEEETTTTEEEETTT  H
```

```
            301                                               350
Sequence    NTVVEGSDSKVALAPLPVAEVNPPAPPVKASWVHKTGSTGGFGSYVAFIP

CHOU        HHHHH TTHHHHHHHHHHHHH TT    HHHHHH TTTTTTTEEEEEEEE
GOR            EEE      HHH      H        H   EEEEE     E  EEEEE
ALB         EEEETTTTEEEEE EEEEEE TT                       EEEEET
JAMSEK      TT      TTT             TTT                      EEEE
PHD            TTT        TTTTT    TTTTT      E    TT  T    EEEEE
DSSP        HHHHHHH HHHHH  EE EEEEEE  TTTEEEEEEEEEETTEEEEEEEE
```

```
            351                         381
Sequence    EKQIGIVMLANTSYPNPARVEAAYHILEALQ

CHOU         EEEEEEEEEETTTTTTTHHHHHHHHHHHHHHHH
GOR         HHHHHEEEE          HHHHHHHHHHHHHH
ALB         TTEEEEEEE        TT HHHHHHHHHH
JAMSEK       EEEEE    TTT
PHD          EEEEE   TTTTTHHHHHHHHHHHHHH T
DSSP         EEEEEEE       HHHHHHHHHHHHHH
```

H = α-Helix

E = β-Strand

T = Turn

CHOU: Chou–Fasman [64]

GOR: Garnier–Osguthorpe–Robson [65]

ALB: Ptitsyn–Finkelstein [87]

JAMSEK: Mrazek–Kypr [88]

PHD: Rost–Sander [70]

DSSP: Kabsch–Sander [66]

Figure 4.3.3   (*Continued*).

number of possibilities to be considered. Many of the scoring functions used in threading programs are potentials of mean force which are also referred to as *knowledge-based potentials* [83, 85]. They are quite different from the traditional force fields (which were described in general in Section 2.2.1). The basic idea of knowledge-based force fields is that molecular structures observed from X-ray analysis or NMR contain a wealth of information on the stabilizing forces within macromolecules. Using statistical methods, the underlying rules governing the 3D structure of proteins have been revealed. It is the basic assumption of the Boltzmann principle that frequently observed states correspond to low-energy states of a system. Thus, the potentials of mean force are compiled by extracting relative frequencies of particular atom pair interactions from a data base of protein structures [85]. The potentials of mean force consist usually of interactions among particular atom pairs and protein–solvent interactions. They incorporate all kind of forces (electrostatic, dispersion, etc.) acting between particular protein atoms as well as the influence of the surrounding solvent on the interaction and therefore, can be used to predict the structure of a macromolecule from its primary sequence. Potentials of mean force have been applied for the prediction of protein folds and even for the detection of errors in protein models and experimentally determined structures [83–85].

The utility of a protein model, either generated by comparative modeling or threading, depends upon the use to which the model is put. The accuracy of ab initio and threading models is too low for problems requiring high-resolution structure information, such as the traditional drug design. Instead, the low-resolution model produced by these methods can reveal structural and functional relationships between proteins not apparent from their amino acid sequence and provide a framework for analyzing spatial relationships between evolutionary conserved residues or between residues shown experimentally to be functionally important. To evaluate the different approaches for protein structure prediction, a competition called 'Critical Assessment of techniques for protein Structure Prediction (CASP)' was organized in 1994–95 [86]. In these CASP experiments the scientific community was invited to predict the 3D structures of novel proteins from their amino acid sequences. The 3D structures of the proteins were solved by X-ray crystallography but were not published. Until now four CASP competitions have taken place. In CASP4 it was concluded that modeling of protein structures has matured now into a practical technology. It is possible, in principle, to produce useful models for more than half of the sequences entering the general sequence databases [82]. The CASP competitions provide a solid basis for assessing the reliability of protein models and their underlying modeling approaches.

References

1. Blundell, T.L. and Johnson, L.N. (1976) *Protein Crystallography*, Academic Press, New York.
2. Bashford, D., Chothia, C., and Lesk, A.M. (1987) Determinants of a protein fold. Unique features of the globin amino acid sequences. *Journal of Molecular Biology*, **196**, 199–216.
3. Greer, J. (1981) Three-dimensional structure of abnormal human haemoglobins Chesapeake and J Capetown. *Journal of Molecular Biology*, **153**, 1027–42.
4. Chothia, C. and Lesk, A.M. (1982) Evolution of proteins formed by beta-sheets. I. Plastocyanin and

azurin. *Journal of Molecular Biology*, **160**, 309–42.

5. Johnson, M.S., Srinivasan, N., Sowdhamini, R., and Blundell, T.L. (1994) Knowledge-based protein modeling. *Critical Reviews in Biochemistry and Molecular Biology*, **29**, 193–316.

6. Sali, A., Overington, J.P., Johnson, M.S., and Blundell, T.L. (1990) From comparisons of protein sequences and structures to protein modeling and design. *Trends in Biochemical Sciences*, **15**, 235–40.

7. Jones, T.A. and Thirup, S. (1986) Using known substructures in protein model building and crystallography. *EMBO Journal*, **5**, 819–22.

8. Dudek, M.J. and Scheraga, H.A. (1990) Protein structure prediction uses a combination of sequence homology and global energy minimization. *Journal of Computational Chemistry*, **11**, 121–51.

9. Levin, R. (1987) When does homology mean something else? *Science*, **237**, 1570.

10. Thornton, J.M. and Gardner, S.P. (1989) Protein motifs and database searching. *Trends in Biochemical Sciences*, **14**, 300–4.

11. Orengo, C.A., Brown, N.P., and Taylor, W.R. (1992) Fast structure alignment for protein databank searching. *Proteins Structure Function and Genetics*, **14**, 139–46.

12. Pearson, W.R. (1990) Rapid and sensitive sequence comparison with FASTP and FASTA. *Methods in Enzymology*, **183**, 63–98.

13. Altschul, S.F., Madden, T.L., Schaffer, A.A. *et al.* (1997) Gapped BLAST and PSI-BLAST: a new generation of protein database search programs. *Nucleic Acids Research*, **25**, 3389–402.

14. HOMOLOGY and MODELLER, Accelrys, San Diego. http://www.accelrys.com.

15. Sali, A. and Blundell, T.L. (1993) Comparative protein modeling by satisfaction of spatial restraints. *Journal of Molecular Biology*, **234**, 779–815.

16. SYBYL BIOPOLYMER, Tripos Associates, St. Louis. http://www.tripos.com.

17. Vriend, G. (1990) What If: A molecular modeling and drug design program. *Journal of Molecular Graphics*, **8**, 52–56. http://www.swift.cmbi.ru.hl/whatif.

18. Thompson, J.D., Higgins, D.G., and Gibson, T.J. (1994) CLUSTAL-W – improving the sensitivity of progressive multiple sequence alignment through sequence weighting, position-specific gap penalties and weight matrix choice. *Nucleic Acids Research*, **22**, 4673–80.

19. Barton, G.J. (1990) Protein multiple sequence alignment and flexible pattern matching. *Methods in Enzymology*, **183**, 403–28.

20. Needleman, S.B. and Wunsch, C.D. (1970) A general method applicable to the search for similarities in the amino acid sequence of two proteins. *Journal of Molecular Biology*, **48**, 443–53.

21. Sander, C. and Schneider, R. (1996) The HSSP database of protein structure-sequence alignments. *Nucleic Acids Research*, **24**, 201–5.

22. Schuler, G.D., Altschul, S.F., and Lipman, D.J. (1991) A workbench for multiple alignment construction and analysis. *Proteins Structure Function and Genetics*, **9**, 180–90.

23. Vingron, M., Argos, P., and Vogt, G. (1991) Protein sequence comparison: methods and significance. *Protein Engineering*, **4**, 375–83.

24. Boswell, D.R. and McLachlan, A.D. (1984) Sequence comparison by exponentially-damped alignment. *Nucleic Acids Research*, **12**, 457–65.

25. Dayhoff, M.O., Schwartz, R.M., and Orcutt, B.C. (1978) A Model of evolutionary change in proteins, in *Atlas of Protein Sequence and Structure* (ed. M.O. Dayhoff), National Biomedical Research Foundation, Washington, DC, Vol. 5, Suppl. 3, pp. 345–52.

26. Gribskov, M., McLachlan, A.D., and Eisenberg, D. (1987) Profile analysis: detection of distantly related proteins. *Proceedings of the National Academy of*

Sciences of the United States of America, **84**, 4355–58.

27. Landes, C., Risler, J.L., and Henaut, A. (1992) A comparison of several similarity indices used in the classification of protein sequences: a multivariate analysis. *Nucleic Acids Research*, **20**, 3631–37.

28. Engelman, D.M., Steitz, T.A., and Goldman, A. (1986) Identifying nonpolar transbilayer helices in amino acid sequences of membrane proteins. *Annual Review of Biophysics and Biophysical Chemistry*, **15**, 321.

29. Gonnet, G.H., Cohen, M.A., and Benner, S.A. (1992) Exhaustive matching of the entire protein sequence database. *Science*, **256**, 1443–45.

30. Doolittle, R. (1990) Searching through sequence databases. *Methods in Enzymology*, **183**, 736–72.

31. Barton, G.J. and Sternberg, M.J.E. (1990) Flexible protein sequence patterns. A sensitive method to detect weak structural similarities. *Journal of Molecular Biology*, **212**, 389–402.

32. Chothia, C. and Lesk, A.M. (1986) The relation between the divergence of sequence and structure in proteins. *EMBO Journal*, **5**, 823–26.

33. Wang, Y. and Li, K.B. (2004) An adaptive and iterative algorithm for refining multiple sequence alignment. *Computational Biology and Chemistry*, **28**, 141–48.

34. Thompson, J.D., Thierry, J.C., and Poch, O. (2003) RASCAL: rapid scanning and correction of multiple sequence alignments. *Bioinformatics*, **19**, 1155–61.

35. Wallace, I.M., O'Sullivan, O., and Higgins, D.G. (2005) Evaluation of iterative alignment algorithms for multiple alignment. *Bioinformatics*, **21**, 1408–14.

36. Perutz, M.F., Bolton, W., Diamond, R. *et al.* (1964) Structure of haemoglobin. An X-ray examination of reduced horse haemoglobin. *Nature*, **203**, 687–90.

37. Maggiora, G.M., Rohrer, D.C., and Mestres, J. (2001) Comparing protein structures: a Gaussian-based approach

to the three-dimensional structural similarity of proteins. *Journal of Molecular Graphics & Modelling*, **19**, 168–78.

38. Reinhardt, A. and Eisenberg, D. (2004) DPANN: improved sequence to structure alignments following fold recognition. *Proteins*, **56**, 528–38.

39. Eisenberg, D., Marcotte, E., McLachlan, A.D., and Pellegrini, M. (2006) Bioinformatic challenges for the next decade(s). *Philosophical Transactions of the Royal Society of London. Series B: Biological Sciences*, **361**, 525–27.

40. Ilyin, V.A., Abyzov, A., and Leslin, C.M. (2004) Structural alignment of proteins by a novel TOPOFIT method, as a superimposition of common volumes at a topomax point. *Protein Science*, **13**, 1865–74.

41. Jung, J. and Lee, B. (2000) Protein structure alignment using environmental profiles. *Protein Engineering*, **13**, 535–43.

42. Vriend, G. and Sander, C. (1991) Detection of common three-dimensional substructures in proteins. *Proteins Structure Function and Genetics*, **11**, 52–58.

43. Matthews, B.W. and Rossmann, M.G. (1985) Comparison of protein structures. *Methods in Enzymology*, **115**, 397–420.

44. Kabsch, W. and Sander, C. (1983) Dictionary of protein secondary structure: pattern recognition of hydrogen-bonded and geometrical features. *Biopolymers*, **22**, 2577–637.

45. Frishman, D. and Argos, P. (1995) Knowledge-based protein secondary structure assignment. *Proteins Structure Function and Genetics*, **23**, 566–79.

46. Sanchez, R., Pieper, U., Melo, F. *et al.* (2000) Protein structure modeling for structural genomics. *Nature Structural Biology Supplement*, **7**, 986–90.

47. Bruccoleri, R.E. and Karplus, M. (1990) Conformational sampling using high-temperature molecular dynamics. *Biopolymers*, **29**, 1847–62.

48. Novotny, J., Bruccoleri, R.E., Davis, M., and Sharp, K.A. (1997)

Empirical free energy calculations: a blind test and further improvements to the method. *Journal of Molecular Biology*, **268**, 401–11.

49. Pellequer, J.L. and Chen, S.W. (1997) Does conformational free energy distinguish loop conformations in proteins? *Biophysical Journal*, **73**, 2359–75.

50. Aloy, P., Pichaud, M., and Russell, R.B. (2005) Protein complexes: structure prediction challenges for the 21st century. *Current Opinion in Structural Biology*, **15**, 15–22.

51. Blundell, T.L., Sibanda, B.L., and Pearl, L. (1983) Three-dimensional structure, specificity and catalytic mechanism of renin. *Nature*, **304**, 273–75.

52. Allen, S.C., Acharya, K.R., Palmer, K.A. *et al.* (1994) A comparison of the predicted and X-ray structures of angiogenin. Implications for further studies of model building of homologous proteins. *Journal of Protein Chemistry*, **13**, 649–58.

53. Ponder, J. and Richards, F.M. (1987) Tertiary templates for proteins. Use of packing criteria in the enumeration of allowed sequences for different structural classes. *Journal of Molecular Biology*, **193**, 775–91.

54. Summers, N.L., Carlson, W.D., and Karplus, M. (1987) Analysis of side-chain orientations in homologous proteins. *Journal of Molecular Biology*, **196**, 175–98.

55. Islam, S.A. and Sternberg, M.J. (1989) A relational database of protein structures designed for flexible enquiries about conformation. *Protein Engineering*, **2**, 431–42.

56. Nayeem, A. and Scheraga, H.A. (1994) A statistical analysis of side-chain conformations in proteins: comparison with ECEPP predictions. *Journal of Protein Chemistry*, **13**, 283–96.

57. Dean, C.M. and Blundell, T.L. (2001) CODA: a combined algorithm for predicting the structurally variable regions of protein models. *Protein Science*, **10**, 599–612.

58. Schrauber, H., Eisenhaber, F., and Argos, P. (1993) Rotamers: to be or not to be? An analysis of amino acid side-chain conformations in globular proteins. *Journal of Molecular Biology*, **230**, 592–612.

59. Canutescu, A.A., Shelenkov, A.A., and Dunbrack, R.L. Jr. (2003) A graph-theory algorithm for rapid protein side-chain prediction. *Protein Science*, **12**, 2001–14.

60. Havel, T.F. and Snow, M. (1991) A new method for building protein conformations from sequence alignments with homologues of known structure. *Journal of Molecular Biology*, **217**, 1–7.

61. Srinivasan, S., March, C.J., and Sudarsanam, S. (1993) An automated method for modeling proteins on known templates using distance geometry. *Protein Science*, **2**, 277–89.

62. Jang, J.X. and Havel, T.F. (1993) SESAME: a least-squares approach to the evaluation of protein structures computed from NMR data. *Journal of Biomolecular NMR*, **3**, 355–60.

63. Fasman, G.D. (1989) Protein conformational prediction. *Trends in Biochemical Sciences*, **14**, 295–99.

64. Chou, P.Y. and Fasman, G.D. (1974) Prediction of protein conformation. *Biochemistry*, **13**, 211–45.

65. Garnier, J., Osguthorpe, D.J., and Robson, B. (1978) Analysis of the accuracy and implications of simple methods for predicting the secondary structure of globular proteins. *Journal of Molecular Biology*, **120**, 97–120.

66. Kabsch, W. and Sander, C. (1983) How good are predictions of protein secondary structure. *FEBS Letters*, **155**, 179–82.

67. Lim, V.I. (1974) Algorithms for prediction of alpha-helical and beta-structural regions in globular proteins. *Journal of Molecular Biology*, **88**, 873–94.

68. Rees, D.C., DeAntonio, L., and Eisenberg, D. (1989) Hydrophobic organization of membrane proteins. *Science*, **245**, 510–13.

69. Kyte, J. and Doolittle, R.F. (1982) A simple method for displaying the

hydropathic character of a protein. *Journal of Molecular Biology*, **157**, 105–32.

70. Rost, B. and Sander, C. (1993) Prediction of protein secondary structure at better than 70% accuracy. *Journal of Molecular Biology*, **232**, 584–99.

71. Rost., B. and Sander, C. (1994) Combining evolutionary information and neural networks to predict protein secondary structure. *Proteins Structure Function and Genetics*, **19**, 55–72.

72. Rost, B. and Eyrich, V.A. (2001) EVA: large-scale analysis of secondary structure prediction. *Proteins Structure Function and Genetics Supplement*, **5**, 192–99.

73. Bryson, K., McGuffin, L.J., Marsden, R.L. *et al.* (2005) Protein structure prediction servers at University College London. *Nucleic Acids Research*, **33**, W36–W38.

74. Cuff, J.A., Clamp, M.E., Sidddiqui, A.S. *et al.* (1998) JPred: a consensus secondary structure prediction server. *Bioinformatics*, **14**, 892–93.

75. Geer, L.Y., Domrachev, M., Lipman, D.J., and Bryant, S.H. (2002) CDART: protein homology by domain architecture. *Genome Research*, **12**, 1619–23.

76. Jones, D.T. (1999) Protein secondary structure prediction based on position-specific scoring matrices. *Journal of Molecular Biology*, **287**, 797–815.

77. Wilmanns, M. and Eisenberg, D. (1995) Inverse protein folding by the residue pair preference profile method: estimating the correctness of alignments of structurally compatible sequences. *Protein Engineering*, **8**, 626–35.

78. Skolnick, J. (2006) In quest of an empirical potential for protein structure prediction. *Current Opinion in Structural Biology*, **16**, 166–71.

79. Panchenko, A.R., Marchler-Bauer, A., and Bryant, S.H. (2000) Combination of threading potentials and sequence profiles improves fold recognition. *Journal of Molecular Biology*, **296**, 1319–31.

80. Kolinski, A., Betancourt, M.R., Kihara, D. *et al.* (2001) Generalized comparative modeling (GENECOMP): a combination of sequence comparison, threading, and lattice modeling for protein structure prediction and refinement. *Proteins Structure Function and Genetics*, **44**, 133–49.

81. Xu, Y. and Xu, D. (2000) Protein threading using PROSPECT: design and evaluation. *Proteins Structure Function and Genetics*, **40**, 343–54.

82. Moult, J. (1999) Predicting protein three-dimensional structure. *Current Opinion in Biotechnology*, **10**, 583–88.

83. Sippl, M.J. (1990) Calculation of conformational ensembles from potentials of mean force. An approach to the knowledge-based prediction of local structures in globular proteins. *Journal of Molecular Biology*, **213**, 859–83.

84. Jones, D.T. and Thornton, J.M. (1996) Potential energy functions for threading. *Current Opinion in Structural Biology*, **6**, 210–16.

85. Sippl, M.J. (1993) Recognition of errors in three-dimensional structures of proteins. *Proteins Structure Function and Genetics*, **17**, 355–62.

86. Mosimann, S., Meleshko, S., and James, M.N.G. (1995) A critical assessment of comparative molecular modeling of tertiary structures of proteins. *Proteins Structure Function and Genetics*, **23**, 301–17.

87. Ptitsyn, O.B. and Finkelstein, A.V. (1983) Theory of proteins secondary and algorithm of its prediction. *Biopolymers*, **22**, 15–25.

88. Mrazek, J. and Kypr, J. (1988) Computer program Jamsek combining statistical and stereochemical rules for the prediction of protein secondary structure. *Computer Applications in the Biosciences*, **4**, 297–302.

4.4
Optimization Procedures – Model Refinement – Molecular Dynamics

4.4.1
Force Fields for Protein Modeling

Protein models derived from either comparative modeling or crystal structures need further refinement. In the course of generation of protein models, the loop and side-chain conformations, in general, are chosen arbitrarily. Therefore, the conformations do not correspond to energetically reasonable structures. Also, crystal structures have to be relaxed to remove the internal strain resulting from the crystal packing forces or to remove close contacts between hydrogen atoms or amino acid residues that may have been added to the crystal coordinates after structure determination.

As protein models consist of hundreds or thousands of atoms, the only feasible method to compute systems of such size are molecular mechanics calculations. The common force fields used in molecular mechanics calculations are based, in principal, on the equations for the potential energy function as described in Section 2.2.1. However, force fields for protein modeling differ in some respect from small molecule force fields. Besides the specific parameterization for proteins and DNA, some simplifications are frequently introduced. In some force fields, nonpolar hydrogen atoms are not represented explicitly, but are included in the description of the heavy atoms to which they are bonded. In contrast, polar hydrogens, which may act as potential partners in hydrogen bonding, are treated explicitly. This procedure is denoted as united-atom model. In the AMBER [1, 2] force field, both the united-atom model and an all-atom representation can be applied, while the GROMOS force field [3] offers only the united-atom model. Other simplifications can be made by introducing cutoff radii [4] to reduce the time-consuming part of calculating nonbonded interactions between atoms separated by distances larger than a defined cutoff value.

An additional variation is made with respect to the treatment of the electrostatic interactions. As the explicit inclusion of solvent is still a problem, some force fields try to simulate the solvent effect by introducing a distance-dependent dielectric constant [1, 2]. Especially in case of macromolecules, the electrostatic field in the environment of the system cannot be considered to be continuous. A differentiating procedure in calculating the particular properties is necessary to reflect the electrostatic effects, which depend on the local situation, for example, in the binding pocket or on the surface of the protein. A detailed discussion of this subject and a description of methods handling the complex situation are given in Section 4.6.1.

The modifications established in protein force fields are numerous and cannot be discussed in detail here. A comprehensive description of potential simplifications is given in Ref. [5]. It should be kept in mind that each simplification applied can result in a loss of accuracy. The decision on the

Molecular Modeling. Basic Principles and Applications. 3rd Edition
H.-D. Höltje, W. Sippl, D. Rognan, and G. Folkers
Copyright © 2008 Wiley-VCH Verlag GmbH & Co. KGaA, Weinheim
ISBN: 978-3-527-31568-0

force field to be chosen strongly depends on the problem to be investigated and always the most accurate force field that is applicable for the whole study has to be selected. The use of different force fields within a molecular modeling study should generally be avoided.

There are several common force fields for protein modeling implemented in software programs. The following list is not complete but comprises some of the most frequently used methods: AMBER [1, 2], CVFF [6], CHARMM [7] and GROMOS [3] force fields.

4.4.2
Geometry Optimization

The algorithms used in the minimization procedures for proteins are the same as for small molecules and have been discussed in detail in Section 2.2.3. The minimization algorithms applied to optimize the geometry usually find only the local minimum on the potential energy surface that is closest to the initial coordinates. In the case of a well-resolved crystal structure, the minimization will directly yield one energetically favorable conformation. The relaxation of a crystal structure is usually a straightforward procedure. However, crystal coordinates sometimes have, even if highly resolved, several unfavorable atomic interactions. These disordered atomic positions cause large initial forces, resulting in artificial movements away from the original structure when starting the minimization process. A general approach to avoid these large deviations is to relax the protein model gradually.

A more profound solution would be to assign tethering forces to all heavy atoms of the crystal structure in the first stage of minimization. The tethering constant is a force applied to fix atomic coordinates on predefined positions. The strength of the tethering force can be selected by the user and affects the extent of movement of the atoms measured by the root mean square (rms) deviation from the initial coordinates. When tethering the heavy atoms, the hydrogen atoms and perhaps solvent molecules are allowed to adjust their positions in order to minimize the total potential energy. A suitable minimization method for this purpose is the steepest descent algorithm. For this initial relaxation step, a crude convergence criterion can be applied or the process can be completed by defining a maximum number of permissible minimization steps.

Subsequently, it is recommendable to tether only the well-defined main chain atoms. Now the side chains are allowed to move and to adjust their orientations. The steepest descent method is suitable in this case as well.

Ultimately the restraints are removed in the last step so that the final minimum represents a totally relaxed conformation. The minimization algorithm should be changed to conjugate gradient to reach convergence in an effective way.

The application of tethering forces can also be useful and necessary in the modeling of incomplete systems. Incomplete systems may result in an X-ray study if certain parts of the crystals or included solvent molecules cannot be resolved adequately. Also, active-site models of enzymes or binding pockets of proteins used for the investigation of potential ligand–protein interactions are examples of typical incomplete systems.

Owing to the absence of neighboring amino acids or solvent molecules, the atom positions at the surface of a protein are mobile. As a consequence, large deviations from the initial positions will result after minimization and the final geometry has to be regarded as an artifact. Therefore atoms or the ends of side chains are tethered at their original positions to avoid unrealistic atom movements at the surface of the protein.

With the objective of confirming the accuracy of the relaxed protein model, the deviations from the experimental structure should be examined. For this purpose, the initial structure and the final geometry are superimposed using least-squares fit methods. As fitting points, normally either all backbone atoms or only backbone atoms of the well-refined secondary structural elements are used. The quality of the fit can be judged by the rms deviation of the optimized form from the initial geometry. The value of the rms deviation strongly dependent on the number and localization of atoms that are considered for the fit. Naturally, a fit of all heavy atoms would result in a much higher rms value than a fit that is confined to backbone atoms only, mainly owing to the greater mobility of side chains.

If the generated model is based merely on comparative modeling techniques, the loop and side-chain conformations need further refinement. It is necessary to carefully investigate their conformational behavior and analyze the potential energy surface for other possible low energy conformations. A valuable tool for this purpose is molecular dynamics simulation. The relaxed geometry obtained as a result of the minimization procedure can be used as starting point for molecular dynamics simulations.

4.4.3
The Use of Molecular Dynamics Simulations in Model Refinement

As mentioned above, the refinement of models derived from comparative modeling studies is a must. Loop and side-chain conformations of the derived protein model represent only one possible conformation and the minimum structure found by the minimization algorithms represents only one local minimum. To detect the energetically most favored 3D structure of a system, a modified strategy is needed for searching the conformational space more thoroughly.

Molecular dynamics simulations offer an effective means to solve this problem, especially for molecules containing hundreds of rotatable bonds. A molecular dynamics simulation is performed by integrating the classical equations of motion over a period of time for the molecular system. The resulting trajectory for the molecule can be used to compute the average and time-dependent properties of the system. The theory of the molecular dynamics method and its application in conformational search of small molecules have been discussed and illustrated on some impressive examples in Section 4.3.3. Here we focus on the utilization of this technique in the refinement of 3D macromolecular structures.

The use of molecular dynamics has made an essential contribution to the understanding of dynamical processes in proteins at the atomic level. However, there are some basic limitations and problems arising with increasing size and associated with the immense number of degrees of freedom of large molecular systems.

Although the computer resources have become sufficiently powerful to enable handling of quite large systems, it still is necessary to introduce some modifications to reduce the computation time demanded [5]. A very useful side effect of the simplifications used is the fact that they open a possibility for longer time periods to be chosen for the sampling of the dynamical simulation. This offers a way to observe the dynamical behavior of large molecular systems more completely.

Before discussing the various possibilities in detail, it must be mentioned again that each modification and reduction of the number of degrees of freedom can cause a lack of accuracy, and whether a respective simplification can be tolerated has to be checked carefully.

One basic and very common simplified procedure is the use of united-atom potential energy functions. The underlying theory of this methodology has already been described above. Most of the force fields for protein modeling like AMBER [1, 2] and GROMOS [3] are based on these algorithms. Omission of the nonpolar hydrogens in a united-atom force field does significantly reduce the number of particles in a large biomolecule. A further possibility to reduce the demand for computational time is provided by application of the SHAKE [8] algorithm. In the SHAKE procedure, additional forces are assigned to the atoms aiming to keep bond lengths fixed at equilibrium values. This is very useful for several reasons. Most importantly, bond stretching energy terms must not be calculated for the frozen bonds. The magnitude of the integration step depends on the fastest occurring vibrations in a molecule. This is usually the high-frequency vibration of the C–H bond stretching. The value of this period is on the order of 10^{-14} seconds, and therefore the integration step should be chosen to be 10^{-15} seconds (1 femtosecond). Applying the SHAKE algorithm on this type of C–H bonds allows a larger integration step with the effect of reducing the necessary computational expense, thereby offering the possibility of simulating the system over a longer time period. Also, the definition of cutoff radii leading to a neglect of nonbonded interactions beyond the defined distance yields the same effect.

In addition, the application of a well-balanced computational protocol may save computational time. In this respect, several parts of a protein can be kept rigid and the molecular dynamics simulations are then carried out only for flexible parts such as loops or side chains while well-defined secondary structures like α-helices or β-strands in the core of the protein are not taken into account. The availability of NMR data can also be a reason to fix atoms, side chains or parts of the protein at their initial coordinates to impede their movement away from the experimentally derived positions. Again, it has to be cautioned that restraining parts of flexible molecules leads to a reduction in the number of degrees of freedom. Without any doubt, a more comprehensive exploration of the conformational space and hence better results are achieved when no positional restraints on parts of the protein structure are applied.

All the methods that have been mentioned enhance the efficiency of the molecular dynamics simulations. Nevertheless, for some problems, the feasible timescale is still too short. If, for example, the binding of a ligand to an enzyme or receptor protein as well as the thereby triggered conformational change are to be studied, the time required for this process can be in the order of picoseconds or even nanoseconds [9]. The same timescale would be indispensable for a simulation of protein folding. Both types of problems are still not solvable.

Several modifications of high-temperature molecular dynamics simulations have been successfully applied in conformational analysis of peptides and in the refinement of protein models. In this respect, two important methods, the high-temperature annealed molecular dynamics simulation [10] and simulated annealing [11] have been discussed in Section 2.3.3. They are valuable tools and are also widely used for investigating peptides and proteins [12–16].

A sensitive point in all molecular dynamics protocols is the choice of the suitable simulation temperature. Usually the simulation is performed in the range between 300 and 400 K. On one hand, it has to be sufficiently high to prevent the system from getting stuck in one particular region of the conformational space, but on the other, it should not be too high because this could result in distorted high-energy conformations even after minimization [16]. Another commonly observed problem in the application of high-temperature molecular dynamics simulations of proteins and peptides is the appearance of trans–cis interconversions of peptide bonds. These artifacts can be avoided by using lower temperatures at the expense of conformational search efficiency or by introducing torsional restraints onto the peptide bonds.

4.4.4
Treatment of Solvated Systems

The conformational flexibility of a protein especially at the surface and in loop regions is affected strongly by the surrounding environment. The nonexistence of neighboring atoms at the surface of protein leading to *in vacuo* conditions for these regions of the protein, and the problems associated therewith have already been mentioned in the discussion of the minimization process. Of course, the accuracy of molecular dynamics simulations increases by including explicit solvent. Unfortunately, this is still an unresolved problem. One possibility to mimic the effect of solvent and to account for the boundary phenomena is the use of distance-dependent dielectric constants.

Enwrapping the molecule with a sphere of solvent molecules can improve the accuracy of the molecular dynamics simulations, because by doing so, at least parts of the effect of solvation are imitated. At this point, it is important to note that there is a decisive difference between simple solvent water and structural water. Structural water is very important for the functionality of the protein and can influence the conformation even in the core of the protein. Therefore, structural water has to be always included explicitly in the calculations.

The next level of improvement is the embedding of the protein in a complete solvent box containing thousands of water molecules in order to simulate a natural solvent environment. This is not always possible because the required computational effort is immense. However, owing to the dramatic increase in computational power during the last few years, more and more simulations are carried out, where explicit solvent systems have been used [17–22]. A comprehensive review on molecular dynamics simulations of proteins in different environments is given in Ref. [23].

In several cases, the use of realistic water models with thousands of molecules is too time consuming. For this reason, specific methods using a simplified representation of solvent molecules have been developed [18]. Solvent molecules, for example, can be substituted by neutral spherical atoms. This kind of proceeding significantly reduces the computational effort. A detailed discussion of all procedures used in this context is beyond the scope of this book. Nevertheless, it is important to consider that inclusion of the solvent environment at any level of complexity into the calculations is an important means for improving the accuracy and the reliability of molecular dynamics simulations, especially for large biomolecular systems.

Like all computational branches of science, molecular dynamics benefits from the seemingly never-ending improvements in computer hardware; simulations that were hard work for yesterday's supercomputers can be carried out today using standard office workstations. The gain in computational efficiency now offers the simulations of ever bigger systems using more realistic boundary conditions and better sampling due to longer sampling times. Recently, realistic simulations of systems as complex as transmembrane channels and receptors have become feasible [24]. Sophisticated simulation systems with 'natural' lipid bilayers (e.g. dipalmitoylphosphatidylcholine (DPPC), palmitoyloleoylphosphatidylcholine (POPC) or dimyristoylphosphatidylcholin (DMPC)) now represent the state of the art. A membrane model is characterized by a set of force-field parameters that has to be carefully adapted in order to observe realistic physical properties during a molecular dynamics simulation [24].

4.4.5
Ligand-binding Site Complexes

Generated protein models are often used for studying ligand–protein interactions. Small molecules that are mostly new drugs of pharmaceutical interest can be docked into the active site of the protein. A variety of molecular docking programs have been developed in the last decade, which can be used for this purpose. A detailed description of the available techniques is given in Chapter 5.

As the natural binding process is not static and most docking programs till date do not include protein flexibility, molecular dynamics simulations are necessary to simulate the dynamical properties of the ligand–protein complex. Valuable information like hydrogen–bonding pattern, rms deviations and positional fluctuations can be deduced from the simulation to discriminate between binding and nonbinding ligands.

Several prerequisites have to be fulfilled for a meaningful molecular dynamics simulation of ligand–protein complexes. The initial coordinates, both of the ligand and the protein, have to represent an energetically reasonable conformation. The simulated system must include all regions of interest and has to be large enough to sample all forces contributing to the total energy of the system correctly. Truncated active-site complexes can only be studied if all possible ligand–protein interactions are reflected during the molecular dynamics simulation. In addition, the necessary parameters for all ligand and protein atoms must be available in the selected force field. Since most of the force fields have been historically developed either for small organic molecules or macromolecules, the collection of the parameters needed is sometimes difficult. Last but not the least, the simulation time has to be sufficiently long to generate a representative ensemble of data.

In spite of the known limitations, molecular dynamics simulations have also become a powerful tool for investigating dynamical processes of biopolymers like peptides, proteins, enzymes, receptors and membranes. The combination of experimental results like NMR measurements, photoaffinity-labeling or crystal data with theoretical methods can provide a route for gaining a detailed 3D atomic picture of the molecular system, and to study the hitherto experimentally inaccessible processes in proteins.

References

1. Weiner, S.J., Kollman, P.A., Case, D.A. *et al.* (1984) New force field for molecular mechanical simulation of nucleic acids and proteins. *Journal of the American Chemical Society*, **106**, 765–84.
2. Weiner, S.J., Kollman, P.A., Nguyen, D.T., and Case, D.A. (1986) An all atom force field for simulations of proteins and nucleic acids. *Journal of Computational Chemistry*, **7**, 230–52.
3. van Gunsteren, W.F. and Berendsen, H.J.C. (1985) Molecular dynamics simulations: techniques and applications to proteins, in: *Molecular Dynamics and Protein Structure*

(ed. J. Hermans), Polycrystal Books Service, Western Springs, pp. 5–14.

4. Brooks, C.L. III, Pettitt, B.M., and Karplus, M. (1985) The effects of terminating long-ranged forces in fluids. *Journal of Chemical Physics*, **83**, 5897–908.

5. Van Gunsteren, W.F., Bakowies, D., Baron, R. *et al.* (2006) Biomolecular modeling: goals, problems, perspectives. *Angewandte Chemie (International ed. in English)*, **45**, 4064–92.

6. Dauber-Osguthorpe, P., Roberts, V.A., Osguthorpe, D.J. *et al.* (1988) Structure and energetics of ligand binding to proteins: *E. coli* dihydrofolate reductase-trimethoprim, a drug-receptor system. *Proteins Structure Function and Genetics*, **4**, 31–47.

7. Brooks, B.R., Bruccoleri, R.E., Olafson, B.D. *et al.* (1983) CHARMM: a program for macromolecular energy, minimization, and dynamics calculations. *Journal of Computational Chemistry*, **4**, 187–217.

8. Ryckaert, J.P., Ciccotti, G., and Berendsen, H.J.C. (1977) Numerical integration of the cartesian equations of motion of a system with constraints:molecular dynamics of n-alkanes. *Journal of Computational Physics*, **23**, 327.

9. Lybrand, T.P. (1990) Computer simulation of biomolecular systems using molecular dynamics and free energy perturbation methods, in *Reviews in Computational Chemistry* (eds K.B. Lipkowitz and D.B. Boyd), VCH, New York, Vol. 1, pp. 295–320.

10. Mohamadi, F., Richards, N.G.J., Guida, W.C. *et al.* (1990) MacroModel-an integrated software system for modeling organic and biorganic molecules using molecular mechanics. *Journal of Computational Chemistry*, **11**, 440–67.

11. Kirkpatrick, S., Gelatt, C.D., and Vecchi, M.P. (1983) Optimization by simulated annealing. *Science*, **220**, 671–80.

12. Salvino, J.M., Seoane, P.R., and Dolle, R.E. (1993) Conformational analysis of bradykinin by annealed molecular dynamics and comparison to NMR-derived conformations. *Journal of Computational Chemistry*, **14**, 438–44.

13. Wilson, S.R. and Cui, W. (1990) Applications of simulated annealing to peptides. *Biopolymers*, **29**, 225–35.

14. Mackey, D.H.J., Cross, A.J., and Hagler, A.T. (1989) The Role of energy minimization in simulation strategies of biomolecular systems, in *Prediction of Protein Structure and the Principles of Protein Conformation* (ed. G. Fasman), Plenum Press, New York, pp. 317–58.

15. Kerr, I.D., Sankararamakrishnan, R., Smart, O.S., and Sansom, M.S.P. (1994) Parallel helix bundles and ion channels: molecular modeling via simulated annealing and restrained molecular dynamics. *Biophysical Journal*, **67**, 1501–15.

16. Bruccoleri, R.E. and Karplus, M. (1990) Conformational sampling using high-temperature molecular dynamics. *Biopolymers*, **29**, 1847–62.

17. Vijayakumar, S., Ravishanker, G., Pratt, R.F., and Beveridge, D.L. (1995) Molecular dynamics simulation of a class A β-lactamase: structural and mechanistic implications. *Journal of the American Chemical Society*, **117**, 1722–30.

18. Antes, I., Thiel, W., and van Gunsteren, W.F. (2002) Molecular dynamics simulations of photoactive yellow protein (PYP) in three states of its photocycle: a comparison with X-ray and NMR data and analysis of the effects of Glu46 deprotonation and mutation. *European Biophysics Journal*, **31**, 504–20.

19. Schlegel, B., Sippl, W., and Höltje, H.-D. (2005) Molecular dynamics simulations of bovine rhodopsin: influence of protonation states and different membrane-mimicking environments. *Journal of Molecular Medicine*, **12**, 49–64.

20. Jöhren, K. and Höltje, H.-D. (2005) Different environments for a realistic simulation of GPCRs-application to the M2 muscarinic receptor. *Archiv der Pharmazie*, **338**, 260–67.

21. Zhang, Y., Sham, Y.Y., Rajamani, R. *et al.* (2005) Homology modeling and molecular dynamics simulations of the mu opioid receptor in a membrane-aqueous system. *Chembiochem*, **6**, 853–59.

22. Haider, S., Grottesi, A., Hall, B.A. *et al.* (2005) Conformational dynamics of the ligand-binding domain of inward rectifier K channels as revealed by molecular dynamics simulations: toward an understanding of kir channel gating. *Biophysical Journal*, **88**, 3310–20.

23. Karplus, M. and McCammon, J.A. (2002) Molecular dynamics simulations of biomolecules. *Nature Structural Biology*, **9**, 646–52.

24. Hansson, T., Oostenbrink, C., and van Gunsteren, W. (2002) Molecular dynamics simulations. *Current Opinion in Structural Biology*, **12**, 190–96.

4.5
Validation of Protein Models

Once a protein model has been built using comparative Modeling Methods and subsequently optimized by molecular mechanics (MM) or Molecular Dynamics (MD), it is important to assess its quality and reliability. The question arises now, as to how protein models can be tested for correctness and accuracy. This is a very difficult task, because the quality of a homology based protein model depends on a huge number of properties at different levels of structural organization. This is summarized in Figure 4.5.1.

4.5.1
Stereochemical Accuracy

The quality of the 3D structure of a protein model is strongly dependent on the accuracy of the template structure used, that is the quality of the crystal structure [1]. Naturally, the modeled protein cannot show higher accuracy than the crystal structure which has been used as a template. Protein structures derived from X-ray diffraction can contain errors, both experimental ones and in the interpretation of the results [1–3]. The general measures for the quality of crystal structures are the resolution and the R-factor. The better the resolution of the protein crystal, the greater the number of independent experimental observations derived from the diffraction data and hence, the greater the accuracy of the protein structure [4]. The resolution of protein structures contained in the PDB usually is found to be in the range of 1–4 Å. The R-factor is a measure for the agreement between the derived 3D structure of a protein crystal (the 3D structure which fits the best to the electron density map) and the 'real' crystal structure. The R-factor can be determined by comparing the experimentally obtained amplitudes of the X-ray reflections and the amplitudes calculated from the protein structure which shows the best fit to the electron density map (for a detailed discussion about the accuracy of protein X-ray crystallography, the reader is referred to the literature [5]). The better the agreement between observed and calculated amplitudes (resulting in a low R-factor), the better the agreement between the derived and the real crystal structure. The R-factor can be artificially reduced in a number of ways and therefore, might sometimes be misleading [2]. It is commonly accepted to consider structures with a resolution of 2.0 Å or better to be reliable. If in addition the R-factor is below 20%, it can be safely assumed that the protein structure is essentially correct.

To verify the stereochemical quality of a model-built structure, the accuracy of parameters like bond lengths, bond angles, torsion angles and correctness of the amino acid chirality have to be proved. It has been observed in 3D structures of proteins that the bond lengths and angles cluster mainly around the 'ideal values'. Thus, the mean values detected in crystal structures can be regarded

Molecular Modeling. Basic Principles and Applications. 3rd Edition
H.-D. Höltje, W. Sippl, D. Rognan, and G. Folkers
Copyright © 2008 Wiley-VCH Verlag GmbH & Co. KGaA, Weinheim
ISBN: 978-3-527-31568-0

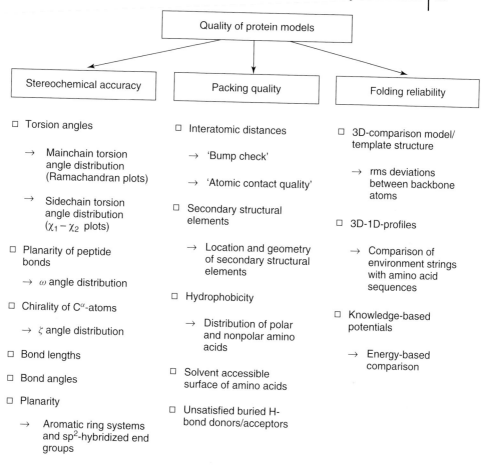

Figure 4.5.1 Quality questionnaire for protein models.

as good indicators of the stereochemical quality and must be compared with the actual values in the generated protein model (see Table 4.5.1) [6] in order to discover stereochemical irregularities which would disclose a bad structure.

Since a manual inspection of all stereochemical parameters of a protein will be tedious and time-consuming, programs have been developed which automatically check all stereochemical properties. Examples are PROCHECK [7], WHAT CHECK [8] and VADAR [9] which are accessible through the internet (e.g. http://www.embl.org, www.biochem.ucl.ac.uk).

One important indicator of stereochemical quality is the distribution of the main chain torsion angles ϕ and ψ. The distribution of all ϕ and ψ torsion angles in a protein can be examined in a Ramachandran plot. As we have described in Section 4.2.1, the favored and unfavored regions of the classical Ramachandran plot have been determined by studying the conformational

Table 4.5.1 Stereochemical parameters derived from high resolution protein structures by Morris *et al.* [6].

Stereochemical parameters	Mean value	Standard deviation
$\phi-\psi$ in most favored regions of Ramachandran plots	>90%	–
χ_1 Torsion angle gauche minus	64.1°	15.7°
trans	183.6°	16.8°
gauche plus	−66.7°	15.0°
χ_2 Torsion angle	177.4°	18.5°
Proline ϕ torsion angle	−65.4°	11.2°
α-helix ϕ torsion angle	−65.3°	11.9°
α-helix ψ torsion angle	−39.4°	11.3°
Disulfide bond separation	2.0 Å	0.1 Å
ω torsion angle	180.0°	5.8°
C^α tetrahedral distortion: ζ torsion angle (virtual torsion angle $C^\alpha-N-C-C^\beta$)	33.9°	3.5°

behavior of isolated dipeptides. Very conveniently the $\phi-\psi$ torsion angles observed in hundreds of well-refined protein structures generally lie within the same regions as determined for the isolated dipeptides. It is one of the remarkable properties of repetitive secondary structures in proteins that the observed ϕ,ψ-values are very close to the optimal dipeptide conformations, as calculated by Ramachandran. Also the ϕ and ψ torsion angles of nonrepetitive structures, like loops or turns, are found within the favored regions of the Ramachandran plot, but are more widely distributed over these areas.

As an example, the Ramachandran plot of a protein crystal structure (cephalosporinase of *Enterobacter cloacae*) is shown in Figure 4.5.2. The torsion angles of all residues, except those for proline residues and those at the chain termini, are presented. Glycine residues are separately identified by triangles (as those are not restricted to any particular region of the plot). The shading represents the different major regions of the plot: the darker the region the more favored is the ϕ,ψ combination. The white region is the disallowed region for normal amino acids and any residue found in this region has to be carefully inspected. Usually, amino acids lying in less favored regions are especially labeled (in Figure 4.5.2 shown in red) with residue name and residue number for easy identification and inspection.

Unfavorable stereochemistry, becoming visible by disallowed ϕ,ψ torsion angles, seems to occur in natural proteins exclusively, if the special geometry is required for function or stability, for example, when residues in the core of the protein are involved in hydrogen bonds or salt-bridges. Residues which are allowed to lie outside the major regions of the Ramachandran plot are proline and glycine. Because glycine and proline have, due to their different stereochemistry, other favored and unfavored regions, it is more convenient to mark these amino acid types particularly, or to exclude them from the

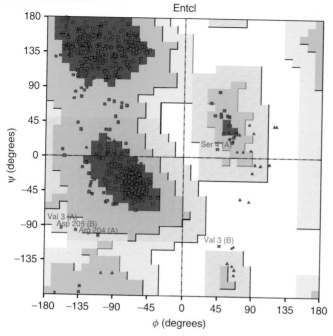

Figure 4.5.2   Ramachandran plot (PROCHECK).

normal Ramachandran plot. Therefore, very often separate Ramachandran plots for all glycines, all prolines and all other amino acids are created. The percentage of residues lying in the favored regions of a Ramachandran plot is one of the best guides to check stereochemical quality of a protein model. Ideally, one would hope to have more than 90% of the residues in the allowed regions [7].

The same check as described above for main chain torsion angles can be applied in the case of the side chain torsion angles χ_i. The χ_1 torsion angles, observed in well-refined structures of proteins [6], are generally close to one of the three possible staggered conformations, the most favored conformation being the one where the bulkiest groups are most remote (see Table 4.5.1: gauche plus, trans and gauche minus torsion angles for χ_1). For the χ_2 torsion angles, a preference for the trans conformation has been found. A similar distribution for the side chain torsion angles in protein crystal structures has been detected by Ponder and Richard [10]. The distribution of the side chain torsion angles of all amino acid types in protein models can be inspected in greater detail in graphs, where side chain torsion angles χ_1 are usually plotted versus χ_2. Examples for this kind of graph are shown in Figure 4.5.3 for a cephalosporinase of *E. cloacae*. Every single plot shows the $\chi_1 - \chi_2$ angle distribution for a particular amino acid type. The green shading on each plot indicates the favorable regions which have been determined from a

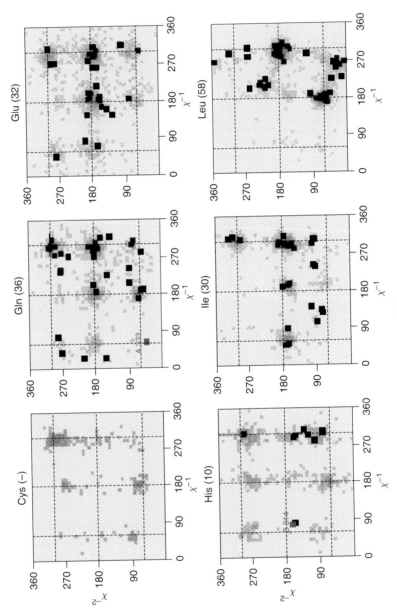

Figure 4.5.3 $\chi_1 - \chi_2$ Plot for different amino acids (PROCHECK).

data set of well-resolved protein crystal structures [7]. Black marks label the corresponding values found in the cephalosporinase. The red marks denote outliers.

Some of the stereochemical parameters of protein structure have been found to be constant in all known proteins. Of course, these properties are a very sensitive measure of the quality of protein models and have to be carefully checked for consistency.

This list contains:

The peptide bond planarity. This property is usually measured by calculating the mean value and the standard deviation of all ω angles in the investigated protein. The smaller the standard deviation, the tighter the clustering around the normal value of 180°, which represents the planar trans configuration (see also Table 4.5.1 for the distribution of ω angles in crystal structures). Also all cis peptide bonds are separately listed and have to be inspected. Cis peptide bonds occur in proteins at about 5% of the bonds that precede proline residues. In the case of all peptide bonds, which do not involve proline residues, the number of cis configurations is observed to be less than 0.05% [11, 12].

The chirality of the C^α-atoms. One of the general principles of protein structure is the preference for one handedness over the other (e.g. the preference for the right-handed conformation of an α-helix). The basis for this is the presence of an asymmetric center at the C^α-atom which is L-configurated in all naturally occurring amino acids. A protein model therefore, must be examined for correct chirality. A parameter, which provides a measure for the correctness of chirality is the ζ (zeta) torsion angle. This is a virtual torsion angle which is not defined by any actual bond in a protein. Rather this torsion angle is determined by the $C^\alpha-N-C'-C^\beta$ atoms of each amino acid residue. The numerical values of the ζ torsion angle should reside between 23° and 45°. A negative value signifies the appearance of an incorrect D-amino acid [7].

Main chain bond lengths and angles. The distribution of each of the different main chain bond lengths and angles in a protein is compared with the distribution observed in well-resolved crystal structures. Usually, deviations more than 0.05 Å for bond lengths and 10° for bond angles are regarded as distorted geometries which have to be inspected in detail [3].

Aromatic ring systems (Phe, Tyr, Trp, His) and sp²-hybridized end groups (Arg, Asn, Asp, Glu, Gln) have to be checked for planarity. The deviation of these parameters, that is the distorted geometry is often the result of bad interatomic contacts. Removing the steric constraints and subsequently optimizing the model yields a relaxed structure with ideal geometrical parameters in most cases.

4.5.2
Packing Quality

Specific packing interactions within the interior are assumed to play an important role for the structural specificity of proteins [13–15]. It has been observed that globular proteins are tightly packed with packing densities comparable to those found in crystals of small organic molecules [13]. The interior of globular proteins contains side chains that fit together in a manner which is strikingly complementary, like pieces of a 3D jigsaw puzzle. The high packing densities observed in proteins are the consequence of the fact that segments of secondary structure are packed together closely; helix against helix, helix against strands of a β-sheet and strands against strands of different β-sheets [15–18]. The interior packing of globular proteins is a major contribution to the stability of the overall conformation. Therefore the packing quality of a protein model can be used to estimate its reliability. It can be judged using a variety of methods, which will be described in detail in this section.

The first step, is to verify that the generated and refined protein model includes no bad van der Waals contacts. Therefore, all interatomic distances have to be examined to check that they reside within ranges which have been observed in well-refined crystal structures. Several procedures exist for this distance check. In the simplest procedure, all interatomic distances are measured and those with distances below a determined threshold are defined as bad contacts which have to be inspected in detail (for example, 2.6 Å is used as threshold in the PROCHECK program [7]). A more accurate judgment of interatomic distances is performed by programs like WHATCHECK [8]. For all well-refined protein crystal structures stored in the Protein Data Bank all interatomic distances shorter than the sum of their van der Waals radii +1.0 Å are determined and stored. The distance that subdivides the collected values such that 5% of all observed distances are shorter and 95% are longer than this measure, is defined as 'short normal distance'. As there are 163 different atom types in the naturally occurring amino acids 163 × 163 'short normal distances' are defined. All distances occurring in the protein model which are more than 0.25 Å shorter than the short normal distances are reported by the program.

The next step involves the examination of the secondary structural elements of the protein model. As we have already mentioned in Section 4.3.2, the secondary structural elements are the most conserved regions in highly homologous proteins. Thus, it has to be proved that the secondary structural elements observed in the template protein can also be detected in the protein model, that is, the secondary structure has been maintained during the building and optimization process. Programs which can be applied for this purpose are the DSSP [19] or the STRIDE program [20] (see Section 4.3.2). These programs allow a more sophisticated assignment of secondary structure than the manual inspection of α-helices and β-sheets.

A variety of methods exist, which use the huge amount of information derived from protein crystal structures to estimate the packing quality of model-built structures [21–24]. From the assumption that atom–atom interactions are the primary determinant of protein conformation, Vriend and Sander have developed a program that checks the packing quality of a protein model by calculating a so called 'contact quality index' [21]. This index is a measure of the agreement between the distributions of atoms around an amino acid side chain in the protein model and equivalent distributions observed in well-resolved protein structures. For this reason, a data base which contains a contact probability distribution for all amino acid side chains, has been generated. This magnitude describes the probability for a certain atom type to occur in a particular region around the side chain. These probability values are used to check the contact quality in the protein model. The better the agreement between the distributions in the model and in the crystal structures the higher the contact quality index, and the more favorable the residue packing.

The distribution of polar and nonpolar residues between the interior and the surface of proteins has been found to be a general principle of the architecture of globular proteins. At a simple level, a globular protein can be considered to consist of a hydrophobic interior surrounded by a hydrophilic external surface which interacts with the solvent molecules. These building principles have been identified in most 3D structures of globular proteins and can be summarized as follows:

- The interior of globular proteins is densely packed without large empty space and is generally hydrophobic. Nonpolar side chains predominate in the protein interior; Val, Leu, Ile, Phe, Ala, and Gly residues comprise 63% of the interior amino acids [11]. Ionized pairs of acidic and basic groups hardly occur in the interior, even though such pairs might be expected to have no net charge due to the formation of salt-bridges.
- Charged and polar groups are located on the surface of globular proteins accessible to the solvent. On an average, Asp, Glu, Lys, and Arg residues comprise 27% of the protein surface and only 4% of the interior residues [11]. (Integral membrane proteins differ from globular proteins primarily in having extremely nonpolar surfaces which are in contact with the hydrophobic membrane core.)

These features make a major contribution to the stability of folded proteins [15, 25, 26]. The underlying principle for this distribution is the hydrophobic effect, that is, the removal of hydrophobic residues from contact with water. It has been observed that the free energies, associated with the transfer from water to organic solvent, of polar, neutral and nonpolar residues are correlated with the extent to which they occur in the interior and exterior of proteins [27]. Therefore, the distribution of hydrophobic and hydrophilic residues in proteins, can be used to estimate the reliability of protein models [27–30]. Several programs have been developed which use this feature as a measure of the packing quality [8, 29, 30] of a protein model.

It has been also observed that the hydrophobicity of an amino acid (defined as free energy of transfer from water to organic solvent), is related linearly to its surface area, that is, the more hydrophobic the residue the more completely buried it will be [31]. The buried surface area of a particular amino acid is the difference between the solvent-accessible surface of the residue in an extended polypeptide chain (usually defined as the *standard state* in the tripeptide Gly–XXX–Gly) and the solvent-accessible surface of the residue in the folded protein. It has been demonstrated that the buried surface area, that is, the area which is lost when a residue is transferred from the standard state to a folded protein is proportional to its hydrophobicity.

Additionally, the total surface buried within globular proteins has been found to correlate with their molecular weights. In other words, upon folding, globular proteins bury a constant fraction of their available surface [27]. Several programs have been developed which use the general properties of amino acid surfaces in order to provide an estimation of the packing quality of globular proteins [8, 29, 31]. For a detailed review of the topic of molecular surfaces and their contributions to protein stability the reader is referred to the literature [15, 32].

Although the residues that form the protein interior are usually nonpolar or neutral, there are rare cases of buried polar residues. It has been observed in many investigations of protein crystal structures that virtually all polar groups in the protein interior are paired in hydrogen bonds. Many of these polar groups form hydrogen bonds within their own secondary structure (i.e. α-helices and β-sheets). Others are involved in binding cofactors, metal ions or are located in the active site of proteins. Buried ionizable groups, which occur rarely inside globular proteins, are usually always involved in salt-bridges. Sometimes, the positive and negative charges are bridged by water molecules. Due to these observations, it is necessary to check whether all polar buried residues are paired in hydrogen bonds and whether all charged residues are involved in salt-bridges, in the protein model. Salt-bridges and hydrogen bonds are usually identified on the basis of their interatomic distances [33].

4.5.3
Folding Reliability

Proteins with homologous amino acid sequences generally have similar folds. Therefore the overall 3D structure of the protein model and its template should be similar. The homologous proteins have to possess the same conformation, especially in the structurally conserved regions. In cases where the originally constructed protein model contains large regions of steric strain (due to the incorrect architecture), the protein may undergo correspondingly large movements in its 3D structure during the refinement process. The resulting protein conformation is not reliable, because it shows only little agreement with the 3D structure of the template protein. When checking protein conformations, one normally measures the similarity in 3D structure by the rms deviations of the C^α-atomic or the backbone coordinates after optimal rigid body superposition of the two structures (for details see [34]). A very large rms deviation means the two structures are dissimilar, a value of zero means that they are identical in conformation. Homologous proteins generally show low rms deviations for their C^α-atoms, but no general value exists which can be used as indicator to show whether two protein structures are similar or dissimilar. Chothia *et al.* have performed an investigation on structural similarity of homologous proteins [35]. The overall extent of the structural divergence of two homologous proteins was measured by optimally superposing the common conserved regions (the so called common core) and calculating the rms difference in the positions of their backbone atoms. For a test set of 32 homologous pairs of proteins they have found rms differences for the common cores which vary between 0.62 and 2.31 Å (see Figure 4.5.4).

When the overall structural similarity of the protein model and the template protein have been evaluated, the question arises if the generated conformation for an unknown protein is the correct native fold. How can one prove whether the constructed model is correct in its overall conformation? In the search for criteria that discriminate between the correct conformation and incorrectly folded models, Novotny *et al.* have performed an interesting

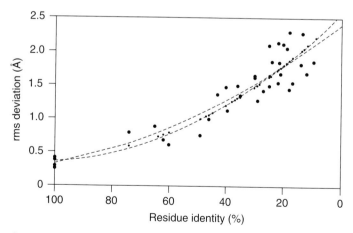

Figure 4.5.4 rms/Sequence identity plot.

investigation [36]. They have studied two structurally dissimilar but identically large proteins, hemerythrin (1HMQ) and the variable domain of mouse immunoglobulin κ-chain (1MCP-L). The two proteins have been modified by placing the amino acid sequence of one protein on to the backbone structure of the other and vice versa, in order to obtain incorrect models. The model structures were optimized to remove steric overlaps of side chains. After minimization, the total energies of native folds and incorrect protein models were approximately the same. The authors concluded that the energies obtained from standard force-field calculations cannot be used to distinguish between correct and incorrect protein conformations. On the other hand, the investigation has shown that the packing criteria of the incorrect models were different from those normally found in native proteins. The incorrect structures clearly violated the general principles of close packing, hydrogen bonding, minimum exposed nonpolar surface area and solvent accessibility of charged groups. Examination of the interior showed that the packing of side chains at the secondary structure interfaces also differed from the characteristics observed in natural proteins (e.g. side chain ridges and grooves spirally wound on α-helices, predominantly flat surfaces of β-sheets). This analysis has distinctly clarified that the validity of model-built structures can only be assessed by a careful inspection of the structural features of a protein model.

For this reason, several methods have been developed which try to distinguish between correct and incorrect folded protein structures [37–45]. One of those approaches is the 3D-Profiles method [37–40]. This method is based on the general principle that the 3D structure of a protein must be compatible with its own amino acid sequence. It measures this compatibility by reducing the 3D structure of the protein model to a simplified one-dimensional representation, the so called *environment string*. The environment string has the same length as the corresponding amino acid sequence. This one-dimensional string then can be compared with the respective amino acid sequence which is also a one-dimensional parameter.

In the first step, the 3D structure of the protein model has to be converted into a one-dimensional parameter. For this reason, the program determines several features of the environment of each residue: the area of the side chain that is buried in the protein, the fraction of the side chain area that is exposed to polar regions and the secondary structure to which the particular amino acid belongs. Based on these characteristics, each residue position is categorized into an environment class. A total of 18 distinct environment classes are implemented in the program [39]. In this manner the 3D structure is translated into a one-dimensional string which represents the environment class of each residue in the protein model.

Although the environment string is one-dimensional, it cannot be aligned with an amino acid sequence without some measure of compatibility for each of the distinct environment classes with each of the twenty natural occurring amino acids. For that reason the program includes a compatibility scoring matrix (comparable to the scoring matrices described in Section 4.3.1), which has been derived from sets of known protein structures [40]. Applying this compatibility matrix, the environment string and the amino acid sequence are aligned and a so called 3D–1D score is obtained for the particular alignment. For obvious reasons, it is more convenient to calculate local 3D–1D scores for small and medium sized regions of about 5 to 30 residues length, than a global score for the complete alignment. The local scores are then plotted against residue positions to reveal local regions of relatively high or low compatibility between the 3D structure and the amino acid sequence [39]. Regions showing unusually low scores are likely to be regions where the protein conformation is incorrect, or where structural refinement is necessary.

The folding reliability can be also tested using knowledge-based force-field methods [43–45]. These methods are based on the compilation of potentials of mean forces from a data base of known 3D-protein structures. The basic idea of these approaches is that atom–atom interactions in proteins are the primary determinant of proper protein folding.

A program, named PROSA-II has been developed, which uses the mean force potentials to calculate the total energy of amino acid sequences in a number of different folds [43]. The calculated total energy of a particular protein conformation is a qualitative criterion for the confidence or quality of a predicted protein model. This is in contrast to investigations where the total energies, derived from standard molecular mechanic force fields, have been used to estimate the reliability of different protein conformations [36]. To test the predictivity of PROSA-II different native and incorrectly modified protein conformations have been used as test set. It has been shown that for a huge number of proteins the derived total energy of the correctly folded protein is much lower than for any alternative (incorrect) protein conformation. Therefore, the program can be successfully applied to recognize erroneous protein folds or to detect faulty parts of structures in protein models.

References

1. Bränden, C.J. and Jones, T.A. (1990) Between objectivity and subjectivity. *Nature*, **343**, 687–89.
2. Jones, T.A., Zou, J.Y., and Cowan, S.W. (1991) Improved methods for building protein models in electron density maps and the location of errors in these models. *Acta Crystallographica*, **A47**, 110–19.
3. Engh, R.A. and Huber, R. (1991) Accurate bond and angle parameters for X-ray protein structure refinement. *Acta Crystallographica*, **A47**, 392–400.
4. Hubbard, T.J.P. and Blundell, T.L. (1987) Comparison of solvent-inaccessible cores of homologous proteins: definitions useful for protein modeling. *Protein Engineering*, **1**, 159–71.
5. Drenth, J. (1994) *Principles of Protein X-ray Crystallography*, Springer-Verlag, New York.

6. Morris, A.L., MacArthur, M.W., Hutchinson, E.G., and Thornton, J.M. (1992) Stereochemical quality of protein structure coordinates. *Proteins Structure Function and Genetics*, **12**, 345–64.

7. Laskowski, R.A., MacArthur, M.W., Moss, D.S., and Thornton, J.M. (1993) PROCHECK: a program to check the stereochemical quality of protein structures. *Journal of Applied Crystallography*, **26**, 283–91.

8. Hooft, R.W.W., Vriend, G., Sander, C., and Abola, E.E. (1996) Errors in protein structures. *Nature*, **381**, 272.

9. Willard, L., Ranjan, A., Zhang, H. *et al.* (2003) VADAR: a web server for quantitative evaluation of protein structure quality. *Nucleic Acids Research*, **31**, 3316–19.

10. Ponder, J. and Richards, F.M. (1987) Tertiary templates for proteins. Use of packing criteria in the enumeration of allowed sequences for different structural classes. *Journal of Molecular Biology*, **193**, 775–91.

11. Creighton, T.E. (1993) *Proteins: Structures and Molecular Properties*, 2nd edn, W. H. Freeman and Company, New York.

12. Stewart, D.E., Sarkar, A., and Wampler, J.E. (1990) Occurrence and role of cis peptide bonds in protein structures. *Journal of Molecular Biology*, **214**, 253–60.

13. Richards, F.M. (1974) The interpretation of protein structures: total volume, group volume distributions and packing density. *Journal of Molecular Biology*, **82**, 1–14.

14. Richards, F.M. (1977) Areas, volumes, packing and protein structure. *Annual Review of Biophysics and Bioengineering*, **6**, 151–76.

15. Chothia, C. (1984) Principles that determine the structure of proteins. *Annual Review of Biochemistry*, **53**, 537–72.

16. Zehfus, M.H. and Rose, G.D. (1986) Compact units in proteins. *Biochemistry*, **25**, 5759–65.

17. Janin, J. and Chothia, C. (1980) Packing of alpha-helices onto beta-pleated sheets and the anatomy of alpha/beta proteins. *Journal of Molecular Biology*, **143**, 95–128.

18. Leszczynski, J.F. and Rose, G.D. (1986) Loops in globular proteins: a novel category of secondary structure. *Science*, **234**, 849–55.

19. Kabsch, W. and Sander, C. (1983) Dictionary of protein secondary structure: pattern recognition of hydrogen-bonded and geometrical features. *Biopolymers*, **22**, 2577–637.

20. Frishman, D. and Argos, P. (1995) Knowledge-based protein secondary structure assignment. *Proteins Structure Function and Genetics*.

21. Hooft, R.W., Sander, C., and Vriend, G. (1997) Objectively judging the quality of a protein structure from a Ramachandran plot. *Computer Applications in the Biosciences*, **13**, 425–30.

22. Hunt, N.G., Gregoret, L.M., and Cohen, F.E. (1994) The origins of protein secondary structure. Effects of packing density and hydrogen bonding studied by a fast conformational search. *Journal of Molecular Biology*, **241**, 214–25.

23. Laskowski, R.A., Thornton, J.M., Humblet, C., and Singh, J. (1996) X-SITE: use of empirically derived atomic packing preferences to identify favourable interaction regions in the binding sites of proteins. *Journal of Molecular Biology*, **259**, 175–201.

24. Privalov, P.L. and Gill, S.J. (1988) Stability of protein structure and hydrophobic interaction. *Advances in Protein Chemistry*, **39**, 191–234.

25. Chothia, C. (1976) The nature of the accessible and buried surfaces in

proteins. *Journal of Molecular Biology*, **105**, 1–12.

26. Wolfenden, R., Anderson, L., Cullis, P.M., and Southgate, C.B. (1983) Affinities of amino acid side chains for solvent water. *Biochemistry*, **20**, 849–55.

27. Miller, S., Janin, J., Lesk, A.M., and Chothia, C. (1987) Interior and surface of monomeric proteins. *Journal of Molecular Biology*, **196**, 641–56.

28. Lee, B. and Richards, F.M. (1971) The interpretation of protein structures: estimation of static accessibility. *Journal of Molecular Biology*, **55**, 379–400.

29. Eisenberg, D. and McLachlan, A.D. (1986) Solvation energy in protein folding and binding. *Nature*, **319**, 199–203.

30. Lijnzaad, P., Berendsen, H.J., and Argos, P. (1996) Hydrophobic patches on the surfaces of protein structures. *Proteins Structure Function and Genetics*, **25**, 389–97.

31. Rose, G.D., Geselowitz, A.R., Lesser, G.L. *et al.* (1985) Hydrophobicity of amino acid residues in globular proteins. *Science*, **229**, 834–38.

32. Rose, G.D. and Dworkin, J.E. (1989) The hydrophobicity profile, in *Prediction of Protein Structure and Function and the Principles of Protein Conformation* (ed. G.D. Fasman), Plenum Press, New York, pp. 625–34.

33. Rashin, A. and Honig, B. (1984) On the environment of ionizable groups in globular proteins. *Journal of Molecular Biology*, **174**, 515–21.

34. Rao, S.T. and Rosman, M.G. (1973) Comparison of super-secondary structures in proteins. *Journal of Molecular Biology*, **76**, 214–28.

35. Chothia, C. and Lesk, A.M. (1986) The relation between the divergence of sequence and structure in proteins. *EMBO Journal*, **5**, 823–26.

36. Novotny, J., Bruccoleri, R. and Karplus, M. (1984) An analysis of incorrectly folded protein models. Implications for structure predictions. *Journal of Molecular Biology*, **177**, 787–818.

37. Fischer, D. and Eisenberg, D. (1999) Predicting structures for genome proteins. *Current Opinion in Structural Biology*, **9**, 208–11.

38. Bowie, J.U., Lüthy, R., and Eisenberg, D. (1991) A method to identify protein sequences that fold into a known three-dimensional structure. *Science*, **253**, 164–70.

39. PROFILES-3D, User Guide, Accelrys, San Diego, http://www.accelrys.com.

40. Lüthy, R., McLachlan, A.D., and Eisenberg, D. (1991) Secondary structure-based profiles: use of structure-conserving scoring tables in searching protein sequence databases for structural similarities. *Proteins Structure Function and Genetics*, **10**, 229–39.

41. Novotny, J., Rashin, J.J., and Bruccoleri, R.E. (1988) Criteria that discriminate between native proteins and incorrectly folded models. *Proteins Structure Function and Genetics*, **4**, 19–25.

42. Hendlich, M., Lackner, P., Weitckus, S. *et al.* (1990) Identification of native protein folds amongst a large number of incorrect models. The calculation of low energy conformations from potentials of mean force. *Journal of Molecular Biology*, **216**, 167–80.

43. Domingues, F.S., Koppensteiner, W.A., Jaritz, M. *et al.* (1999) Sustained performance of knowledge-based potentials in fold recognition. *Proteins Structure Function and Genetics*, **37**, 112–20.

44. Casari, G. and Sippl, M.J. (1992) Structure-derived hydrophobic

potential. Hydrophobic potential derived from x-ray structures of globular proteins is able to identify native folds. *Journal of Molecular Biology*, **224**, 725–32.

45. Watson, J.D., Laskowski, R.A., and Thornton, J.M. (2005) Predicting protein function from sequence and structural data. *Current Opinion in Structural Biology*, **15**, 275–84.

4.6
Properties of Proteins

4.6.1
Electrostatic Potential

As we have already mentioned, electrostatic interactions are among the most important factors in defining the conformation of a molecule in aqueous solution and in determining the energetics of interaction between two approaching molecules. The protein itself, the solvent, cofactors and prosthetic groups are nearly always charged or dipolar, and so there is a broad range of effects due to their electrostatic interactions [1–4]. Unlike dispersion forces, electrostatic interactions are effective over relatively large distances. Due to their strong influence on the structure and function of macromolecules in aqueous solution, it is absolutely necessary to explicitly consider the effect of electrostatic interactions in any theoretical study on proteins [1]. For this purpose, we need theoretical models which are able to describe the electrostatic effects in proteins correctly.

The interaction between any two charges is described by Coulomb's law (see Section 2.2.1). In the simplest form it is only valid for two point charges in vacuum. If the charges are immersed in any other matter, then particles of the surrounding matter are polarized by their presence, and the induced dipoles of the particles interact with the original point charges. Thus, the total resolved force on each of the point charges is altered, and the electrostatic interaction is decreased under the influence of the dielectric medium.

In classical electrostatic approaches, the materials are considered to be homogeneous dielectric media, which can be polarized by charges and dipoles. A dielectric constant is used as a macroscopic measure of the polarizability of a medium rather than explicitly accounting for the polarization of each atom. The portrayed proceeding is called a continuum model.

It must be kept in mind that this view is simplistic and that the concept of dielectric constant, which constitutes a genuine macroscopic property, is valid only for homogeneous media. Less homogeneous environments must be treated explicitly. Special problems arise at the boundaries between regions of very different dielectric properties [5]. The surface of a protein represents such a case, because it divides the molecule into two regions which differ dramatically in composition. The molecular interior possesses a very low dielectric constant and includes a particular number of charges (most of them near the surface). Outside the protein there is a polar aqueous medium, which normally contains a distinct amount of ions. For two point charges separated by a specific distance in a macromolecule in aqueous solution, the electrostatic interaction energy depends on the shape of the macromolecule and the exact positions of the charges (for detailed description of this topic see [5–7]). While using Coulomb's law for the calculation of electrostatic interactions, this fact will not be taken into consideration.

Molecular Modeling. Basic Principles and Applications. 3rd Edition
H.-D. Höltje, W. Sippl, D. Rognan, and G. Folkers
Copyright © 2008 Wiley-VCH Verlag GmbH & Co. KGaA, Weinheim
ISBN: 978-3-527-31568-0

The multiple interactions occurring among the point charges and dipoles of the protein and the solvent are mutually dependent and they change the simple relationship of Coulomb's law into a very complex one. The electrostatic interaction among molecules in a homogeneous environment can be averaged and expressed, as we have seen above, with the help of a simple dielectric constant. This concept is not valid for the inhomogeneous environment of proteins. Their electrostatic properties involve interactions among the multiple charges and dipoles of the proteins, and between these and the surrounding solvent and any ions in it. In this situation interactions between particular charges and dipoles have to be calculated individually. This is impractical in the presence of so many atoms of the protein and the solvent.

The major problem in studying electrostatic effects on proteins is, as we have seen, the treatment of polarization effects [4]. In many electrostatic problems, real materials are treated as simple continua, and the effects of the underlying microscopic structure of the material are only incorporated into the macroscopic dielectric constant. On the microscopic level, the shielding of the charges arises from the polarizability of the individual atoms. Thus, an approach which discards the use of a dielectric constant and considers the individual atoms of the system and their mutual polarizabilities would be the best way to solve the problem. Of course, the exact quantum mechanical treatment would be a suitable solution, but this is at the moment, due to limitations of computer power, not practicable for systems of the size of proteins. Therefore, in general empirical approaches are used for the exact calculation of electrostatic interactions within proteins [4–12].

Most of these approaches make use of the point charge approximation, that is, the charge distribution of a protein is described by locating point charges at the atom centers. Several methods have been developed to obtain corresponding partial charges [13–15]. The procedures used are comparable to those described for the small molecules (see Section 2.4.1). Because the complete protein is too large for a quantum mechanical charge calculation, the charges have been calculated for smaller fragments, like individual amino acids. The point charges for individual atoms of particular amino acids so derived, are then stored in point charge libraries from which they can be retrieved and assigned to each atom in the protein of interest. The often used Kollman charges, for example, have been determined by scaling point charges to fit the ab initio derived molecular electrostatic potential [14]. In case of proteins, the ionization state has to be taken into consideration as well. Therefore, formal charges are assigned to those amino acid residues that are expected to exist in charged state under physiological conditions. These charges are placed on one or two of the atoms of a residue. For example, an aspartic acid residue obtains the formal charge -1, which is assumed to be distributed over the two carboxylic oxygen atoms.

In one of the first approaches to a more reliable consideration of electrostatic interaction within proteins, the use of a distance-dependent dielectric constant was introduced. The mathematical equation used for the corresponding function often has the form $\varepsilon(r) = r$, where r is the distance between the atoms of interest [16]. The distance-dependent dielectric constant is based on plausibility rather than on any experimentally measurable effect. It is assumed that at distances in the order of atomic dimensions, the dielectric constant between two charges is that of vacuum conditions, and that at much larger separations, the dielectric constant of water $\varepsilon = 80$ holds true. For intermediate distances, it is assumed that the dielectric varies with distance in an appropriate way. Distance-dependent dielectric constants can partially mimic the solvent-screening effects on electrostatic energies and are sufficient to stabilize macromolecules in MD simulations. However, they cannot correctly describe properties like the electrostatic forces and the electrostatic potential.

A solution to the electrostatic problem may be provided by using the Poisson–Boltzmann equation. This equation belongs to the class of differential equations that are typical for the description of boundary phenomena. The Poisson–Boltzmann equation provides a rigorous approach to the calculation of the electrostatic effects of proteins, including the electrostatic potential. Several procedures which make use of the Poisson–Boltzmann equation have been developed. Two commercially available programs are DelPhi [17, 18] and UHBD [10, 19].

In the framework of the Poisson–Boltzman approach the macromolecular system is considered to consist of two separate dielectric regions. The solvent-accessible surface of the protein defines the boundary between these two regions. The interior of this surface is defined as the solute and the exterior is defined as the solvent. Water molecules located in the interior of the protein are usually treated as part of the solute rather than of the solvent. The protein is described in terms of its 3D structure with the location of point charges on the atom centers. A low dielectric constant is used for all points inside the solvent-accessible surface. Common values for this parameter range from 2 to 5. The Poisson–Boltzmann equation is also able to consider the electrostatic effects associated with ions embedded in the solvent. Thus the physiological conditions ($0.145 \ \mathrm{mol \ l^{-1}}$) can be incorporated in the calculation.

Use of the Poisson–Boltzmann approach yields the total electrostatic potential of a charged molecule in a solvent according to the following simplified equation:

$$\phi_i^{tot} = \phi_i^{coul} + \phi_i^{self} + \phi_i^{cross} + \phi_i^{own}$$

The solvent molecule responds to the electrostatic field generated by each point charge in the molecule. This response, which consists of two electrostatic effects, the dipolar orientation and the electronic polarization, in turn sets up an electrostatic field at the positions of the original point charges, which is called *the reaction field* [20]. The magnitude of the reaction field is determined by the point charge, its distance from the molecular surface, the shape of the surface and the dielectric constants of molecule interior and solvent. The reaction field exerts a force on all point charges in the system, including the source charge itself. The total electrostatic potential ϕ_i^{tot} is the sum of the interaction of each point charge with its self-reaction field ϕ_i^{self}, the reaction field induced by other point charges ϕ_i^{cross}, the direct coulombic interaction with other point charges ϕ_i^{coul} and the intrinsic electrostatic potential generated by each point charge ϕ_i^{own} (for a detailed description of this topic see [5, 7, 21]).

The Poisson–Boltzmann equation is actually a reliable model for the electrostatic interaction in proteins, because it considers the effect of polarization as well as the ionic strength. Unfortunately, this equation is a very complex differential equation and can be solved analytically only for small regular systems. The alternative to the analytical solution is the use of numerical techniques to find an approximate solution even for large protein systems. For the numerical solution the programs use the so called *finite difference method* (FDPB). Here the protein is mapped onto a 3D cubical grid. The calculated values for the charge density and the electrostatic potential are located on each point of the cubical grid. The numerical solution yields values which are accurate to within 5% in comparison to analytical solutions (which are available for small systems). The most critical regions and thus, the regions of largest errors are usually those located near charged residues on the protein surface. Several procedures have been recently developed to avoid these errors [18].

Calculation of the electrostatic potential of a protein molecule is only one of the possibilities offered by the Poisson–Boltzmann method. In addition, parameters like the total electrostatic energy of the system, the solvation energy and the reaction-field energy of proteins can also be calculated. Nevertheless, the most important parameter is the electrostatic potential, which can be displayed in various ways (as described for the small molecules in Section 2.4.1).

Electrostatic potentials have been shown to play an important role in molecular recognition and binding. For example, the electrostatic potential of the superoxide dismutase enzyme has been shown to be responsible for enhanced external diffusion rates of the substrates to the active site [22]. The investigation of the electrostatic potentials of two trypsin enzymes, rat and cow trypsin, has yielded interesting results [23]. Though these two enzymes although have the same catalytic mechanism, they differ in net charge by 12.5 units. The calculation of the electrostatic potentials, using the Poisson–Boltzmann approach, revealed that both active sites are effectively shielded from the charges located on the surface, resulting in nearly identical electrostatic potentials inside the active sites.

Gramicidin A, a well-known membrane cation transporting protein, is an example for the graphical representation of the electrostatic potential of a protein (see Figure 4.6.1). Gramicidin A forms a dimer in the membrane. The calculation of the electrostatic potential has been performed for the gramicidin A dimer embedded in a low dielectric membrane layer (which is treated as part of the low dielectric solute system), using program DelPhi.

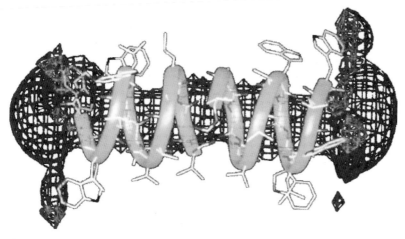

Figure 4.6.1 Representation of the electrostatic potential of a gramicidin A dimer embedded in a membrane environment. Calculations were performed using DelPhi. (Color code: magenta = negative, green = positive potentials.)

4.6.2
Interaction Potentials

Other important features for studying interaction, recognition and binding of possible substrates to a protein are provided by the evaluation of molecular interaction fields. As we have already comprehensively discussed in Section 2.5.4, interaction potentials are useful indicators for the prediction of binding properties of molecules. Programs, like the widely used GRID [24, 25], can be used to map regions within a protein where a water molecule or a substrate is preferentially attracted. The interaction fields, derived with a particular probe, also can be used as starting point for docking studies of a substrate to its active site. The techniques and procedures applied in this context, are the same as described in the case of small molecules in Section 2.4.2. Various examples are given in literature where these programs have been successfully used to predict binding regions [26–28], to dock molecules into active sites [29–32] and to optimize structures of ligands in order to optimize the binding properties [26, 33, 34]. An outstanding collection of articles and reviews on this subject can be found in [35].

4.6.3
Hydrophobicity

In Section 4.5.2 on the packing quality of proteins, we have discussed the important role of hydrophobic properties in the process of protein folding. Protein binding activities are also often determined by hydrophobic interactions. As was discussed for small molecules (Section 2.4.3), several methods are available for the representation of hydrophobic and hydrophilic properties of molecules. The hydrophobicity can be either represented directly on the molecular surface or as hydrophobic field in the space surrounding the molecule. Useful programs in this respect are, for example, GRID [24], HINT [36] as well as MOLCAD [37]. A detailed description of the different methods and a comparison of the results derived in studies on proteins is given in the literature [38].

References

1. MacArthur, M.W., Laskowski, R.A., and Thornton, J.M. (1998) Validation of protein models derived from experiment. *Current Opinion in Structural Biology*, **8**, 631–37.
2. Honig, B., Hubbel, W., and Flewelling, R.F. (1986) Electrostatic interactions in membranes and proteins. *Annual Review of Biophysics and Biophysical Chemistry*, **15**, 163–93.
3. Matthew, J.B. (1985) Electrostatic effects in proteins. *Annual Review of Biophysics and Biophysical Chemistry*, **14**, 387–417.
4. Schutz, C.N. and Warshel, A. (2001) What are the dielectric "constants" of proteins and how to validate electrostatic models? *Proteins Structure Function and Genetics*, **44**, 400–17.
5. Warshel, A. and Aqvist, J. (1991) Electrostatic energy and macromolecular function. *Annual Review of Biophysics and Biophysical Chemistry*, **20**, 267–98.
6. Gilson, M., Rashin, A., Fine, R., and Honig, B. (1985) On the calculation of electrostatic interactions in proteins. *Journal of Molecular Biology*, **183**, 503–16.
7. Harvey, S.C. (1989) Treatment of electrostatic effects in macromolecular modeling. *Proteins Structure Function and Genetics*, **5**, 78–92.
8. Zauhar, R.J. and Morgan, R.S. (1985) A new method for computing the macromolecular electric potential. *Journal of Molecular Biology*, **186**, 815–20.
9. States, D.J. and Karplus, M. (1987) A model for electrostatic. effects in proteins. *Journal of Molecular Biology*, **197**, 122–30.
10. Karplus, M. and McCammon, J.A. (2002) Molecular dynamics simulations of biomolecules. *Nature Structural Biology*, **9**, 646–52.
11. Warwicker, J. and Watson, H.C. (1982) Calculation of the electric potential in the active site cleft due to alpha-helix dipoles. *Journal of Molecular Biology*, **157**, 671–79.
12. Warshel, A. and Papazyan, A. (1998) Electrostatic effects in macromolecules: fundamental concepts and practical modeling. *Current Opinion in Structural Biology*, **8**, 211–17.
13. Jorgensen, W.L. and Tirado-Rives, J. (1988) The OPLS potential functions for proteins. Energy minimization for crystals of cyclic peptides and crambin. *Journal of the American Chemical Society*, **110**, 1657–66.
14. Weiner, P.K. and Kollman, P.A. (1981) AMBER: assisted model building with energy refinement. A general program for modeling molecules and their interactions. *Journal of Computational Chemistry*, **2**, 287–99.
15. Abraham, R.J., Grant, G.H., Haworth, I.S., and Smith, P.E. (1991) Charge calculations in molecular mechanics. Part 8. Partial atomic charges from classical calculations. *Journal of Computer-Aided Molecular Design*, **5**, 21–39.

16. McCammon, J.A., Wolyness, P.G., and Karplus, M. (1979) Picosecond dynamics of tyrosine side chains in proteins. *Biopolymers*, **18**, 927–42.

17. DelPhi User Guide, Accelrys, San Diego, http://www.accelrys.com.

18. Luo, R., David, L., and Gilson, M.K. (2002) Accelerated Poisson-Boltzmann calculations for static and dynamic systems. *Journal of Computational Chemistry*, **23**, 1244–53.

19. Antosiewicz, J., McCammon, J.A., and Gilson, M.K. (1994) Prediction of pH-dependent properties of proteins. *Journal of Molecular Biology*, **238**, 415–36.

20. Bottcher, C.J.F. (1973) *Theory of Electric Polarization*, Elsevier Press, Amsterdam.

21. Gilson, M.K., McCammon, J.A., and Madura, J.D. (1995) Molecular dynamics simulation with a continuum electrostatic model of the solvent. *Journal of Computational Chemistry*, **9**, 1081–95.

22. Sharp, K., Fine, R., and Honig, B. (1987) Computer simulations of the diffusion of a substrate to an active site of an enzyme. *Science*, **236**, 1460–63.

23. Soman, K., Yang, A., Honig, B., and Fletterick, R. (1989) Electrical potentials in trypsin isozymes. *Biochemistry*, **28**, 9918–26.

24. Goodford, P.J. (1985) A computational procedure for determining energetically favourable binding sites on biologically important macromolecules. *Journal of Medicinal Chemistry*, **28**, 849–57.

25. Wade, R.C., Clark, K.J., and Goodford, P.J. (1993) Further development of hydrogen bond functions for use in determining energetically favourable binding sites on molecules of known structure. 1. Ligand probe groups with the ability to form two hydrogen bonds. *Journal of Medicinal Chemistry*, **36**, 140–47.

26. Windshügel, B., Jyrkkarinne, J., Poso, A. *et al.* (2005) Molecular dynamics simulations of the human CAR ligand-binding domain: deciphering the molecular basis for constitutive activity. *Journal of Molecular Modeling*, **11**, 69–79.

27. Von Itzstein, M., Dyason, J.C., Oliver, S.W. *et al.* (1996) A study of the active site of influenza virus sialidase: an approach to the rational design of novel anti-influenza drugs. *Journal of Medicinal Chemistry*, **39**, 388–91.

28. Wade, R.C. (1997) Flu' and structure-based drug design. *Structure*, **5**, 1139–46.

29. Meng, E.C., Shoichet, B.K., and Kuntz, I.D. (1992) Automated docking with grid-based energy evaluation. *Journal of Computational Chemistry*, **13**, 505–24.

30. Byberg, J.R., Jorgensen, F.S., Hansen, S., and Hough, E. (1992) Substrate-enzyme interactions and catalytic mechanism in phospholipase C: a molecular modeling study using the GRID program. *Proteins Structure Function and Genetics*, **12**, 331–38.

31. Stoddard, B.L. and Koshland, D.E. (1993) Molecular recognition analyzed by docking simulations: the aspartate receptor and isocitrate dehydrogenase from Escherichia coli. *Proceedings of the National Academy of Sciences of the United States of America*, **90**, 1146–53.

32. Bitomsky, W. and Wade, R.C. (1999) Docking of glycosaminoglycans to heparin-binding proteins. *Journal of the American Chemical Society*, **121**, 3004–13.

33. Varney, M.D., Marzoni, G.P., Palmer, C.L. *et al.* (1992) Crystal-structure-based design and synthesis of benzcdindole-containing inhibitors of thymidylate synthase. *Journal of Medicinal Chemistry*, **35**, 663–76.

34. Ocain, T.D., Deininger, D.D., Russo, R. *et al.* (1992) New modified heterocyclic phenylalanine derivatives. Incorporation into potent inhibitors of human renin. *Journal of Medicinal Chemistry*, **35**, 823–32.

35. Cruciani, G. (2005) In *Molecular Interaction Fields, Methods and Principles in Medicinal Chemistry, Series* (eds H. Kubinyi, G. Folkers, and R. Mannhold), VCH Publishers, New York.

36. Kellogg, G.E., Semus, S.F., and Abraham, D.J. (1991) HINT: a new method of empirical hydrophobic field calculation for CoMFA. *Journal of Computer-Aided Molecular Design*, **5**, 545–52.

37. Heiden, W., Moeckel, G., and Brickmann, J. (1993) A new approach to analysis and display of local lipophilicity/hydrophilicity mapped on molecular surfaces. *Journal of Computer-Aided Molecular Design*, **7**, 503–14.

38. Folkers, G., Merz, A., and Rognan, D. (1993) CoMFA as a tool for active site modeling, in *Trends in QSAR and Molecular Modeling* (ed. C.G. Wermuth), ESCOM Science Publishers B. V., Leiden, Vol. 92, pp. 233–44.

5
Virtual Screening and Docking

High-throughput screening (HTS) of chemical libraries is a well-established method for finding new lead compounds in drug discovery. However, as the available databases get larger and larger, the costs of such screenings rise, whereas the hit rates decrease. This problem could possibly be avoided by not screening the whole database experimentally, but only a small subset that should be enriched in compounds that are likely to bind to the target. This preselection can be done by virtual screening (VS) [1], a computational method that is used to select the most promising compounds from an electronic database for experimental screening. VS can be carried out by searching databases for molecules fitting user-defined constraints: similarity to a set of known actives [2], pharmacophore [3], three-dimensional (3D) structure of the macromolecular target [4]. This chapter is not aimed at reviewing all computational procedures that are available to screen an electronic database, but will be more focused on protein-based VS. The four essential steps of any VS process (preparation, docking, scoring, postfiltering; see Figure 5.1) will thus be detailed in the following sections.

5.1
Preparation of the Partners

5.1.1
Preparation of the Compound Library

Besides corporate databases, there are numerous commercially available screening collections available both electronically and physically (see a near-exhaustive list of suppliers in Table 5.1). Such compound libraries should generally be first filtered to remove unsuitable compounds that anyway would not reach and pass clinical trials owing to undesired properties. In any case, the filtering level (high, low) should be adapted to the size of the electronic database and the properties that are required for selection (hit, lead, pharmacological reagent).

Molecular Modeling. Basic Principles and Applications. 3rd Edition
H.-D. Höltje, W. Sippl, D. Rognan, and G. Folkers
Copyright © 2008 Wiley-VCH Verlag GmbH & Co. KGaA, Weinheim
ISBN: 978-3-527-31568-0

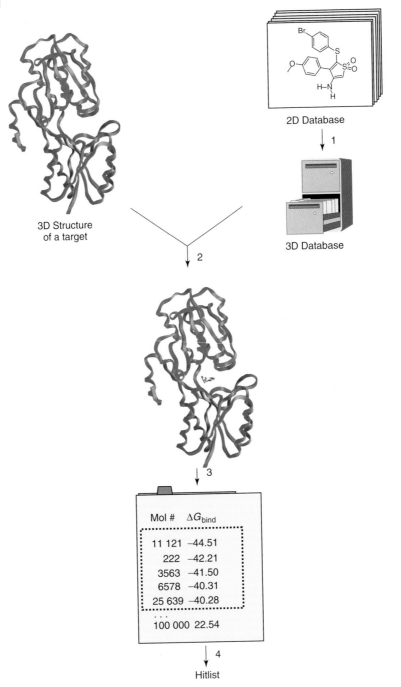

2D Database

↓ 1

3D Structure
of a target

3D Database

↓ 2

↓ 3

Mol #	ΔG_{bind}
11 121	−44.51
222	−42.21
3563	−41.50
6578	−40.31
25 639	−40.28
. . .	
100 000	22.54

↓ 4

Hitlist

Figure 5.1 Protein-based virtual screening flowchart: (1) 3D library setup, (2) docking, (3) scoring, (4) postprocessing and final hit selection.

Table 5.1 Chemical libraries suitable for virtual screening.

Supplier	Size	Web site
Physically available compounds		
AKos screening samples	700 000	http://www.akosgmbh.de/
Ambinter	500 000	http://www.ambinter.com/
AMRI (Comgenex)	241 000	http://www.albmolecular.hu/cgi-bin/index.php
ASDI Biosciences	105 600	http://www.asdibiosciences.com/
Asinex Synergy	11 000	http://www.asinex.com/
Asinex Gold	227 480	http://www.asinex.com/
Asinex Platinium	130 646	http://www.asinex.com/
Asinex building blocks	5908	http://www.asinex.com/
Bionet	44 546	http://www.keyorganics.ltd.uk/
ChemBridge Express-Pick	435 000	http://chembridge.com/chembridge/
ChemDiv new chemistry	28 300	http://www.chemdiv.com/
ChemDiv discovery chemistry	620 200	http://www.chemdiv.com/
ChemStar	60 086	http://www.chemstaronline.com/
CNRS National Library	29 356	http://chimiotheque-nationale.enscm.fr/
Enamine	857 000	http://www.enamine.relc.com/
Exclusive chemistry	1906	http://www.exchemistry.com/
Innovapharm	680 000	
InterBioScreen natural compounds	40 000	http://www.ibscreen.com/
InterBioScreen synthetic compounds	340 000	http://www.ibscreen.com/
LifeChemicals	208 000	http://www.lifechemicals.com/
Maybridge	66 000	http://www.maybridge.com
MDD	33 000	http://www.worldmolecules.com/databases.html
NCI DTP	127 000	http://dtp.nci.nih.gov/webdata.html
Otava	109 000	http://www.otava.com.ua/
Peakdale Molecular	8500	http://www.peakdale.co.uk/
Pharmeks	165 000	http://www.pharmeks.com/
Princeton BioMolecular Research	530 000	http://www.princetonbio.com/
Specs	240 000	http://www.specs.net/
TimTec Stock	225 000	http://www.timtec.net/
TosLab	20 500	http://www.toslab.com/
Tripos Leadquest	52 000	http://www.tripos.com/
Vitas-M STK	234 000	http://www.vitasmlab.com/
Vitas-M Tulip	24 700	http://www.vitasmlab.com/
Web databases		
ChemDB	4 100 000	http://cdb.ics.uci.edu/CHEMDB/Web/index.htm
ChemMine	5 800 000	http://bioweb.ucr.edu/ChemMineV2/
ChemNavigator	24 000 000	http://www.chemnavigator.com/
PubChem	4 000 000	http://pubchem.ncbi.nlm.nih.gov/search/
MDL screening compounds directory	3 500 000	http://www.mdli.com
Zinc	3 381 581	http://blaster.docking.org/zinc/

There are two ways of removing unsuitable molecules from databases. The first possibility is to use a series of different filters, each one excluding compounds with certain properties [5]. Highly reactive and toxic compounds can be removed according to reactive moieties such as acyl-halides, sulfonyl-halides, Michael acceptors, and so on. The best known filter is probably the Lipinski

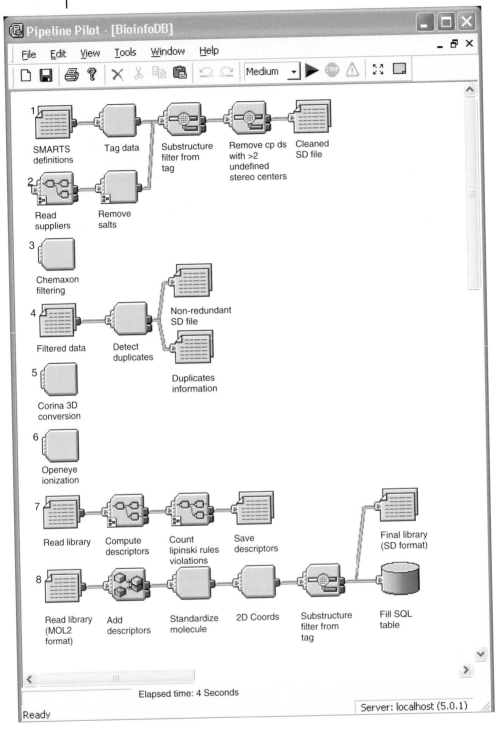

◀ **Figure 5.2** Automated Pipeline Pilot [11] workflow for library setup. (1) Reading raw data from various suppliers; (2) cleaning data (remove erroneous and complex structures, counterions); (3) filtering according to predefined property rules; (4) detection of duplicates; (5) conversion in 3D format with potential enumeration of several conformers, stereoisomers and tautomers; (6) ionization at physiological pH; (7) computation of physicochemical and topological descriptors; (8) conversion in 2D format and storage in relational databases.

'Rule-of-Five' [6], suggesting that poor absorption or permeation are more likely when the molecular weight is over 500, the calculated octanol–water partition coefficient (clogP) is higher than 5, when there are more than 10 hydrogen bond acceptors and more than 5 hydrogen bonds donors. All compounds that fulfill two or more of these conditions are likely to show poor permeability and should be removed from the database. Meanwhile, more elaborate filters [7] for specific absorption/distribution/metabolism/excretion (ADME) properties are being developed, such as filters for prediction of aqueous solubility, membrane permeation and metabolic clearance. A second possibility for filtering out unsuitable molecules is to design a knowledge-based binary classification system (e.g. neural networks, genetic algorithms (GAs), decision trees) that automatically distinguishes between appropriate and inappropriate compounds [8]. In addition to 'non-druglike' compounds, it might be desirable to eliminate molecules with a high propensity to bind to a wide array of protein targets. Such compounds are often called *promiscuous binders* [9] as they can perturb the assay or detection method (e.g. fluorescent molecules) or specifically bind to several macromolecular targets through aggregation or denaturation [10]. Since commercial and corporate libraries are dynamic entities, it is of utmost importance to automate, as much as possible, the library preparation for facilitating future upgrades. Using the above-cited rules, automated workflows can be readily set up to perform the following tasks (Figure 5.2):

- reading raw data from various suppliers (usually in sd file format)
- cleaning data: remove erroneous and complex structures, counterions (usually in 1D SMILES or SMARTS file formats)
- filtering according to predefined property rules
- detecting duplicates
- converting 1D to 3D format with potential enumeration of several conformers, stereoisomers and tautomers
- ionizing at physiological pH
- computing physicochemical and topological descriptors
- storing in relational databases for future browsing purpose.

In some publicly available web resources (e.g. Zinc [12], ChemDB [13]; Table 5.1), the above tasks have already been performed and the resulting multisupplier library is ready to screen, in others (e.g. ChemNavigator), the library preparation still remains to be done.

5.1.2
Representation of Proteins and Ligands

Reproducing the conformational space accessible to a macromolecule is a very difficult task and involves unavoidable approximation. Docking procedures can thus be classified into three categories depending on the approximation level: (i) rigid body docking: both protein and ligand are treated as rigid bodies, (ii) semiflexible docking: only the ligand is considered flexible, (iii) fully flexible docking: both ligand and protein are treated as flexible molecules.

5.1.2.1 Protein Flexibility
Proteins are highly flexible molecules that exist in a range of conformational states, with low-energy barriers separating them. Protein motions can be classified into three groups:

(1) Small-scale fast motions mainly involve side-chain movements, but some movements of backbone atoms are also present. Binding sites frequently display enhanced flexibility compared to other areas of the protein surface.

(2) Large-scale, slow domain motions, for example, hinge-bending movements, are movements where rigid domains are connected by flexible joints that tether them and constrain their movement. Hinge-bending is believed to allow the 'induced fit' of the ligand.

(3) Renaturation upon ligand binding affects many proteins that are in a partially unfolded state owing to either a small hydrophobic core, or the presence of uncompensated buried charges. Ligand binding stabilizes the bound conformation, shifting the equilibrium in its direction. Conformational changes of the protein upon ligand binding can thus extend from only few side-chain movements to large hinge-bending movements.

Obviously, proteins should be treated as flexible in ligand docking, especially when using a ligand-free structure as target. However, for a long time, this has been computationally impossible. The development of docking algorithms taking protein flexibility into account has started only recently, aided by the exponential growth of computer speed, RAM and disk capacity.

One approach to handle protein flexibility is to start from a single 3D structure and further allow defined substructures to move. This task can be achieved by exploring the conformational space of protein side chains using a rotamer library [14] while freezing backbone atoms, allowing hinge-bending movements of protein domains. An alternative approach is to use an ensemble of several possible conformations derived experimentally (X-ray diffraction, nuclear magnetic resonance (NMR)) or computed from molecular dynamics or Monte Carlo simulations. The conformational ensemble is then translated into a composite grid and used to calculate for any probe atom the potential stored in the composite grid. The grids are then used for molecular docking by a simple readout of the interaction energy values previously stored [15]. A last strategy, developed recently, uses a unified protein description from superimposed structures. Similar parts of the structures are merged whereas dissimilar areas are treated as separated alternatives [16]. The different structures of the ensemble can be recombined to form new overall structures during the docking process.

Including protein flexibility in the docking process is especially important while using low-resolution protein models arising from comparative modeling approaches.

5.1.2.2 Ligand Flexibility

Since drugs are much smaller than macromolecules, ligand flexibility is computationally easier to handle and thus today it is standard in docking routines. The simplest approach is to store multiple conformations of the ligands in the database, each conformation being regarded as rigid during the docking process. Another way of introducing ligand flexibility is to store only one conformation per ligand in the database, but to then treat the ligands as flexible molecular entities during the docking process. The incremental construction method, which divides the ligand into fragments and incrementally builds it in the receptor binding site, is commonly used. When placing a fragment, only the most likely dihedral angles are searched for the best solutions. Examples for docking programs using this method are FlexX and Dock (Table 5.2). Another possibility is to encode flexibility around dihedral angles as genes into GAs as in Gold or AutoDock, for example. Genetic operations (crossing over, mutation) that specifically target these genes will address ligand flexibility per se. Last, exhaustive computational techniques for examining conformational space (molecular dynamics, simulated annealing, Monte Carlo) can be used on condition that the pace of the search algorithm does not penalize the VS process.

Most recent docking algorithms handle ligand flexibility during execution, which presents the noticeable advantage of reducing disk space necessary to store the database. However, one can generally rely on pre-generated conformational ensembles by state-of-the art conformer generators (e.g. CATALYST, OMEGA, CORINA), which usually produce among the set of proposed conformers, one that is close to a bioactive conformation [64].

Table 5.2 Main docking programs.

Name	Origin	Web site	Reference
Adam	IMMD[a]	http://www.immd.co.jp/en/index.html	17
AutoDock	Scripps	http://www.scripps.edu/mb/olson/doc/autodock/	18
CDocker	Eli Lilly		19
Clix	CSIRO[b]		20
CombiDock	UCSF[c]		21
Darwin	Wistar Institute		22
Divali	UCSF		23
Dock	UCSF	http://dock.compbio.ucsf.edu/	24
DockIt	Metaphorics L.L.C.	http://www.metaphorics.com/	
DockVision	Alberta University	http://www.dockvision.com/	25
Dream++	UCSF		26
eHiTS	SymBioSys Inc.	http://www.simbiosys.ca/	27
Eudoc	Mayo Clinic Cancer Center		28
FDS	Southampton University		29
FFLD	Zürich University	http://www.biochem-caflisch.unizh.ch/	30
FlexScreen	Forschungszentrum Karlsruhe GmbH	http://iwrwww1.fzk.de/biostruct/RLD/flexscreen.htm	31
FlexX	BioSolveIT	http://www.biosolveit.de/	32
Flog	Merck Research Laboratories		33
Fred	OpenEye	http://www.eyesopen.com/	34
FTDock	Imperial Cancer Research Fund	http://www.sbg.bio.ic.ac.uk/docking/	35
GAsDock	Dalian University of Technology		36
Glide	Schrödinger	http://www.schrodinger.com/	37
Gold	CCDC[d]	http://www.ccdc.cam.ac.uk/	38
Hammerhead	Arris Pharmaceutical Corporation		39
HierDock	California Institute of Technology		40
ICM	Molsoft	http://www.molsoft.com/	41
LibDock	Accelrys	http://www.accelrys.com/	42
LigandFit	Accelrys	http://www.accelrys.com/	43
Ligin	Weizmann Institute of Science	http://swift.cmbi.kun.nl/swift/ligin/	44
MCDock	Georgetown University Medical Center		45
MOE-Dock	Chemical Computing Group	http://www.chemcomp.com/	
MolDock	Molegro ApS	http://www.molegro.com/index.php	46
PAS-Dock	Norwegian University of Science and Technology		47
PhDock	Wyeth Research		48
Ph4Dock	Ryoka Systems Inc.		49
ProDock	Cornell University		50
Pro_Leads	Proteus Molecular Design Ltd.		51

Table 5.2 *(continued).*

Name	Origin	Web site	Reference
ProPose	4SC	http://www.4sc.de/	52
Psi-Dock	Peking University		53
Q-fit	University College, London		54
QXP	Novartis Pharmaceuticals		55
RiboDock	Vernalis		56
Sandock	Edinburgh University		57
SDocker	Lilly Research Labs	http://www.biochem-caflisch.unizh.ch/	58
Seed	Zürich University		59
SFDock	Peking University		60
SkelGen	De Novo Pharmaceuticals Ltd.	http://www.denovopharma.com/	61
Slide	Michigan State University	http://www.bch.msu.edu/~kuhn/ projects/slide/home.html	62
Surflex	Biopharmics	http://www.biopharmics.com/ products.html	63

a Institute of Medical Molecular Diagnostic Ltd.
b Commonwealth Scientific and Industrial Rearch Organisation.
c University of California, San Francisco.
d Cambridge Crystallographic Data Center.

Whatever the strategy used, it is generally advisable to refine the ligand structures by a fast molecular mechanics-based energy-minimization protocol. Usual file formats for storing the ligand coordinates are mol2 (TRIPOS), sd (MDL) or mae files (Schrödinger).

5.2
Docking Algorithms

The numerous currently described docking tools (Table 5.2) differ in aspects such as the description of molecular interactions, the algorithms used to generate ligand structures and the average run time per molecule. The algorithms can be grouped into deterministic and stochastic approaches. Deterministic algorithms are reproducible, whereas stochastic algorithms include a random factor and are thus not fully reproducible.

5.2.1
Incremental Construction Methods

In an incremental construction algorithm the ligand is not docked as a complete molecule at once, but is instead divided into single fragments and incrementally reconstructed inside the active site (Figure 5.3).

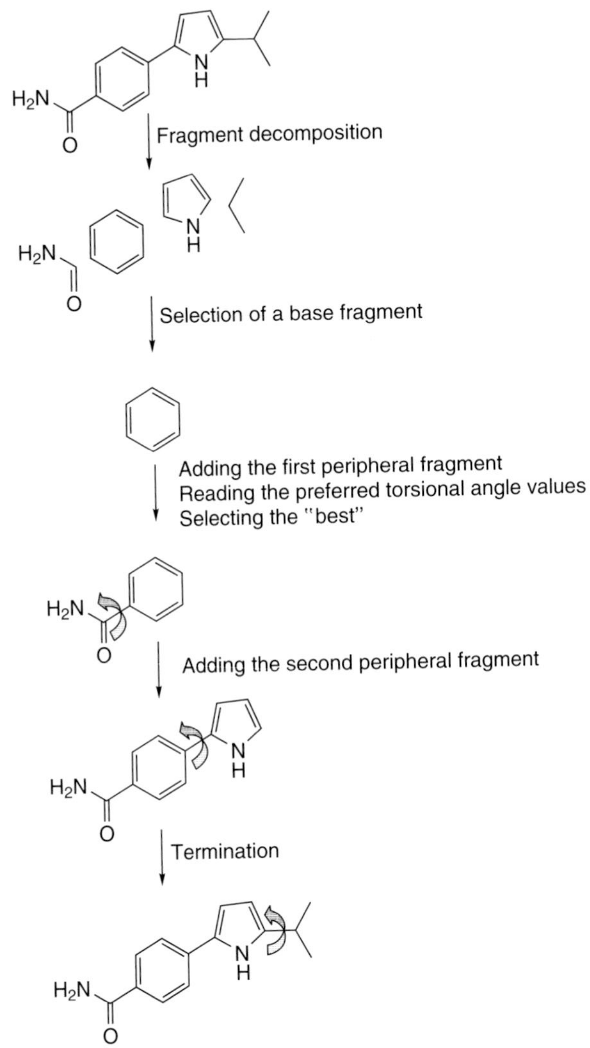

Figure 5.3 Incremental construction of a ligand.

The first incremental construction method was Dock [24]. The first step in Dock is the identification of points in the active site where ligand atoms may be located. These points, called *sphere centers*, are identified by generating a set of overlapping spheres that fill the site. These sphere centers try to capture shape characteristics of the active site with a minimum of points. The ligand is partitioned along each flexible bond to generate rigid segments. An anchor fragment is then selected from the rigid fragments, either manually or automatically. This anchor fragment is orientated within the active site, independently of the rest of the ligand, by 'matching' ligand atoms with sphere centers (Dock is therefore sometimes also called a *fast shape matching algorithm*). All possible anchor placements are scored in terms of their interactions with the protein (for the implementation of the Dock scoring function see Section 5.3), and the best ones are used for subsequent 'growing' of the ligand. Finally, the best-scored poses of the complete ligand are selected.

FlexX [32] also treats the ligands as flexible and the protein as rigid. Similarl to Dock, it divides the ligands along its rotational bonds into rigid fragments, first docks a base fragment into the active site and then reattaches the remaining fragments. It differs, however, significantly from Dock in the method used for determining the placement of the base fragment. Instead of defining points where ligand atoms may be located, FlexX defines interaction sites for each possible interacting group of the active site and the ligand. The interaction sites are assigned an interaction type (hydrogen bond acceptor, hydrogen bond donor, etc.) and are modeled by an interaction geometry consisting of an interaction center and a spherical surface. The base fragment is oriented by searching for placements where three interactions between the protein and the ligand can occur. The remaining ligand components are then incrementally attached to the core. At each growing step, a list of preferred torsional angle values is read and the best conformation in terms of protein–ligand interactions is kept for further 'growing' of the ligand (for the implementation of FlexX scoring function see Section 5.3) [32]. Dock and FlexX are by far the most widely used docking tools using a fragment-based approach. Other programmes (e.g. Propose, Slide, Surflex; Table 5.2) use a rather similar fragmenting approach, although various strategies for adding the peripheral fragments and various scoring functions (cf. Section 5.3) can be used to rank poses.

5.2.2
Genetic Algorithms

A GA is a computer program that mimics the process of evolution by manipulating a collection of data structures called *chromosomes* (Figure 5.4). Each of these chromosomes encodes a possible solution to the problem to be solved. Gold [38] uses such a GA for docking a ligand to a protein. Each chromosome encodes a possible protein–ligand complex conformation. A chromosome is assigned a fitness score on the basis of the relative quality of that solution in terms of protein–ligand interactions. Starting from an initial, randomly generated 'parent' population of chromosomes, the GA repeatedly applies two major genetic operators, crossover and mutation, resulting in 'children' chromosomes that replace the least-fit members of the population. The crossover operator requires two parents and produces two children, whereas the mutation operator requires one parent and produces one child. Crossover thus combines features from two different chromosomes in one, whereas mutation introduces random perturbations. The 'parent'

(1) A set of operators (crossing over, mutation, etc.) is chosen. Each operator is assigned a weight.

(2) An initial population is randomly created and the fitness of all individuals computed.

(3) An operator is chosen using a roulette wheel selection, based on operator weights.

(4) The parents able to reproduce are chosen, using roulette wheel selection based on fitness scores.

(5) The genetic operaor is applied, children chromosomes generated, and the fitness of each child determined.

(6) If not already present in the population, children replace the least-fit individuals.

(7) Goto three unless the maximum number of genetic operations is reached.

Figure 5.4 Main features of a genetic algorithm.

chromosomes are randomly selected from the existing population with a bias toward the best, thus introducing an evolutionary pressure into the algorithm. This emphasis on the survival of the best individuals ensures that, over time, the population should move toward an optimal solution, that is to the correct binding mode.

AutoDock 3.0 [18] uses a Lamarckian genetic algorithm (LGA). The characteristic of an LGA is that environmental adaptations of an individual's phenotype are transcribed into its genotype, on the basis of Jean Baptiste de Lamarck's assertion that phenotypic characteristics acquired during an individual's lifetime can become heritable traits. In AutoDock 3.0, each generation is thus followed by a local search (energy minimization) on a user-defined proportion of the population and resulting ligand coordinates are stored in the chromosome, replacing the parent.

Genetic algorithms are now commonplace for searching conformational place and several GA implementations (e.g. Darwin, Divali, GAsDock, MolDock, Psi-Dock; Table 5.2) have been recently described in combination with local minimization strategies using various force fields.

5.2.3
Tabu Search

Tabu search (TS) algorithms were first implemented in the Pro_Leads docking software [51]. A TS is characterized by imposing restrictions to enable a search process to negotiate otherwise difficult regions (Figure 5.5). These restrictions take the form of a tabu list that stores a number of previously visited solutions.

(1) Create an initial solution at random. Make it the current solution.

(2) Evaluate the current solution. If best so far, record it as best solution.

(3) Update the tabu list

    (a) If Tabu list not full, add current solution to the list.

    (b) Else, replace the oldest member with the current solution.

(4) Generate and evaluate x possible moves (e.g. 1000) from the current solution.

(5) Rank x moves in ascendng order of interaction energy.

(6) Examine the moves in rank order.

    (a) If move has the lowest energy than the best solution so far, accept it and goto seven.

    (b) If move is not tabu, accept is and goto seven.

    (c) If no acceptable move, exit.

(7) If the maximum number of iterations is reached, exit with the best solution found.

    (a) If best solution has not changed for a number of iterations (e.g. 100) goto one.

    (b) Else, goto two.

Figure 5.5 Main features of a tabu search.

By preventing the search from revisiting these regions, the exploration of new search space is encouraged. Only one current solution is maintained during the course of a search. At the start of a run, the current solution is initialized by randomizing the position and orientation of the ligand within a certain box around the active site. From the current solution, a user-defined number of moves is generated by a mutation-like procedure and finally ranked according to a scoring function. The TS maintains a tabu list that stores a number of previously visited solutions, and a move is 'tabu' if it generates a solution that is not different enough (e.g. rms deviation <0.75 Å) from the stored solutions. The highest ranked move is always accepted as the new 'current solution' if its energy is lower than the best energy till then, and it replaces, at the same time, the previous 'best solution'. Otherwise, the algorithm chooses the best non-tabu move. If neither criteria can be met, the algorithm terminates, if a new current solution can be found, it is added to the tabu list. The new current solution is simply added to the end of the list until it is full (e.g. 25 solutions). Thereafter, the current solution replaces an existing solution stored in the tabu list in a 'first-in, first-out' manner, which means it replaces the tabu solution having the longest residence in the list. Once the new current solution has been identified, a new set of moves is generated from it and the search procedure continues with the next iteration. TS and GA algorithms have recently been combined in PAS-Dock, Psi-Dock and SFDock (Table 5.2) for limiting the conformational search space while docking a small molecule to a protein. While GA usually converges quickly at the close proximity of a global minimum, it can be trapped in local minima. Using a tabu list helps in avoiding this drawback.

5.2.4
Simulated Annealing and Monte Carlo Simulations

Simulated annealing is a special molecular dynamics simulation, in which the system is cooled down at regular time intervals by decreasing the simulation temperature. The system thus gets trapped in the nearest local minimum conformation. Disadvantages of simulated annealing are that the result depends on the initial placement of the ligand and that the algorithm does not explore the solution space exhaustively. In a Monte Carlo (MC) search, the conformational space is sampled by random movements. Several MC implementations in docking programmes are possible (Table 5.2). In AutoDock2.4, Monte Carlo simulated annealing (MCSA) protocol is used. During each constant temperature cycle, random changes are made to the ligand's current orientation and conformation. The new state is immediately accepted if its energy is lower than the energy of the preceding state. Otherwise, the configuration is accepted or rejected based upon a probability expression (Boltzmann equation). The probability of acceptance P is given as

$$P = e^{\left(-\frac{\Delta E}{kT}\right)}, \tag{1}$$

where ΔE is the difference in energy from the previous step, T is the absolute temperature in kelvin, and k is the Boltzmann constant. This means, the higher the temperature of the cycle, the higher the probability that the new state is accepted.

In MCDock, Monte Carlo simulation is used in two steps. The first step serves for a pure geometrical optimization of the ligand position by random moves to minimize intermolecular overlap and is followed by MCSA using the CHARMM force field. Both ProDock and ICM (Table 5.2) use internal coordinates to represent molecular structures and couple MC random moves with force-field-based energy minimization. DockVision, a very user-friendly docking tool, operates in three steps (i) random positioning of precalculated ligand conformations in the user-defined binding cavity, (ii) checking for steric clashes with the target using a floating algorithm combining an MC procedure with a grid-based steric score function, (iii) searching for optimal binding modes after relapse of potential clashes by force-field-based MC sampling. QXP combines random MC moves on dihedral angles of the ligand with energy minimization with an intermediate fast template fitting procedure aimed at optimally locating the ligand in the binding cavity. Affinity uses a two-step procedure to dock a flexible ligand to a partially flexible protein: (i) a classical MC procedure is applied to locate the ligand in the binding site, (ii) location is afterward optimized by a simulated annealing protocol using a grid-based force field to treat the bulk (nonmovable part) of the complex and a more sophisticated full force field including implicit solvation effects is applied to the movable part of the system (ligand, amino acids of the binding site). Glide, another commercially available software uses a suite of hierarchical filters to remove unlikely solutions starting from low-level approximations (distance matches) to high-level calculations (full-force-field-based MCSA minimization) with free energy scoring. Glide implements a novel algorithm for rapid conformational generation, minimizing computational costs by clustering the core regions of the generated 3D ligand conformations and treating the positions of the end rotamer groups essentially independently.

In FDS (Table 5.2) a Monte Carlo algorithm is coupled to the AMBER molecular mechanics force field with the generalized born–surface area (GB–SA) continuum model. This solvent model is not only less expensive than an explicit representation, but also yields increased sampling.

5.2.5
Shape-fitting Methods

Shape-fitting methods are fast docking routines for which the steric and electrostatic complementarities of precalculated ligand conformations to the protein target are estimated. FTDock (Table 5.2) uses a grid-based representation of both the ligand and the target with assignment of defined values on the grid that depends on the atom accessibility. Fast Fourier transforms are then used to optimize ligand–target complementarity after global rotation/translation of the ligand. Ligin optimizes ligand orientation through a complementarity function summing up atomic contact surfaces to which a special weight is assigned depending on the favored/disfavored nature of the interaction. Sandock takes into account both the steric and electrostatic complementarity of the ligand, fitted by a distance-matching algorithm to the protein accessible surface. Last, LibDock uses a grid representation of the binding site for matching a triplet of ligand atoms to a triplet of hot spots (polar–apolar interaction sites). After elimination of improper orientation, possible matches are optimized by a soft atom pairwise potential.

5.2.6
Miscellaneous Approaches

Several other conformational sampling and pose scoring methods have been described in docking tools. Molecular dynamics (MD) coupled to simulating annealing (SA) conformationial sampling was implemented in the CDocker approach (Table 5.2), which uses essentially a set of CHARMm [65] scripts with soft-core potentials for enhancing conformational sampling. A distance geometry (DG) approach using sets of inter- and intra-atomic distances is enabled in DockIt. Systematic translation/rotation of the ligand within a predefined docking box has been described in Eudoc (Table 5.2). Last, several methods (SEED, FFLD, eHiTS, Table 5.2) decompose the ligand into rigid fragments that are separately docked and used as constraints to rebuild the full ligand either using a GA (SEED, FFLD) or a graph matching algorithm for selecting the most compatible fragment poses and later fitting flexible side chains between constrained fragments (eHiTS).

5.3
Scoring Functions

The free energy of binding is given by the Gibbs–Helmholtz equation:

$$\Delta G = \Delta H - T\Delta S, \tag{2}$$

with ΔG giving the free energy of binding, ΔH the enthalpy, T the temperature in Kelvin and ΔS the entropy. ΔG is related to the binding constant K_i by the equation

$$\Delta G = -RT \ln K_i, \tag{3}$$

with R being the gas constant.

There is a wide variety of different techniques available for predicting the binding free energy of a small molecule ligand on the basis of the given 3D structure of a protein–ligand complex. These techniques differ significantly in accuracy and speed. If one wants to predict the binding free energy difference between a ligand and a reference molecule, very accurate but time-consuming techniques such as free energy perturbation can be used [66]. If the aim is, however, to compare free energies of hundreds or thousands of protein–ligand complexes as generated by VS, much faster, but consequently, less accurate scoring functions have to be used. Scoring functions can be mainly categorized into three groups (Table 5.3): empirical scoring functions, force-field-based functions and knowledge-based potential of mean force. In VS, scoring functions are used for two purposes: (i) during the docking process, they serve as fitness function in the optimization placement of the ligand; (ii) when the docking is completed, it is used to rank each ligand of the database for which a docking solution has been found. In principle, different scoring functions could be used for these two purposes, although the same function is usually utilized in most docking tools.

5.3.1
Empirical Scoring Functions

Empirical scoring functions use several terms describing properties known to be important in drug binding to construct a master equation for predicting binding affinity. Multilinear regression is used to optimize the coefficients to weight the computed terms using a training set of protein–ligand complexes

Table 5.3 Main scoring functions.

Name	References
Empirical functions	
Chemscore	67
FlexX	32
Fresno	68
Glidescore	37
Hint	69
Ligscore	70
Ludi	71
PLP	72
Screenscore	73
X-Score	74
Force fields	
AutoDock	18
Dock	24
Goldscore	75
Potential of mean force	
Bleep	76
Drugscore	77
Pmf	78
SmoG	79

for which both the binding affinity and an experimentally determined high-resolution 3D structure are known. These terms generally describe polar interactions such as hydrogen bonds and ionic interactions, apolar interactions such as lipophilic and aromatic interactions, loss of ligand flexibility (entropy) and eventually also desolvation effects.

A breakthrough in deriving empirical scoring functions came with the introduction of Boehm's function developed for the *de novo* design program LUDI (Table 5.3), which is now implemented in a modified form in FlexX and represents a typical empirical function:

$$\Delta G = \Delta G_0 + \Delta G_{rot} * N_{rot} + \Delta G_{hb} \Sigma f(\Delta R, \Delta \alpha) + \Delta G_{io} \Sigma f(\Delta R, \Delta \alpha)$$
$$+ \Delta G_{aro} f(\Delta R, \Delta \alpha) + \Delta G_{lipo} f^*(\Delta R). \tag{4}$$

The ΔG coefficients are unknown and are determined by multilinear regression in order to fit the experimental measured binding affinities. The first terms are a fixed ground term and a term taking into account the loss of entropy during ligand binding by burial of rotatable bonds (ΔG_{rot}: energy loss per rotatable bond, N_{rot}: number of rotatable ligand bonds). ΔG_{hb} and ΔG_{io} give the binding energy for each optimal hydrogen bond and salt bridge, respectively. $f(\Delta R, \Delta \alpha)$ is a scaling function penalizing deviations from the ideal interaction geometry in terms of distance (ΔR) and angle ($\Delta \alpha$). An equivalent scaling function is used for the aromatic interactions (ΔG_{aro}). The lipophilic term (ΔG_{lipo}) is calculated as a sum over all pairwise atom–atom contacts. The function $f^*(\Delta R)$ accounts for contacts with a more or less ideal distance and penalizes forbiddingly close contacts.

One major disadvantage of the empirical scoring functions is the need of a training set to derive the weight factors of the individual energy terms. One can therefore expect an empirical scoring function to perform well only for proteins (metalloenzymes, proteases) similar to that used in the training set [80].

5.3.2
Force-field-based Scoring Functions

Force-field-based scoring functions such as the Dock energy score (Table 5.3) are based on the nonbonded terms of a classical molecular mechanics force field (e.g. AMBER, CHARMm, etc.). A Lennard–Jones potential describes van der Waals interactions, whereas the Coulomb energy describes the electrostatic components of the interactions. The nonbonded interaction energy takes the following form:

$$E = \sum_{i=1}^{\text{lig}} \sum_{j=1}^{\text{rec}} \left[\frac{A_{ij}}{r^{12}} - \frac{B_{ij}}{r^6} + 332\frac{q_i q_j}{Dr_{ij}} \right], \tag{5}$$

where A_{ij} and B_{ij} are van der Waals repulsion and attraction parameters between two atoms i and j at a distance r_{ij}, q_i and q_j are the point charges on atoms i and j, D is the dielectric function, and 332 is a factor that converts the electrostatic energy into kilocalories per mole. The main drawback of force-field calculations is the omission of the entropic component of the binding free energy. Therefore, attention should be paid not to overestimate the larger and most polar molecules that usually get the highest enthalpy interaction scores.

5.3.3
Knowledge-based Scoring Functions

A major disadvantage of empirical scoring functions lies in the fact that it is unclear to what extent they can be applied to protein–ligand complexes that were not represented in the training set used for deriving the master equation. Furthermore, empirical scoring functions dissect the protein–ligand binding free energy into all its physically meaningful contributions and try to evaluate them explicitly. Desolvation and entropy terms are, however, especially difficult to quantify.

A more recently developed approach avoiding these disadvantages uses knowledge-based scoring functions (Table 5.3) with potentials of mean force. A potential of mean force simply encodes structural information gathered from protein–ligand X-ray coordinates into Helmholtz free interaction energies of protein–ligand atom pairs. It is assumed that the more often a protein atom of type i and a ligand atom of type j are found at a certain distance r_{ij}, the more favorable this interaction is. Each interaction type between a protein atom of type i and a ligand atom type j at a distance r_{ij} is then assigned a protein–ligand interaction free energy $A(r)$ depending on its frequency.

$$A(r) = -k_B T \ln g_{ij}(r), \tag{6}$$

where k_B is the Boltzmann constant, T the absolute temperature and $g_{ij}(r)$ the atom pair distribution function for a protein–ligand atom pair ij. The distribution function is calculated from the number density of occurrences of that pair ij at a certain distance r in a database of protein–ligand complexes (usually the PDB).

The score is defined as the sum over all interatomic interactions of the protein–ligand complex. Advantages of this approach are that no fitting to experimentally measured binding free energies of the complexes in the training set is needed, and that solvation and entropic terms are treated implicitly.

5.3.4
Critical Overview of Fast Scoring Functions

The scoring function still remains the Achilles' heel of structure-based VS. Several recent and independent studies agree to conclude that many fast scoring functions can indeed distinguish near-native poses (rmsd lower than 2.0 from the X-ray pose) from decoys for about 70% of high-resolution protein–ligand X-ray structures [81]. However, when docking is applied to a large database, the corresponding scoring function should be robust enough to rank putative hits by increasing binding free energy values. Unfortunately, an accurate prediction of absolute binding free energies is still impossible, whatever the method, the treatment of solvation and the charge model [80]. Predicting binding free energy changes is possible on the condition that a customized scoring function is applied to a series of congeneric ligands. However, for a database containing a large diversity of compounds and for targets that have not been traditionally used for calibrating scoring functions, the obtained accuracy is usually limited (about 7 kJ mol^{-1} or 1.5 pK unit). From this observation, two sources of improvement are possible: (i) design more accurate scoring functions, and (ii) design smarter strategies to postprocess docking outputs (see next section). Many computational chemists actually favor the second option. The accuracy of scoring functions leveled off several years ago, for the simple reason that some unknown parameters (e.g. role of bound water, protein flexibility) remain extremely difficult to predict whatever the physical principles used to derive a scoring function.

5.4
Postfiltering Virtual Screening Results

Assuming that scoring functions are far from being able to quantitatively predict binding free energies (binding affinities), smart postdocking strategies are necessary to select virtual hits for experimental validation. Most of them attempt to detect false positives and consequently enhance true positive rates.

5.4.1
Filtering by Topological Properties

In order to eliminate protein–ligand complexes with improper geometries, different filters can be applied that evaluate three-dimensional protein–ligand complexes in terms of their steric complementarity. Stahl *et al.* developed a set of such filters [82] including the fraction of the ligand volume buried inside the binding pocket, the size of lipophilic cavities along the protein–ligand interface, the solvent-accessible surface of nonpolar parts of the ligand, and the number of close contacts between nonhydrogen-bonded polar atoms of the ligand and the protein. In an optimal pose, the buried ligand volume should be as large as possible, whereas the size of lipophilic cavities, the solvent-accessible surface of nonpolar parts of the ligand and the number of close contacts between nonhydrogen-bonded polar atoms should be as small as possible.

5.4.2
Filtering by Consensus Mining Approaches

Since the mean value of repeated samplings tends to be closer to the reality [83], using either several docking approaches [84] or several scoring functions [85, 86] is a possible way to remove false positives. On the basis of this assumption, any possible docking/scoring combination can be undertaken with, however, the risk of dramatically decreasing the size of the hit list, while increasing the level of constraints. As VS basically aims at finding the best possible compromise between hit rate (percentage of true actives in the hit list) and hit coverage (percentage of true actives effectively recovered), one should use consensus approaches with caution and always compare various hit selection strategies on the condition that hit lists of comparable sizes are produced [87].

Along the same line, an investigation of various conformations of the target for docking may contribute to better handling of the binding site flexibility by outputting an averaged docking score [88]. Last, various poses of each ligand (and not necessarily the top-ranked pose) can be analyzed in all consensus postprocessing scenarios [89]. Consensus approaches usually require a substantial amount of preexisting knowledge (several X-ray structure coordinates, many known actives) and are usually customized for a particular target and are thus not applicable to the next one.

5.4.3
Filtering by Combining Computational Procedures

Docking output coordinates can be submitted, at least for the most interesting target–ligand complexes, to more sophisticated computational approaches that are better suited to predict absolute binding free energies. Incorrect treatment of long-range electrostatic effects and desolvation are, for example, clear drawbacks of fast scoring functions. It is therefore possible to apply more powerful scoring methods using either molecular mechanics coupled to continuum solvation models [90, 91] (e.g. Molecular Mechanics-Poisson-Boltzmann solvent accessible surface area (MM-PBSA) or molecular mechanics-generalized born model solvent accessibility (MM-GBSA) approximations) or even quantum mechanics [92, 93]. Such approaches are still limited to a prefiltered set of virtual hits since they are computationally more demanding. However, they are applicable on standard PC clusters/farms at reasonable computing costs.

Machine learning methods able to detect information in very noisy datasets can also help in trapping those hits of interest. Among the most promising methods, Bayesian statistics relies upon the frequency of 2D substructural features among known active and inactive compounds and has been shown to be a powerful postdocking filter [94].

Last, the known binding mode of true actives can be fingerprinted in order to translate 3D information about the protein–ligand complex into a simple bit vector with a finite number of bits coding, for either each atom or each residue of a binding site of interest, the molecular interactions of the ligands (H-bonds, ionic interactions, hydrophobic contacts) [95]. Assuming that true actives usually share similar binding modes (same kind of interactions with well-defined residues), generating molecular interaction fingerprints for all predicted poses and scoring them by similarity to that of known actives, is a promising way to remove false positives showing protein–ligand interaction patterns that are inconsistent with experimental data.

5.4.4
Filtering by Chemical Diversity

In order to decrease the number of virtual hits to be purchased and or tested for biological evaluation (docking 50 000 compounds will typically generate hit lists of 1–2000 molecules), virtual hits are often clustered according to their chemical diversity. Although there is no clear definition of molecular diversity (it is indeed dependent on the molecular descriptors and on the metric used to measure diversity), this strategy presents the advantage of focusing less on individual numerical values (docking scores) and more on how these values are distributed in chemical space (Figure 5.6). By prioritizing scaffolds instead of individual compounds, false negatives may be recovered if they share a common scaffold with true positives.

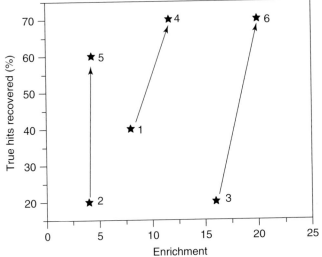

Figure 5.6 Influence of postprocessing strategies in retrieving true vasopressin V_{1a} receptor antagonists by structure-based screening of a database of 990 randomly chosen 'druglike' compounds seeded with 10 true actives [96]. (1) Top 5% ligands as scored by FlexX; (2) top 5% ligands as scored by Gold; (3) hits common to (1) and (2); (4) ClassPharmer [97] prioritization of scaffolds populated by more than two compounds and for which 60% of the representatives have a FlexXscore lower than −22 kJ mol^{-1}; (5) ClassPharmer prioritization of scaffolds populated by more than two compounds for which 60% of the representatives have a Goldscore higher than 37.5; (6) ClassPharmer prioritization of classes populated by more than two compounds for which 60% of the representatives have a FlexX score lower than −22 kJ mol^{-1} and a Goldscore higher than 37.5.

5.4.5
Filtering by Visual Inspection

At the earliest possible stage, one cannot avoid looking at the predicted 3D structure of the protein-hit complex to ascertain that (i) the ligand has been really docked in the expected binding site and not at its periphery, (ii) the bound conformation of the ligand has a physicochemical meaning, (iii) the ligand interacts with key residues of the active site. This step can be very time-consuming and tedious, but ensures that every selected hit presents expected properties.

5.5
Comparison of Different Docking and Scoring Methods

Numerous docking programs based on very different physicochemical approximations have been reported (see Table 5.2). Since any docking tool needs to combine a docking engine with a fast scoring function, recent literature is full of benchmarks addressing three possible issues: (i) the capability of a docking algorithm to reproduce the X-ray pose of selected small molecular-weight ligand [81, 89], (ii) the propensity of fast scoring functions to recognize near-native poses among a set of decoys [98] and to predict absolute binding free energies [80], (iii) the discrimination of known binders from randomly chosen molecules in VS experiments [99, 100]. However, analyzing all these data for a comparative analysis of available docking tools is very difficult. First, many tools are not available. Second, independent studies assessing the relative performance of docking algorithms/scoring functions are still rare and focus on the usage of few methods. Third, the quality of judgment may vary depending on the examined properties (quality of the top-ranked pose, quality of all plausible poses, binding free energy prediction, VS utility). Fourth, most docking programs assume approximation levels that can vary considerably and lead, for example, to very inhomogeneous docking paces ranging from few seconds to few hours. Last, many docking programs have been calibrated and validated on small protein–ligand datasets. Hence, detailed benchmarks (>100 PDB-ligand complexes) are only reported for few docking tools. The most recent validation studies on different datasets agree to conclude that the accuracy of a docking tool is largely target-dependant and should be examined on a case-by-case basis. Glide and Gold seem to be the most robust programs for their propensity to generate near-native poses in about 75–80% of the cases [81], provided that several solutions are stored. A major problem is that the scoring function does not always (only in about 40–50% of the cases) predict the correct solution as the most probable one, which considerably complicates the analysis of docking results. Numerous reasons explain this limited accuracy. Some are easy to correct (e.g. incorrect atom type for either the ligand or the protein), some are more difficult (e.g. accuracy of the protein 3D structure, flexibility of the ligand, accuracy of the scoring function), and some are really tricky to overcome (protein flexibility, role of bound water). However, predicting which docking program will be the most suited for a research project is still unrealistic. If known ligands are available, the best possible option is to try a systematic combination of docking/scoring parameters and select for productive screening the one that best segregates true actives from true inactives. If no or very few ligands are available, some guides may be followed to choose the tool that seems the most appropriate to the physicochemical properties of the protein cavity.

5.6
Examples of Successful Virtual Screening Studies

Some recent reports from the literature (2003–2006) are reviewed in Table 5.4. Most of them still make use of high-resolution X-ray structures [73–79, 101–127]. However, encouraging data begin to emerge from homology models

Table 5.4 Successful structure-based screening data from the recent literature (2003–2006).

Target	Docking	Library	Size	Hit rate[a]	References
Aldose reductase	FlexX	ACD	260 000	55%@ 20 μM	101
L-Xylulose reductase	Dock	NCI[b]	249 071	5%@ 100 μM	135
IMPDH[c]	FlexX	Roche reagents	3425	8%@ 100 μM	136
DHFR[d]	FlexX	Roche reagents library	9448	21%@ 25 μM	102
DHFR	Dock	ACD[e]	n.a.	33%@ 20 mM	103
Thymidine phosphorylase	Dock	NCI	250 000	7%@ 20 μM	104
t-RNA guanine transglycosylase	FlexX	Seven collections	827 000	55%@ 10 μM	105
Chk-1 kinase	FlexX	AstraZeneca	550 000	36%@ 68 μM	106
	rDock	Vernalis	700 000	0.8%@ 50 μM	107
Casein kinase II	Dock	Novartis	450 000	33%@ 10 μM	108
	Moe/Glide/ Fred/Gold	MMS database	2000	n.a.	109
BCR-ABL[m]	Dock	ChemDiv	200 000	13%@ 30 μM	110
P56 Lck	Dock	n.a.[f]	2 000 000	17%@ 100 μM	111
EphB2	Gold	ChemDiv	50 452	5%@ 10 μM	112
Protein kinase B	FlexX	ChemBridge	50 000	10%@ 20 μM	113
Thymidine kinase	FlexX	CMC[g] + KEGG[h]	7986	10%@ 20 μM	114
Cdk2	Lidaeus	Commercial compounds	50 000	5%@ 20 μM	115
Acetylcholinesterase	ADAM & EVE	ACD + Maybridge	160 000	11%@ 10 μM	116
Acetylcholinesterase	AutoDock	Combinatorial library	271	50%@ 10 μM	117
Phosphodiesterase 4D	FlexX	Combinatorial library	320	55%@ 100 nM	137
Rac GTPase	FlexX	NCI	140 000	6%@ 50 μM	118
SARS CoV 3C-like protease	Dock	ACD + MDDR + NCI	630 000	7%@ 200 μM	119
HIV-1 integrase	Gold	ChemBridge	1700	43%@ 100 μM	120
AICAR transformylase[i]	AutoDock	NCI	1990	51%@ 20 μM	121
XIAP[n]	Dock	TCM[j]	8000	3%@ 5 μM	138
Stat3β	Dock	Four collections	429 000	1%@ 20 μM	139
CytP450 2D6	Gold	NCI subset	111	39%@ 10 μM	122
CytP450 2D6	Gold	VU Amsterdam	5760	62%@ 5 μM	123
SHBG[k]	Glide	Natural compounds	23 836	7%@ 25 μM	124
Thyroid hormone β receptor	ICM	ACD	190 000	14%@ 30 μM	125
Edema factor	Dock	ACD	205 226	8%@ 100 μM	126

Table 5.4 (continued).

Target	Docking	Library	Size	Hit rate[a]	References
RmIc[o]	FlexX	Combinatorial library	3888	32%@ 20 μM	127
Ribosomal A-site	rDock	Vernalis	1 000 000	26%@ 500 μM	128
Hsp90	rDock	Vernalis	700 000	n.a.	129
β-lactamase	Dock	Zinc	33 000	30%@ 120 μM	130
11βHSD-1[l]	Gold/ Glide	Natural compounds	114 000	13%@ 20 μM	131
Falcipain-2	Gold	ChemBridge	50 000	28%@ 60 μM	132
15-lipoxygenase	ChemBridge	Glide	50 000	3%@ 10 μM	133
5-HT$_{1A}$ receptor	Dock	>20 suppliers	1 600 000	21%@ 5 μM	134
NK$_1$ receptor	Dock	>20 suppliers	1 600 000	15%@ 5 μM	134
D$_2$ receptor	Dock	>20 suppliers	1 600 000	17%@ 5 μM	134
CCR$_3$ receptor	Dock	>20 suppliers	1 600 000	12%@ 5 μM	134
5-HT$_4$ receptor	Dock	>20 suppliers	1 600 000	21%@ 5 μM	134
A$_{1a}$ receptor	Gold	Aventis	n.a.	30%@ 1 μM	140
NK$_1$ receptor	FlexX	Seven collections	827 000	14%@ 1 μM	141
D$_3$ receptor	LigandFit	NCI	250 000	40%@ 1 μM	142

a   Hit rate at a concentration threshold. The hit rate is the ratio of the number of active compounds to the total number of compounds tested.
b   National Cancer Institute database.
c   Inosine-5'-monophosphate dehydrogenase.
d   Dihydrofolate reductase.
e   Available chemicals directory.
f   Not available.
g   CMC: comprehensive medicinal chemistry database.
h   KEGG database (http://www.genome.jp/kegg/ligand.html).
i   Aminoimidazole carboxamide ribonucleotide transformylase.
j   Traditional Chinese medicine database (http://www.tcm3d.com).
k   Sex hormone binding globulin.
l   11β-Hydroxysteroid dehydrogenase.
m   BCR-ABL tyrosine Kinase.
n   X-linked inhibitor of apoptosis.
o   dDTP-6-deoxy-D-xylo-4-hexulose-3,5-epimerase.

[128–134] and thus broaden the applicability of structure-based screening methods to a wider array of pharmaceutically interesting targets.

Macromolecular targets (e.g. kinases, reductases, esterases) presenting a well-defined hydrophilic pocket for which the directionality of intermolecular interactions play a key role in ligand recognition are particularly well suited for VS for the simple reason that most docking tools and scoring functions have been calibrated for such situations. Thus, it is no surprise the latter protein families are overrepresented in targets for which true inhibitors have been discovered by database docking [73–79, 101–115] (Table 5.4). A second reason for these successes is that both database prefiltering and postdocking analysis can be constrained by existing knowledge about preferential chemotypes and binding modes [111].

Nevertheless, first-in-class micromolar inhibitors [138, 139] have already been discovered using high-throughput docking techniques, thus illustrating the power of VS methods for difficult targets. In most of the cases, druglike and or leadlike compounds are identified. However, VS may also be applied in fragonomic projects [135, 136] for which low-molecular-weight fragments (typically bellow 300) are typically screened by structural biology methods (X-ray diffraction, NMR). A very difficult aspect of *in silico* fragment screening remains scoring since many docking programs will typically produce a continuum of energetically comparable poses for which a final selection is difficult. Ranking such fragments by similarity to known molecular interaction fingerprints is a promising way to remove all bias associated with scoring functions [143].

A large majority of structure-based screening projects are aimed at identifying hits to start medicinal chemistry with. However, lead optimization might be possible on the condition that the binding mode of the starting lead can be unambiguously recovered and that a rational strategy can be designed for selecting the next compounds to synthesize [137].

5.7
Outlook

Structure-based VS is currently mature enough to be applied as a first-choice screening method notably in academic labs where HTS infrastructures are missing. Experimental screening and VS are not mutually exclusive but complementary methods that optimally work in different situations. Whereas HTS requires industrial settings (library deck storage and maintenance, screening robots, numerous readout techniques), VS can be applied at relatively low costs in academic settings. It is clear that VS will generate many more false positives than HTS, which is not a problem per se on the condition that hit coverage is large enough (false negatives) and that a new medicinal chemistry project can be started from experimentally validated virtual hits. A clear positioning of VS with respect to HTS is that *in silico* techniques might be applied for targets where an experimental HTS assay is either cumbersome or even impossible. Moreover, VS can also be applied to library collections that are not available to experimental screening and thus maximize the chances of get interesting hits.

A clear danger in VS, notably to beginners in the field, is the temptation to screen larger and larger libraries to enhance hit rates. Most, if not all previously reported success stories indicate that library setup and postdocking analyses are probably the key factors to success. It is now common practice and strongly advised to first filter the starting compound library by experimentally driven constraints (substructure search, 2D or 3D pharmacophore), rigorously set up a docking/scoring strategy aimed at discriminating known actives from known inactives and carefully analyze docking results even for compounds/poses that are not highly prioritized at first sight. One should also not be totally biased by the hit rate in VS analysis since this number is strongly target and library dependent. The number of new validated chemotypes amenable to medicinal chemistry optimization is therefore a better descriptor than the simple hit rate, which considerably varies with regard to the current knowledge on a particular target. VS is a natural complement to traditional medicinal chemistry and particularly well suited for proposing new molecular scaffolds that can be easily converted into focussed ligand libraries of higher values. Both methodological improvements (docking, scoring, hit triage, prediction of absorption, distribution, metabolism, excretion and toxicity (ADMET) properties) and better screening collections (focused and targeted libraries) should contribute to improve the value of this powerful tool.

References

1. Walters, W.P., Stahl, M.T., and Murcko, M.A. (1998) Virtual screening – an overview. *Drug Discovery Today*, **3**, 160–78.
2. Stahura, F.L. and Bajorath, J. (2004) Virtual screening methods that complement HTS. *Combinatorial Chemistry & High Throughput Screening*, **7**, 259–69.
3. Guner, O., Clement, O., and Kurogi, Y. (2004) Pharmacophore modeling and three dimensional database searching for drug design using catalyst: recent advances. *Current Medicinal Chemistry*, **11**, 2991–3005.
4. Klebe, G. (2006) Virtual ligand screening: strategies, perspectives and limitations. *Drug Discovery Today*, **11**, 580–94.
5. Rishton, G.M. (1997) Reactive compounds and in vitro false positives in HTS. *Drug Discovery Today*, **2**, 382–84.
6. Lipinski, C.A., Lombardo, F., Dominy, B.W., and Feney, P.J. (1997) Experimental and computational approaches to estimate solubility and permeability in drug discovery and development settings. *Advanced Drug Delivery Reviews*, **23**, 3–25.
7. Roche, O. and Guba, W. (2005) Computational chemistry as an integral component of lead generation. *Mini Reviews in Medicinal Chemistry*, **5**, 677–83.
8. Muresan, S. and Sadowski, J. (2005) "In-house likeness": comparison of large compound collections using artificial neural networks. *Journal of Chemical Information and Modeling*, **45**, 888–93.
9. Pearce, B.C., Sofia, M.J., Good, A.C. *et al.* (2006) An empirical process for the design of high-throughput screening deck filters. *Journal of Chemical Information and Modeling*, **46**, 1060–68.
10. Shoichet, B.K. (2006) Screening in a spirit haunted world. *Drug Discovery Today*, **11**, 607–15.
11. SciTegic, Accelrys Inc., San Diego, USA. sales@scitegic.com.
12. Irwin, J.J. and Shoichet, B.K. (2005) ZINC. A free database of commercially available compounds for virtual screening. *Journal of Chemical Information and Computer Sciences*, **45**, 177–182.
13. Chen, J., Swamidass, S.J., Dou, Y. *et al.* (2005) ChemDB: a public database of small molecules and

related chemoinformatics resources. *Bioinformatics*, **21**, 4133–39.

14. Leach, A.R. and Lemon, A.P. (1998) Exploring the conformational space of protein side chains using dead-end elimination and the A* algorithm. *Proteins*, **33**, 227–39.

15. Osterberg, F., Morris, G.M., Sanner, M.F. *et al.* (2002) Automated docking to multiple target structures: incorporation of protein mobility and structural water heterogeneity in AutoDock. *Proteins*, **46**, 34–40.

16. Claussen, H., Buning, C., Rarey, M., and Lengauer, T. (2001) FlexE: efficient molecular docking considering protein structure variations. *Journal of Molecular Biology*, **308**, 377–95.

17. Mizutani, M.Y., Takamatsu, Y., Ichinose, T. *et al.* (2006) Effective handling of induced-fit motion in flexible docking. *Proteins*, **63**, 878–91.

18. Morris, G.M., Goodsell, D.S., Halliday, R.S. *et al.* (1998) Automated docking using a Lamarckian genetic algorithm and an empirical binding free energy function. *Journal of Computational Chemistry*, **19**, 1639–62.

19. Wu, G., Robertson, D.H., Brooks, C.L. 3rd, and Vieth, M. (2003) Detailed analysis of grid-based molecular docking: A case study of CDOCKER-A CHARMM-based MD docking algorithm. *Journal of Computational Chemistry*, **24**, 1549–62.

20. Lawrence, M.C. and Davis, P.C. (1992) CLIX: a search algorithm for finding novel ligands capable of binding proteins of known three-dimensional structure. *Proteins*, **12**, 31–41.

21. Sun, Y., Ewing, T.J., Skillman, A.G., and Kuntz, I.D. (1998) CombiDOCK: structure-based combinatorial docking and library design. *Journal of Computer-Aided Molecular Design*, **12**, 597–604.

22. Taylor, J.S. and Burnett, R.M. (2000) DARWIN: a program for docking flexible molecules. *Proteins*, **41**, 173–91.

23. Clark, K.P. (1995) and Ajay Flexible ligand docking without parameter adjustment across four ligand-receptor complexes. *Journal of Computational Chemistry*, **16**, 1210–26.

24. Ewing, T.J., Makino, S., Skillman, A.G., and Kuntz, I.D. (2001) DOCK 4.0: search strategies for automated molecular docking of flexible molecule databases. *Journal of Computer-Aided Molecular Design*, **15**, 411–28.

25. Hart, T.N. and Read, R.J. (1992) A multiple-start Monte Carlo docking method. *Proteins*, **13**, 206–22.

26. Makino, S., Ewing, T.J.A., and Kuntz, I.D. (1999) DREAM++: flexible docking program for virtual combinatorial libraries. *Journal of Computer-Aided Molecular Design*, **13**, 513–32.

27. Zsoldos, Z., Reid, D., Simon, A. *et al.* (2007) eHiTS: A new fast, exhaustive flexible ligand docking system. *Journal of Molecular Graphics & Modelling*, **26**, 198–212 .

28. Pang, Y.P., Perola, E., Xu, K., and Prendergast, F.G. (2001) EUDOC: a computer program for identification of drug interaction sites in macromolecules and drug leads from chemical databases. *Journal of Computational Chemistry*, **22**, 1750–71.

29. Taylor, R.D., Jewsbury, P.J., and Essex, J.W. (2003) FDS: flexible ligand and receptor docking with a continuum solvent model and soft-core energy function. *Journal of Computational Chemistry*, **24**, 1637–56.

30. Cecchini, M., Kolb, P., Majeux, N., and Caflisch, A. (2004) Automated docking of highly flexible ligands by genetic algorithms: a critical assessment. *Journal of Computational Chemistry*, **25**, 412–22.

31. Merlitz, H., Herges, T., and Wenzel, W. (2004) Fluctuation analysis and accuracy of a large-scale in silico screen. *Journal of Computational Chemistry*, **25**, 1568–75.

32. Rarey, M., Kramer, B., Lengauer, T., and Klebe, G. (1996) A fast flexible docking method using an incremental construction algorithm. *Journal of Molecular Biology*, **261**, 470–89.

33. Miller, M.D., Sheridan, R.P., Kearsley, S.K., and Underwood, D.J. (1994) Advances in automated docking applied to human immunodeficiency virus type 1 protease. *Methods in Enzymology*, **241**, 354–70.

34. OpenEye Scientific Software, Santa Fe, USA. business@eyesopen.com.

35. Gabb, H.A., Jackson, R.M., and Sternberg, M.J. (1997) Modelling protein docking using shape complementarity, electrostatics and biochemical information. *Journal of Molecular Biology*, **272**, 106–20.

36. Li, H., Li, C., Gui, C. *et al.* (2004) GAsDock: a new approach for rapid flexible docking based on an improved multi-population genetic algorithm. *Bioorganic & Medicinal Chemistry Letters*, **14**, 4671–76.

37. Friesner, R.A., Banks, J.L., Murphy, R.B. *et al.* (2004) Glide: a new approach for rapid, accurate docking and scoring. 1. Method and assessment of docking accuracy. *Journal of Medicinal Chemistry*, **47**, 1739–49.

38. Verdonk, M.L., Cole, J.C., Hartshorn, M.J. *et al.* (2003) Improved protein-ligand docking using GOLD. *Proteins*, **52**, 609–23.

39. Welch, W., Ruppert, J., and Jain, A.N. (1996) Hammerhead: fast, fully automated docking of flexible ligands to protein binding sites. *Chemistry & Biology*, **3**, 449–62.

40. Floriano, W.B., Vaidehi, N., Zamanakos, G., and Goddard, W.A. III. (2004) HierVLS hierarchical docking protocol for virtual ligand screening of large-molecule databases. *Journal of Medicinal Chemistry*, **47**, 56–71.

41. Totrov, M. and Abagyan, R. (1997) Flexible protein-ligand docking by global energy optimization in internal coordinates. *Proteins*, 1(Suppl.), 215–20.

42. Diller, D.J. and Merz, K.M. Jr. (2001) High throughput docking for library design and library prioritization. *Proteins*, **43**, 113–24.

43. Venkatachalam, C.M., Jiang, X., Oldfield, T., and Waldman, M. (2003) LigandFit: a novel method for the shape-directed rapid docking of ligands to protein active sites. *Journal of Molecular Graphics & Modelling*, **21**, 289–307.

44. Sobolev, V., Moallem, T.M., Wade, R.C. *et al.* (1997) CASP2 molecular docking predictions with the LIGIN software. *Proteins*, 1(Suppl.), 210–14.

45. Liu, M. and Wang, S. (1999) MCDOCK: a Monte Carlo simulation approach to the molecular docking problem. *Journal of Computer-Aided Molecular Design*, **13**, 435–51.

46. Thomsen, R. and Christensen, M.H. (2006) MolDock: a new technique for high-accuracy molecular docking. *Journal of Medicinal Chemistry*, **49**, 3315–21.

47. Tondel, K., Anderssen, E., and Drablos, F. (2006) Protein Alpha Shape (PAS) Dock: a new gaussian-based score function suitable for docking in homology modelled protein structures. *Journal of Computer-Aided Molecular Design*, **20**, 131–44.

48. Joseph-McCarthy, D. and Alvarez, J.C. (2003) Automated generation of MCSS-derived pharmacophoric DOCK site points for searching multiconformation databases. *Proteins*, **51**, 189–202.

49. Goto, J., Kataoka, R., and Hirayama, N. (2004) Ph4Dock: pharmacophore-based protein-ligand docking. *Journal of Medicinal Chemistry*, **47**, 6804–11.

50. Trosset, J.Y. and Scheraga, H.A. (1999) Prodock: software package for protein modeling and docking. *Journal of Computational Chemistry*, **20**, 412–27.

51. Baxter, C.A., Murray, C.W., Clark, D.E. *et al.* (1998) Flexible docking using Tabu search and an empirical estimate of binding affinity. *Proteins*, **33**, 367–82.

52. Seifert, M.H. (2005) ProPose: steered virtual screening by simultaneous protein-ligand docking and ligand-ligand alignment. *Journal of Chemical Information and Modeling*, **45**, 449–60.

53. Pei, J., Wang, Q., Liu, Z. *et al.* (2006) PSI-DOCK: towards highly efficient and accurate flexible ligand docking. *Proteins*, **62**, 934–46.

54. Jackson, R.M. (2002) Q-fit: a probabilistic method for docking molecular fragments by sampling low energy conformational space. *Journal of Computer-Aided Molecular Design*, **16**, 43–57.

55. McMartin, C. and Bohacek, R.S. (1997) QXP: powerful, rapid computer algorithms for structure-based drug design. *Journal of Computer-Aided Molecular Design*, **11**, 333–44.

56. Morley, S.D. and Afshar, M. (2004) Validation of an empirical RNA-ligand scoring function for fast flexible docking using Ribodock. *Journal of Computer-Aided Molecular Design*, **18**, 189–208.

57. Burkhard, P., Taylor, P., and Walkinshaw, M.D. (1998) An example of a protein ligand found by database mining: description of the docking method and its verification by a 2.3 A X-ray structure of a thrombin-ligand complex. *Journal of Molecular Biology*, **277**, 449–66.

58. Wu, G. and Vieth, M. (2004) SDOCKER: a method utilizing existing X-ray structures to improve docking accuracy. *Journal of Medicinal Chemistry*, **47**, 3142–48.

59. Majeux, N., Scarsi, M., Apostolakis, J. *et al.* (1999) Exhaustive docking of molecular fragments with electrostatic solvation. *Proteins*, **37**, 88–105.

60. Hou, T., Wang, J., Chen, L., and Xu, X. (1999) Automated docking of peptides and proteins by using a genetic algorithm combined with a tabu search. *Protein Engineering*, **12**, 639–48.

61. Firth-Clark, S., Willems, H.M., Williams, A., and Harris, W. (2006) Generation and selection of novel estrogen receptor ligands using the de novo structure-based design tool, SkelGen. *Journal of Chemical Information and Modeling*, **46**, 642–47.

62. Zavodszky, M.I., Sanschagrin, P.C., Korde, R.S., and Kuhn, L.A. (2002) Distilling the essential features of a protein surface for improving protein-ligand docking, scoring, and virtual screening. *Journal of Computer-Aided Molecular Design*, **16**, 883–902.

63. Jain, A.N. (2003) Surflex: fully automatic flexible molecular docking using a molecular similarity-based search engine. *Journal of Medicinal Chemistry*, **46**, 499–511.

64. Kirchmair, J., Wolber, G., Laggner, C., and Langer, T. (2006) Comparative performance assessment of the conformational model generators omega and catalyst: a large-scale survey on the retrieval of protein-bound ligand conformations. *Journal of Chemical Information and Modeling*, **46**, 1848–61.

65. Brooks, B.R., Bruccoleri, R.E., Olafson, B.D. *et al.* (1983) CHARMM: a program for macromolecular energy, minimization, and dynamics calculations. *Journal of Computational Chemistry*, **4**, 187–217.

66. Miyamoto, S. and Kollman, P.A. (1993) Absolute and relative binding free energy calculations of the interaction of biotin and its analogs with streptavidin using molecular dynamics/free energy perturbation approaches. *Proteins*, **16**, 226–45.

67. Eldridge, M., Murray, C.W., Auton, T.A. *et al.* (1997) Empirical scoring functions: I. The development of a fast empirical scoring function to estimate the binding affinity of ligands in receptor complexes. *Journal of Computer-Aided Molecular Design*, **11**, 425–45.

68. Rognan, D., Laumoeller, S.L., Holm, A. *et al.* (1999) Predicting binding affinities of protein ligands

from three-dimensional coordinates: application to peptide binding to class I major histocompatibility proteins. *Journal of Medicinal Chemistry*, **42**, 4650–58.

69. Cozzini, P., Fornabaio, M., Marabotti, A. *et al.* (2002) Simple, intuitive calculations of free energy of binding for protein-ligand complexes. 1. Models without explicit constrained water. *Journal of Medicinal Chemistry*, **45**, 2469–83.

70. Krammer, A., Kirchhoff, P.D., Jiang, X. *et al.* (2005) LigScore: a novel scoring function for predicting binding affinities. *Journal of Molecular Graphics & Modelling*, **23**, 395–407.

71. Böhm, H.J. (1994) The development of a simple empirical scoring function to estimate the binding constant for a protein-ligand complex of known three-dimensional structure. *Journal of Computer-Aided Molecular Design*, **8**, 243–56.

72. Gehlhaar, D.K., Verkhivker, G.M., Rejto, P.A. *et al.* (1995) Molecular recognition of the inhibitor AG-1343 by HIV-1 protease: conformationally flexible docking by evolutionary programming. *Chemistry & Biology*, **2**, 317–24.

73. Stahl, M. and Rarey, M. (2001) Detailed analysis of scoring functions for virtual screening. *Journal of Medicinal Chemistry*, **44**, 1035–42.

74. Wang, R., Lai, L., and Wang, S. (2002) Further development and validation of empirical scoring functions for structure-based binding affinity prediction. *Journal of Computer-Aided Molecular Design*, **16**, 11–26.

75. Jones, G., Wilett, P., Glen, R.C. *et al.* (1997) Development and validation of a genetic algorithm for flexible docking. *Journal of Molecular Biology*, **267**, 727–48.

76. Mitchell, J.B., Laskowski, R.A., Alex, A., and Thornton, J.M. (1999) BLEEP: potential of mean force describing protein-ligand interactions. II. Calculation of binding energies and comparison with experimental data. *Journal of Computational Chemistry*, **20**, 117–1185.

77. Gohlke, H., Hendlich, M., and Klebe, G. (2000) Knowledge-based scoring function to predict protein-ligand interactions. *Journal of Molecular Biology*, **295**, 337–56.

78. Muegge, I. and Martin, Y.C. (1999) A general and fast scoring function for protein-ligand interactions: a simplified potential approach. *Journal of Medicinal Chemistry*, **42**, 791–804.

79. Ishchenko, A.V. and Shakhnovich, E.I. (2002) SMall molecule growth 2001 (SMoG2001): an improved knowledge-based scoring function for protein-ligand interactions. *Journal of Medicinal Chemistry*, **45**, 2770–80.

80. Ferrara, P., Gohlke, H., Price, D.J. *et al.* (2004) Assessing scoring functions for protein-ligand interactions. *Journal of Medicinal Chemistry*, **47**, 3032–47.

81. Kellenberger, E., Rodrigo, J., Muller, P., and Rognan, D. (2004) Comparative evaluation of eight docking tools for docking and virtual screening accuracy. *Proteins*, **57**, 225–42.

82. Stahl, M. and Bohm, H.J. (1998) Development of filter functions for protein-ligand docking. *Journal of Molecular Graphics & Modelling*, **16**, 121–32.

83. Wang, R. and Wang, S. (2001) How does consensus scoring work for virtual library screening? An idealized computer experiment. *Journal of Chemical Information and Computer Sciences*, **41**, 1422–26.

84. Paul, N. and Rognan, D. (2002) ConsDock: a new program for the consensus analysis of protein-ligand interactions. *Proteins*, **47**, 521–33.

85. Charifson, P.S., Corkery, J.J., Murcko, M.A., and Walters, W.P. (1999) Consensus scoring: A method for obtaining improved hit rates from docking databases of three-dimensional structures into proteins. *Journal of Medicinal Chemistry*, **42**, 5100–9.

86. Bissantz, C., Folkers, G., and Rognan, D. (2000) Protein-based virtual screening of chemical databases. 1. Evaluation of different docking/scoring combinations. *Journal of Medicinal Chemistry*, **43**, 4759–67.

87. Xing, L., Hodgkin, E., Liu, Q., and Sedlock, D. (2004) Evaluation and application of multiple scoring functions for a virtual screening experiment. *Journal of Computer-Aided Molecular Design*, **18**, 333–44.

88. Vigers, G.P. and Rizzi, J.P. (2004) Multiple active site corrections for docking and virtual screening. *Journal of Medicinal Chemistry*, **47**, 80–89.

89. Kontoyianni, M., McClellan, L.M., and Sokol, G.S. (2004) Evaluation of docking performance: comparative data on docking algorithms. *Journal of Medicinal Chemistry*, **47**, 558–65.

90. Lyne, P.D., Lamb, M.L., and Saeh, J.C. (2006) Accurate prediction of the relative potencies of members of a series of kinase inhibitors using molecular docking and MM-GBSA scoring. *Journal of Medicinal Chemistry*, **49**, 4805–8.

91. Kuhn, B., Gerber, P., Schulz-Gasch, T., and Stahl, M. (2005) Validation and use of the MM-PBSA approach for drug discovery. *Journal of Medicinal Chemistry*, **48**, 4040–48.

92. Khandelwal, A., Lukacova, V., Comez, D. *et al.* (2005) A combination of docking, QM/MM methods, and MD simulation for binding affinity estimation of metalloprotein ligands. *Journal of Medicinal Chemistry*, **48**, 5437–47.

93. Ferrara, P., Curioni, A., Vangrevelinghe, E. *et al.* (2006) New scoring functions for virtual screening from molecular dynamics simulations with a quantum-refined force-field (QRFF-MD). Application to cyclin-dependent kinase 2. *Journal of Chemical Information and Modeling*, **46**, 254–63.

94. Klon, A.E., Glick, M., and Davies, J.W. (2004) Combination of a naive Bayes classifier with consensus scoring improves enrichment of high-throughput docking results. *Journal of Medicinal Chemistry*, **47**, 4356–59.

95. Deng, Z., Chuaqui, C., and Singh, J. (2004) Structural interaction fingerprint (SIFt): a novel method for analyzing three-dimensional protein-ligand binding interactions. *Journal of Medicinal Chemistry*, **47**, 337–44.

96. Bissantz, C., Bernard, P., Hibert, M., and Rognan, D. (2003) Protein-based virtual screening of chemical databases. II. Are homology models of G-Protein Coupled Receptors suitable targets? *Proteins*, **50**, 5–25.

97. Simulations Plus, Inc., Lancaster, USA. info@simulationplus.com.

98. Wang, R., Lu, Y., Fang, X., and Wang, S. (2004) An extensive test of 14 scoring functions using the PDBbind refined set of 800 protein-ligand complexes. *Journal of Chemical Information and Computer Sciences*, **44**, 2114–25.

99. Perola, E., Walters, W.P., and Charifson, P.S. (2004) A detailed comparison of current docking and scoring methods on systems of pharmaceutical relevance. *Proteins*, **56**, 235–49.

100. Cummings, M.D., DesJarlais, R.L., Gibbs, A.C. *et al.* (2005) Comparison of automated docking programs as virtual screening tools. *Journal of Medicinal Chemistry*, **48**, 962–76.

101. Kraemer, O., Hazemann, I., Podjarny, A.D., and Klebe, G. (2004) Virtual screening for inhibitors of human aldose reductase. *Proteins*, **55**, 814–23.

102. Wyss, P.C., Gerber, P., Hartman, P.G. *et al.* (2003) Novel dihydrofolate reductase inhibitors. Structure-based versus diversity-based library design and high-throughput synthesis and screening. *Journal of Medicinal Chemistry*, **46**, 2304–12.

103. Rastelli, G., Pacchioni, S., Sirawaraporn, W. *et al.* (2003)

Docking and database screening reveal new classes of Plasmodium falciparum dihydrofolate reductase inhibitors. *Journal of Medicinal Chemistry*, **46**, 2834–45.

104. McNally, V.A., Gbaj, A., Douglas, K.T. *et al.* (2003) Identification of a novel class of inhibitor of human and Escherichia coli thymidine phosphorylase by in silico screening. *Bioorganic & Medicinal Chemistry Letters*, **13**, 3705–9.

105. Brenk, R., Naerum, L., Gradler, U. *et al.* (2003) Virtual screening for submicromolar leads of tRNA-guanine transglycosylase based on a new unexpected binding mode detected by crystal structure analysis. *Journal of Medicinal Chemistry*, **46**, 1133–43.

106. Lyne, P.D., Kenny, P.W., Cosgrove, D.A. *et al.* (2004) Identification of compounds with nanomolar binding affinity for checkpoint kinase-1 using knowledge-based virtual screening. *Journal of Medicinal Chemistry*, **47**, 1962–68.

107. Foloppe, N., Fisher, L.M., Howes, R. *et al.* (2006) Identification of chemically diverse Chk1 inhibitors by receptor-based virtual screening. *Bioorganic & Medicinal Chemistry*, **14**, 4792–802.

108. Vangrevelinghe, E., Zimmermann, K., Schoepfer, J. *et al.* (2003) Discovery of a potent and selective protein kinase CK2 inhibitor by high-throughput docking. *Journal of Medicinal Chemistry*, **46**, 2656–62.

109. Cozza, G., Bonvini, P., Zorzi, E. *et al.* (2006) Identification of ellagic acid as potent inhibitor of protein kinase CK2: a successful example of a virtual screening application. *Journal of Medicinal Chemistry* **49**, 2363–66.

110. Peng, H., Huang, N., Qi, J. *et al.* (2003) Identification of novel inhibitors of BCR-ABL tyrosine kinase via virtual screening. *Bioorganic & Medicinal Chemistry Letters*, **13**, 3693–99.

111. Huang, N., Nagarsekar, A., Xia, G. *et al.* (2004) Identification of non-phosphate-containing small molecular weight inhibitors of the tyrosine kinase p56 Lck SH2 domain via in silico screening against the pY + 3 binding site. *Journal of Medicinal Chemistry*, **47**, 3502–11.

112. Toledo-Sherman, L., Deretey, E., Slon-Usakiewicz, J.J. *et al.* (2005) Frontal affinity chromatography with MS detection of EphB2 tyrosine kinase receptor. 2. Identification of small-molecule inhibitors via coupling with virtual screening. *Journal of Medicinal Chemistry*, **48**, 3221–30.

113. Forino, M., Jung, D., Easton, J.B. *et al.* (2005) Virtual docking approaches to protein kinase B inhibition. *Journal of Medicinal Chemistry*, **48**, 2278–81.

114. Douguet, D., Munier-Lehmann, H., Labesse, G., and Pochet, S. (2005) LEA3D: a computer-aided ligand design for structure-based drug design. *Journal of Medicinal Chemistry*, **48**, 2457–68.

115. Wu, S.Y., McNae, I., Kontopidis, G. *et al.* (2003) Discovery of a novel family of CDK inhibitors with the program LIDAEUS: structural basis for ligand-induced disordering of the activation loop. *Structure*, **11**, 399–410.

116. Mizutani, M.Y. and Itai, A. (2004) Efficient method for high-throughput virtual screening based on flexible docking: discovery of novel acetylcholinesterase inhibitors. *Journal of Medicinal Chemistry*, **47**, 4818–28.

117. Dickerson, T.J., Beuscher, A.E. IV, Rogers, C.J. *et al.* (2005) Discovery of acetylcholinesterase peripheral anionic site ligands through computational refinement of a directed library. *Biochemistry*, **44**, 14845–53.

118. Gao, Y., Dickerson, J.B., Guo, F. *et al.* (2004) Rational design and characterization of a Rac GTPase-specific small molecule inhibitor. *Proceedings of the National Academy of Sciences of the United States of America*, **101**, 7618–23.

119. Liu, Z., Huang, C., Fan, K. *et al.* (2005) Virtual screening of novel noncovalent inhibitors for SARS-CoV 3C-like proteinase. *Journal of Chemical Information and Modeling*, **45**, 10–17.

120. Dayam, R., Sanchez, T., Clement, O. *et al.* (2005) Beta-diketo acid pharmacophore hypothesis. 1. Discovery of a novel class of HIV-1 integrase inhibitors. *Journal of Medicinal Chemistry*, **48**, 111–20.

121. Li, C., Xu, L., Wolan, D.W. *et al.* (2004) Virtual screening of human 5-aminoimidazole-4-carboxamide ribonucleotide transformylase against the NCI diversity set by use of AutoDock to identify novel nonfolate inhibitors. *Journal of Medicinal Chemistry*, **47**, 6681–90.

122. Kemp, C.A., Flanagan, J.U., van Eldik, A.J. *et al.* (2004) Validation of model of cytochrome P450 2D6: an in silico tool for predicting metabolism and inhibition. *Journal of Medicinal Chemistry*, **47**, 5340–46.

123. de Graaf, C., Oostenbrink, C., Keizers, P.H. *et al.* (2006) Catalytic site prediction and virtual screening of cytochrome P450 2D6 substrates by consideration of water and rescoring in automated docking. *Journal of Medicinal Chemistry*, **49**, 2417–30.

124. Cherkasov, A., Shi, Z., Fallahi, M., and Hammond, G.L. (2005) Successful in silico discovery of novel nonsteroidal ligands for human sex hormone binding globulin. *Journal of Medicinal Chemistry*, **48**, 3203–13.

125. Schapira, M., Raaka, B.M., Das, S. *et al.* (2003) Discovery of diverse thyroid hormone receptor antagonists by high-throughput docking. *Proceedings of the National Academy of Sciences of the United States of America*, **100**, 7354–59.

126. Soelaiman, S., Wei, B.Q., Bergson, P. *et al.* (2003) Structure-based inhibitor discovery against adenylyl cyclase toxins from pathogenic bacteria that cause anthrax and whooping cough. *The Journal of Biological Chemistry*, **278**, 25990–97.

127. Babaoglu, K., Page, M.A., Jones, V.C. *et al.* (2003) Novel inhibitors of an emerging target in mycobacterium tuberculosis; substituted thiazolidinones as inhibitors of dTDP-rhamnose synthesis. *Bioorganic & Medicinal Chemistry Letters*, **13**, 3227–30.

128. Foloppe, N., Chen, I.J., Davis, B. *et al.* (2004) A structure-based strategy to identify new molecular scaffolds targeting the bacterial ribosomal A-site. *Bioorganic & Medicinal Chemistry*, **12**, 935–47.

129. Barril, X., Brough, P., Drysdale, M. *et al.* (2005) Structure-based discovery of a new class of Hsp90 inhibitors. *Bioorganic & Medicinal Chemistry Letters*, **15**, 5187–91.

130. Irwin, J.J., Raushel, F.M., and Shoichet, B.K. (2005) Virtual screening against metalloenzymes for inhibitors and substrates. *Biochemistry*, **44**, 12316–28.

131. Miguet, L., Zhang, Z., Barbier, M., and Grigorov, M.G. (2006) Comparison of a homology model and the crystallographic structure of human 11beta-hydroxysteroid dehydrogenase type 1 (11betaHSD1) in a structure-based identification of inhibitors. *Journal of Computer-Aided Molecular Design*, **20**, 67–81.

132. Desai, P.V., Patny, A., Sabnis, Y. *et al.* (2004) Identification of novel parasitic cysteine protease inhibitors using virtual screening. 1. The ChemBridge database. *Journal of Medicinal Chemistry*, **47**, 6609–15.

133. Kenyon, V., Chorny, I., Carvajal, W.J. *et al.* (2006) Novel human lipoxygenase inhibitors discovered using virtual screening with homology models. *Journal of Medicinal Chemistry*, **49**, 1356–63.

134. Becker, O.M., Marantz, Y., Shacham, S. *et al.* (2004) G protein-coupled receptors: in silico drug discovery in 3D. *Proceedings of the National Academy of Sciences of the United States of America*, **101**, 11304–9.

135. Carbone, V., Ishikura, S., Hara, A., and El-Kabbani, O. (2005)

Structure-based discovery of human L-xylulose reductase inhibitors from database screening and molecular docking. *Bioorganic & Medicinal Chemistry*, **13**, 301–12.

136. Pickett, S.D., Sherborne B.S., Wilkinson, T. *et al.* (2003) Discovery of novel low molecular weight inhibitors of IMPDH via virtual needle screening. *Bioorganic & Medicinal Chemistry Letters*, **13**, 1691–94.

137. Krier, M., Araujo-Junior, J.X., Schmitt, M. *et al.* (2005) Design of small-sized libraries by combinatorial assembly of linkers and functional groups to a given scaffold: application to the structure-based optimization of a phosphodiesterase 4 inhibitor. *Journal of Medicinal Chemistry*, **48**, 3816–22.

138. Nikolovska-Coleska, Z., Xu, L., Hu Z. *et al.* (2004) Discovery of embelin as a cell-permeable, small-molecular weight inhibitor of XIAP through structure-based computational screening of a traditional herbal medicine three-dimensional structure database. *Journal of Medicinal Chemistry*, **47**, 2430–40.

139. Song, H., Wang, R., Wang, S., and Lin, J. (2005) A low-molecular-weight compound discovered through virtual database screening inhibits Stat3 function in breast cancer cells. *Proceedings of the National Academy of Sciences of the United States of America*, **102**, 4700–5.

140. Evers, A. and Klabunde, T. (2005) Structure-based drug discovery using GPCR homology modelling: successful virtual screening for antagonists of the alpha1A adrenergic receptor. *Journal of Medicinal Chemistry*, **48**, 1088–97.

141. Evers, A. and Klebe, G. (2004) Successful virtual screening for a submicromolar antagonist of the neurokinin-1 receptor based on a ligand-supported homology model. *Journal of Medicinal Chemistry*, **47**, 5381–92.

142. Varady, J., Wu, X., Fang, X. *et al.* (2003) Molecular modeling of the three-dimensional structure of dopamine 3 (D3) subtype receptor: discovery of novel and potent D3 ligands through a hybrid pharmacophore- and structure-based database searching approach. *Journal of Medicinal Chemistry*, **46**, 4377–92.

143. Marcou, G. and Rognan, D. (2007) Optimizing fragment and scaffold docking by use of molecular interaction fingerprints. *Journal of Chemical Information and Modeling*, **47**, 195–207.

6
Scope and Limits of Molecular Docking

The aim of this chapter is to demonstrate that different docking problems do need different docking procedures. Docking scenarios typically fall into two categories.

First, if the active site of the protein is not known, the search for both the binding site and, subsequently, the binding mode of the ligand is called a 'blind docking'. Methods of blind docking are also important for investigating protein–protein interactions. A special subcase of blind docking is the docking to homology models of a target where the position of the active site is assumed to be similar to one in a template protein. Docking to models of transmembranal proteins such as G-protein coupled receptors (GPCRs) fall into this category.

Second, if the site of the binding is known from X-ray diffraction or from nuclear magnetic resonance (NMR) studies, the docking into the known active site is called a 'direct docking'. In this case, the focus of interest is the binding mode of the ligand. Several issues have to be addressed before direct docking can be performed. Figure 6.1 summarizes these problems and docking categories.

The following are some of the issues:

(1) The active site contains discrete crystal molecules of water mediating the binding between the ligand and the target. Are they to be considered or ignored in the model? This issue is addressed in the first section.

(2) The binding site of the ligand is overlapping with the binding site of a cofactor. Should the cofactor be kept in the active site or not? Is the cofactor necessary for the binding of the inhibitor, knowing that for substrate binding the cofactor is essential? This issue is addressed in the second section.

(3) The active site contains catalytic metal ions. How are they to be taken into account? Metals are usually essential for the enzymatic catalysis. While considering metals, an important factor is their parameterization.

Molecular Modeling. Basic Principles and Applications. 3rd Edition
H.-D. Höltje, W. Sippl, D. Rognan, and G. Folkers
Copyright © 2008 Wiley-VCH Verlag GmbH & Co. KGaA, Weinheim
ISBN: 978-3-527-31568-0

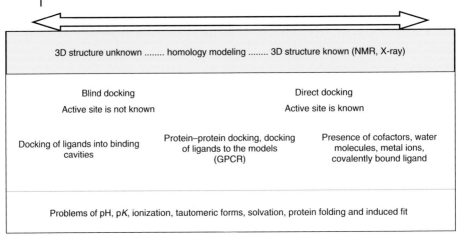

Figure 6.1 Categories of molecular docking.

How are they coordinated to amino acid residues of proteins, water molecules and ligands and what charge is located on the metal ion? Studies on metals in docking and screening are numerous (see two examples in Refs 1 and 2) but are not presented in this chapter.

(4) Recent developments in docking deal with many new factors that have so far been omitted in most of the docking approaches, namely, (i) the consideration (or rather nonconsideration) of tautomeric states of the compounds under scrutiny, (ii) the effect of the pH or pK microenvironment of the active site on ligand binding, (iii) simulation of the induced fit and conformational changes of proteins during the ligand binding and finally (iv) the situation when it is known that the putative inhibitor is expected to bind to the protein covalently. The impact of tautomerism on docking (i) as an example of these factors is brought out in the last section of this chapter.

6.1
Docking in the Polar Active Site that Contains Water Molecules

In this section, we focus on the role of water molecules in the active site. One of the characteristic features of an active site is the conformational change that occurs upon ligand binding and the subsequent rearrangement of water molecules. Discrete water molecules may mediate the binding of the ligand to the enzyme, and may even directly participate in the catalysis [3]. When running virtual screening assays, it is always a dilemma to decide whether water molecules should be considered. For the screening of large databases, the detailed binding mode of different compounds is usually unknown and hence this question is not easily decided. There are attempts, however, to take into account all the water molecules that contribute positively to the binding [4−6]. Recently, screening applications have begun to include water molecules (see references in [7]). If several water molecules are present, it may be hard to correctly predict how many of them participate in binding of the ligand and where their exact positions in the site are. Therefore, docking procedures usually recommend the removal of water molecules from the binding site and the screening is thus performed for an empty active site devoid of water. However, in cases such as the heat-labile cholera toxin [6], HIV-1 protease [8] or Herpes simplex virus 1 thymidine kinase (HSV1 TK), explicit water molecules play an important role in ligand binding.

The following example shows the positive contribution of water molecules in the active site to an increased accuracy of docking, better prediction of the binding mode and consequently, better results in screening. The target is the HSV1 TK, which is an important biomedical target that has been used in virus-directed enzyme-prodrug gene therapy of cancer [9, 10]. A specific characteristic of this protein is that it phosphorylates not only its natural substrate thymidine (dT) and some pyrimidine analogs but also certain analogs of purine such as the antiherpetic prodrug acyclovir (ACV) [11, 12]. Analogs studied in HSV1 TK are mainly pyrimidine and purine nucleic acid bases with known, experimentally measured affinities [13, 14], which can be compared with the calculated affinities predicted by the docking program. From crystal structure information, it is known that the HSV1 TK active site is small and polar, and can be formally divided into two contiguous moieties. The first moiety accommodates the substrate and water molecules and the second moiety contains the cofactor adenosine triphosphate (ATP).

First, in order to reproduce the crystallographically determined binding modes, known ligands are docked. The docking of the two main substrates, dT and ACV, is presented. The coordinates of TK with dT and ACV were taken from Protein Data Bank (PDB ID: 2VTK and 2KI5) [13, 15]. To determine the binding mode of dT and ACV, the program AutoDock (version 3.0) was used [16]. As described in detail in Chapter 5, AutoDock contains an efficient Lamarckian genetic algorithm for docking of flexible ligands into a rigid binding site, which generates different low-energy docking poses. In AutoDock, as in many other docking programs, water molecules can be kept and positioned at the exact place in the binding site, thus considering them as part of the protein. The user can define the parameters for water (position of hydrogen atoms and partial charges of the atoms) (e.g. taken from AMBER [17]). Thus, docking poses of both ligands were generated, while water molecules in the active site were either kept or removed.

In addition, docking of the database of 26 known ligands was performed using program FlexX [18]. In contrast to AutoDock, this program contains an incremental construction algorithm, which first places a core fragment and then peripheral fragments of the ligand into a rigid binding site (see Chapter 5). On the basis of this algorithm, only one individual minimum energy pose is produced. FlexX uses formal charges and does not require the time-consuming ab initio charge calculation. In FlexX, water molecules can be defined in the active site either by keeping concrete water molecules during the docking procedure from the FlexX menu 'customize' or by selecting the 'particle concept' option [8], whereby water molecules are positioned within the site during the screening. In the study reported here, water molecules were kept at discrete, crystallographic positions.

In the third and final part of this project, a database of 80 000 druglike compounds from available chemical directory (ACD) [19] was prepared and enriched by the 26 known enzyme ligands. The screening was performed first by using the program DOCK 4.0 [20] and then by the FlexX program. DOCK matches ligands to the inverse image of the active site. Using DOCK, water can also be treated as a part of the protein at a given position. Knowing about the importance of the two discrete molecules of water (see later in this section), two runs were performed: one with water present in the active site and the other one without water. DOCK is suitable for the screening of large databases and may serve as a prescreening tool (see Chapter 5). After the screening by DOCK, 1000 energetically top-ranked compounds from both runs were screened again using FlexX. For the ranking, the 26 known binders served as a reference. The programs AutoDock, FlexX and DOCK were widely used in the studies of this chapter, and lead to plausible results.

Learning from the results.   Thymidine binds via a hydrogen-bond network as shown in Figure 6.2. The presence of two water molecules (W1 and W2) mediating hydrogen bonds between the O(2) of the dT base and Arg176 is also characteristic for other pyrimidine derivatives [15, 21, 22]. ACV – a purine analog with an acyclic side chain mimicking the sugar moiety – binds in the active site in a manner similar to dT, yet via a different H-bond network (Figure 6.3a). ACV does not interact via water-mediated hydrogen bonds. In analogy, the X-ray structures of other purine analogs reveal a binding pattern similar to that of ACV [14].

Before docking, the importance of the two discrete water molecules present in the dT–active site complex has been addressed. Approximately half of the crystal structures of viral thymidine kinase water molecules mediate the binding of the ligand to the protein. If pyrimidines are bound, those two water molecules are always present. It was decided, therefore, to run the docking twice: with the water molecules in the active site (water site) and without them (empty site).

Docking of dT to the water site accurately reproduced the crystallographic binding mode with root mean square deviation (rmsd) values ranging from 0.36 to 0.72 Å for 50 runs. In contrast, docking to the empty site was less accurate, resulting in a fuzzy picture of the adopted docking poses (rmsd

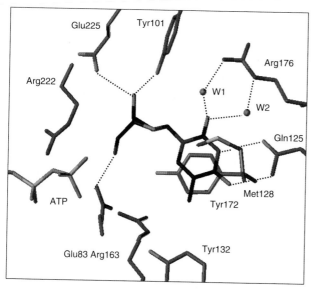

Figure 6.2 Active site of HSV1 TK with bound thymidine. Thymidine is sandwiched between Tyr172 and Met128, creating five direct hydrogen bonds (dotted lines) and two hydrogen bonds mediated via two molecules of water, W1 and W2 (balls).

range increases to 0.74–1.23 Å) (Figure 6.3b). On the other hand, docking of ACV into the water site revealed several possible orientations, and rmsd values when compared with the X-ray orientation were not satisfying (rmsd > 1.2 Å). When the docking for ACV was run in absence of water, the prediction of the binding geometry improved, and the docking procedure led to a single unique orientation (the rmsd value for the lowest-energy orientation decreased to 1.03 Å) (Figure 6.3c). It can be concluded that, in contrast to dT, which prefers docking to the water site, docking of ACV into the empty site leads to a more accurate prediction of the binding geometry.

As already mentioned, a set of 26 ligands were docked using FlexX. In line with the preceding observations, the ligands were both docked reciprocally into the dT-like active site (water site) and into the ACV-like active site (empty site). Lowest-energy conformers have been selected and their affinities constants K_i have been calculated from Equation 1:

$$\Delta G_{\text{FlexX}} = RT \ln K_i \tag{1}$$

where ΔG_{FlexX} stands for the free energy of binding calculated by FlexX. Table 6.1 shows in detail the results of K_i predictions for six ligands of the data set of 26 molecules. Comparison between the predicted and the experimentally determined K_i values shows that for pyrimidines, these values improved by docking into the dT-like active site. A similar improvement can be observed for

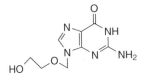

(a)   Acyclovir

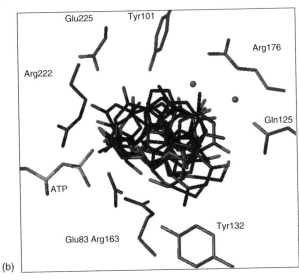

(b)

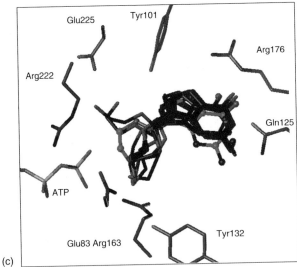

(c)

Figure 6.3 Docking to aciclovir to HSV TK. (c) into the water-empty active site.
(a) Chemical structure of acyclovir.   Lowest-energy cluster of acyclovir is
(b) Docking clusters of acyclovir (C atoms in overlaying the ligand in its crystal structure
black) into the HSV1 TK active site in the   binding mode (ball and sticks, C atoms in
presence of water molecules (red balls) and   green).

Table 6.1 Prediction of affinity constants K_i of HSV1 TK ligands from the docking to water-containing or water-free active sites.

	Ligand	K_i experimental (μmol l^{-1})	Docking in dT-like site (water is present)		Docking in ACV-like site (no presence of waters)	
			ΔG (kJ mol^{-1})	K_i predicted (μmol l^{-1})	ΔG (kJ mol^{-1})	K_i predicted (μmol l^{-1})
Purine analogs	ACV	200	−19.8	341.0	−20.7	239.1
	GCV	47.6	−15.9	1631.3	−20.1	302.1
Pyrimidine analogs	dT	0.2	−35.3	0.7	−24.4	54.2
	BVDU[a]	0.1	−33.1	1.6	−19.7	357.9
	IDU[b]	0.09	−33.7	1.2	−23.9	64.7
	N-MCT[c]	11.54	−26.4	23.5	−12.1	7586.7

a   Bromovinyldeoxyuridine.
b   Iododeoxyuridine.
c   North-methanocarbathymidine.

purine analogs from docking into the ACV-like active site. Hence, the presence of crystal water positively influences the prediction of binding constants of pyrimidines. In contrast, yet in full agreement with the known relevant crystal structures, the same behavior is observed in the case of purine analogs, ACV and ganciclovir (GCV), but for the water-free situation.

Docking purine derivatives into the water site obviously is disfavored, and energy is gained if they are docked into the empty site. This is in full agreement with the predictions from docking dT and ACV using AutoDock. It is encouraging to observe that the docking results unequivocally recognize the discrimination between the two main classes of analogs – pyrimidine and purine derivatives.

Application to virtual screening.   Finally, the knowledge acquired about the impact of discrete water molecules on the binding of the ligands was transferred to the screening of 80 000 druglike compounds from the ACD database. The results led to diverse and interesting observations when the 1000 top-ranked compounds were screened using the water site. One particular hit, which was ranked among top 100 in the water site, was not found at all in top 1000 in the screening assay with the empty site. Experimental affinity assays, however, revealed that this compound indeed binds to HSV1 TK with a binding affinity in the submicromolar range, whereas dT and other pyrimidine inhibitors exhibit significantly lower affinities. This again shows the importance of taking water molecules into consideration for the screening, and highlights the fact that discrete water molecules in the active site, treated as a part of the docking target, may lead to unique and powerful hits.

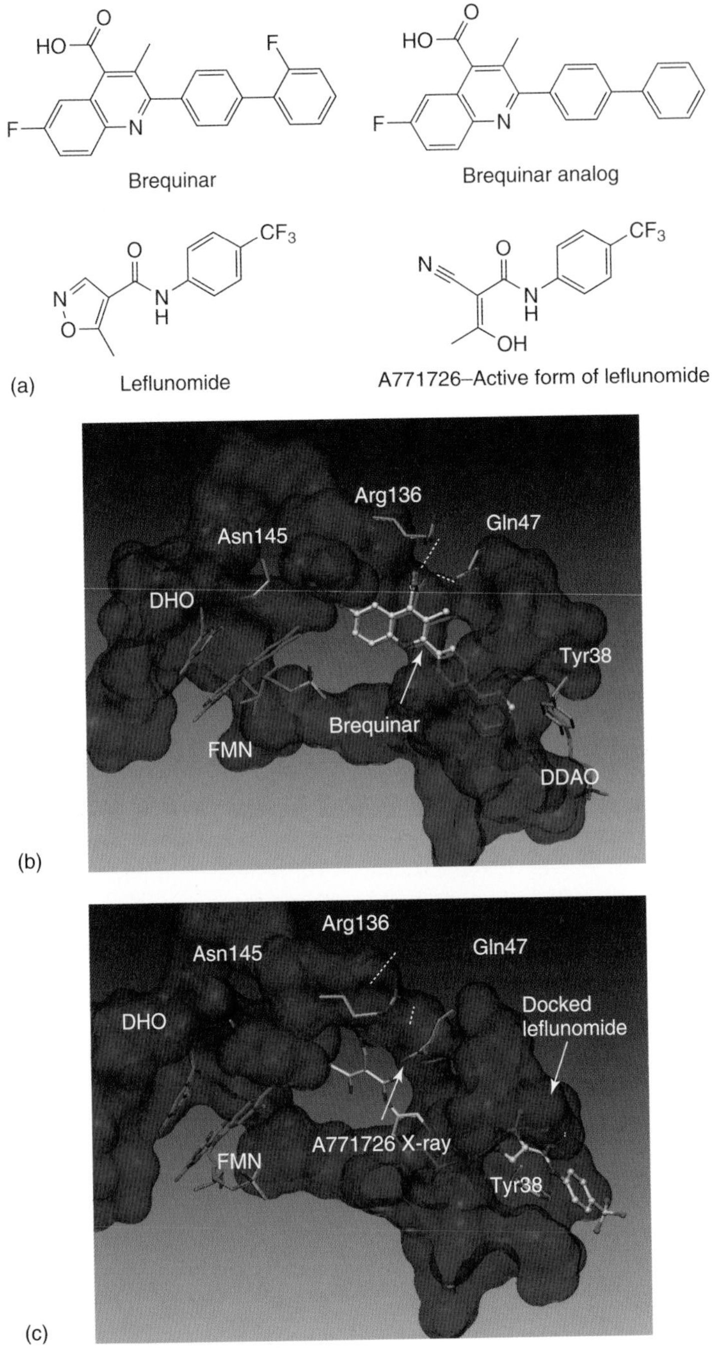
Brequinar
Brequinar analog
Leflunomide
A771726–Active form of leflunomide
(a)
Arg136
Gln47
Asn145
DHO
Tyr38
Brequinar
FMN
DDAO
(b)
Arg136
Asn145
Gln47
DHO
Docked leflunomide
A771726 X-ray
FMN
Tyr38
(c)

◀ **Figure 6.4** Chemical structures and docking of DHODH inhibitors. (a) Chemical structures of brequinar, brequinar analog, leflunomide and leflunomide active form A771726. (b and c): Protein surface represents hydrophobic and other residues of the active site. For seek of clarity, residues above and under docked brequinar are not shown. Hydrogen bonds and the electrostatic interaction are shown in dotted lines. (b) Docked brequinar analog (ball and stick) is overlaying its bound X-ray structure (sticks) and the result is also identical when DDAO is not present in the docking site. (c) Leflunomide (ball and sticks) is docked at the entrance to the protein if DDAO is omitted. Active form of leflunomide A771726 (sticks, X-ray structure) binds nearby FMN, deeply in the ubiquinone-binding pocket. In the case when DDAO is included in the docking site, neither A771726 nor leflunomide was successfully docked into the protein.

6.2
Including Cofactor in Docking?

Docking into human dihydroorotate dehydrogenase (DHODH) falls into the category that shows the importance of considering the cofactor in the binding site during the docking procedure. Moreover, this example shows that it is important to study the known experimental data such as *in vitro* binding assays or conditions of the crystallization.

Human DHODH catalyzes the fourth committed step in the de novo biosynthesis of pyrimidines. DHODH active site, which catalyzes the hydrogenation of the dihydroorotate (DHO), is large and mainly hydrophobic. It contains the cofactor flavin-mononucleotide (FMN) and ubiquinone. The enzyme is associated via ubiquinone with the mitochondrial inner membrane. As the rapidly proliferating human T cells have an exceptional requirement for de novo pyrimidine biosynthesis, inhibitors of DHODH constitute an attractive therapeutic approach to autoimmune diseases, immunosuppression and cancer [23–25].

The DHODH case clearly demonstrates the effect of the presence of cofactors within the binding site and their influence on docking accuracy. There are two solved crystal structures deposited in Protein Data Bank: DHODH in complex with an analog of the inhibitor brequinar (resolved without one fluorine atom that is missing in the structure) (PDB ID: 1D3G) [24, 26], and DHODH in complex with the active form of the inhibitor leflunomide called A771726 (PDB ID: 1D3H) [25, 26] (Figure 6.4a). The peculiarity of docking for this type of enzyme has already been demonstrated on these two complexes. Both crystal structures show a small subpocket, in which the substrate DHO is bound, and next to which FMN is situated. FMN accepts a hydride and proton from DHO, and further reduces the second cofactor ubiquinone. Both inhibitors are located in the hydrophobic binding pocket of ubiquinone and block the electron transfer. The two crystal structures contain a detergent molecule N,N-dimethyldecylamine-N-oxide (DDAO), which imitates ubiquinone, and is located on the other edge of the site where the ubiquinone enters the protein active site. On the basis of the structural information, the question arises as to which of the cofactors should be considered in the docking procedure.

The study of McLean *et al.* [27] presents data indicating that brequinar and A771726 are uncompetitive inhibitors of DHO. These authors conclude that none of the inhibitors affect DHO-mediated flavin reduction. For this reason, it was appropriate to keep DHO (substrate) and FMN (cofactor) in the docking site and to train on the ubiquinone-binding site. Secondly, brequinar is a competitive inhibitor as against ubiquinone, whereas the behavior of A771726 as opposed to ubiquinone is noncompetitive. McLean *et al.* [27] claimed that the binding locations of brequinar or A771726 might not represent the relevant physiological binding site. Their arguments are based on the fact that crystallization of both inhibitors was done at concentrations 4–6 orders of magnitude above the K_i (K_i of A771726 $\sim 100 \, nmol^{-1}$ and brequinar $\sim 1 \, nmol^{-1}$) and 200-fold higher than the lower limit of K_d obtained by isothermal calorimetry. Thus, these authors conclude that the observed geometries present low affinity binding modes. They also state that both inhibitor sites are overlapping, but that the site of A771726 is not in the proximity of FMN. They suggest that A771726 binds close to the ubiquinone protein entrance, and that, this part of the protein may retain sufficient plasticity to allow binding of the inhibitor together with the hydrophobic tail of ubiquinone. This is in contrast to the structure resolved by Liu *et al.* [26] We have, therefore, carried out this docking twice by removing or keeping the detergent molecule DDAO in the ubiquinone-binding site and thus imitating the experimental observations.

In the crystal structure, the carboxyl group of brequinar exhibits electrostatic interaction with amino acids Gln47 and Arg136, whereas its phenyl rings lie in the hydrophobic tunnel. Docking of the brequinar analog into the DDAO-empty site was accurate and reproduced the crystallographically determined position (rmsd = 0.63 Å) (Figure 6.4b). Interestingly, the same docking result was also obtained in the case when the detergent DDAO mimicking the second cofactor ubiquinone was included; its presence did not affect the binding of brequinar.

In contrast to this, the docking of leflunomide and its active form A771726 (noncompetitive inhibitor versus ubiquinone) showed a dependency on the presence or absence of DDAO. When DDAO was kept in the docking site, neither the inhibitor A771726 nor the leflunomide could be docked within the active site. In the case that DDAO was omitted, only leflunomide (and not its active form A771726) could be docked to the protein (Figure 6.4c).

Docking reflected the crystallographic binding only in case of brequinar, which also binds at lower affinities ($K_i = \sim 1 \, nmol^{-1}$), and is competitive to ubiquinone. Leflunomide docking indicates a relevant binding mode of the inhibitor at the entrance to the protein, and supports the hypothesis that the crystallographically determined binding mode represents a low affinity mode.

This example shows the importance of omitting or keeping cofactors in the docking site and that it has a direct impact on the accuracy of docking. It also shows that docking can lead to the proposition of alternate binding modes, correct or incorrect, each of which might initiate different, new experimental studies. Actually, recent studies including an X-ray structure by Baumgartner *et al.* present a novel series of DHODH inhibitors that have a dual mode of binding, and that are competitive to the ubiquinone-binding site [28].

The presented case demonstrates the potential importance of docking including cofactors. In similar situations, the effect of cofactors on docking poses should always be carefully considered.

6.3
Impact of Tautomerism on Docking

Tautomers are often disregarded in computer-aided molecular modeling applications [29, 30]. There are numerous studies of tautomers in gas or aqueous solution; however, little is known about tautomerism of ligands in the binding site of proteins. Tautomeric forms of a molecule differ in shape, functional groups, surface and hydrogen-bonding pattern, which may have a crucial effect on the molecular recognition.

Let us imagine the simple case of prototropy. Prototropy can alter the skeleton of a given molecule, which in principle can be seen as a new distinct molecule with different complementarity to the target. Considering tautomery in ligand–protein interactions therefore has a significant impact on the prediction of the ligand binding using various docking techniques. This section points to the hitherto unaddressed issue of tautomerism in docking.

The general view is that the binding environment within a protein is a very specific one. It is different from the environment of the aqueous solution or the vacuum. Apolar or polar, acidic or basic side chains create local pHs, shifting pK values of nearby polar amino acids and subsequently influence the functional groups of the ligand. Presence of a ligand, metal cations and water also influence the pH (and the pK) conditions in close proximity of amino acid side chains and the process of catalysis [31]. In such a context, the ligand may be ionized or can achieve its excited tautomeric state.

Many enzymes of therapeutical relevance, such as nucleoside kinases, telomerases or DNA-polymerases, are able to accommodate purine or pyrimidine derivatives in their active sites. Hence, nucleic acid bases are molecules with well-known occurrence of tautomery. For example, there is the phenomenon of tautomer-base mispairing in the DNA strand as a source of DNA replication errors [32, 33]. Nucleobases always reveal one tautomer, which is, for stability reasons, incorporated into the nucleic acid. The ability of pyrimidine and purine structures to form hydrogen bonds is linked intimately with the potential existence of tautomeric structures. Similarly, the phenomenon of tautomerism can be observed on histamines and in proteins on histidine side chains [34, 35].

Tautomers have different molecular shapes and different hydrogen-bond donor and acceptor properties resulting in a significant impact on molecular recognition. Nevertheless, up to now, tautomers have been neglected in automated molecular docking. Several questions arise: Does a molecule bind preferably in one distinct tautomer? Is the most stable tautomeric form in aqueous solution also the most stable form in the active site of a protein? What can be the binding contribution of a ligand in its excited tautomeric state in contrast to its 'normal' tautomeric state, for example, its low-energy configuration? How should a compound whose proton shift induces different stereoisomerism be treated?

The impact of tautomerism on docking can be found only rarely in publications [36]. One example is the evidence of tautomer-bound state

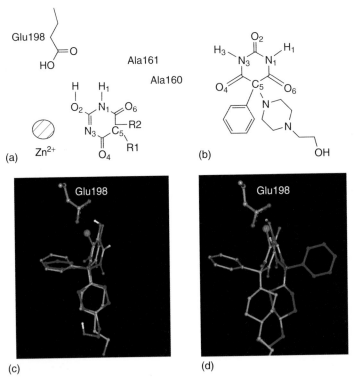

(a) (b) (c) (d)

Figure 6.5 Docking of tautomers of barbiturate inhibitor RO200-1770 to MMP8. (a) Barbiturate derivative Tautomer 1 as bound in the crystal structure to the active site of MMP8. (b) Barbiturate derivative Tautomer 2. Atom C_5 is, due to the tautomerism of H on N_1 or N_3 of the barbiturate ring, pseudochiral, having the S or R stereoisomerism depending on the tautomer. (c) Docking of the Tautomer 1 (green) overlaying the crystal structure position (orange). (d) Docking of the Tautomer 2 (green) with the 180°-flipped phenyl ring compared to the X-ray pose (orange). In both C and D, Zn^{2+} is present during docking (magenta).

reported by Brandstetter *et al.* [37]. Barbiturate inhibitor (RO200–1770) was bound as its enol isomer (let us call it Tautomer 1) to the active site of a matrix metalloproteinase (MMP-8) (PDB ID: JJ9) (Figure 6.5a) [37]. The lone pair of oxygen, O_2, contributes to the coordination of Zn^{2+}, while the hydrogen of O_2 hydroxyl is involved in H bond with Glu198. Thus, it is the tautomeric enol form of the barbiturate (Tautomer 1) that is favored by the protein matrix over the keto form (Tautomer 2), which dominates in solution. The H atom on N_1 is bound to the carbonyl of Ala161 and the ketone O_6 to the adjacent amides of Ala160 and Ala161. The phenyl and the piperidyl rings occupy the hydrophobic subpockets of the binding site.

The program FlexX was used to dock the two tautomers of the barbiturate inhibitor RO200–1770. The docking results for both tautomers of RO200–1770 show very different docking poses (Figure 6.5b and c). Tautomer 1 is docked as in the crystallized complex (rmsd < 1A), whereas Tautomer 2 is docked in a pose where the aromatic ring is flipped 180° (Figure 6.5c). This result shows the importance of the correct addition of hydrogens on the ligand structure.

Compounds in chemical databases are stored in their canonical forms to which the tautomeric or ionic form of a compound can be reduced using strongly defined chemical rules. Many commercial and unregistered databases often contain pairs of tautomers registered under different names and even prices [38, 39]. Trepalin *et al.* estimated that up to 0.5% of commercially available compounds, meant for bioscreening, contain tautomers [38]. Conversely, a large amount of tautomers are missing.

Most algorithms accept chemical structures just like they are imported by the user from databases without consideration of potentially existing tautomers. Thus, including tautomers could be seen as enhancing the number of degrees of freedom to be considered by docking programs. It can be a fast and relatively precise method to cover for the incertitude caused by the variability of the effective pK in different parts of a receptor-binding site. If a database is used for computer-aided lead finding, enriching the database by energetically similar tautomers may significantly improve the success rates in computer-aided drug design.

References

1. Irwin, J.J., Raushel, F.M., and Shoichet, B.K. (2005) Virtual screening against metalloenzymes for inhibitors and substrates. *Biochemistry*, **44**, 12316–28.
2. Gresh, N. (2005) Development, validation, and applications of anisotropic polarizable molecular mechanics to study ligand and drug-receptor interactions. *Current Pharmaceutical Design*, **12**, 2121–58.
3. Gohlke, H. and Klebe, G. (2002) Approaches to the description and prediction of the binding affinity of small-molecule ligands to macromolecular receptors. *Angewandte Chemie (International ed. in English)*, **41**, 2644–76.
4. Pospisil, P., Scapozza, L., and Folkers, G. (2001) The role of water in drug design: thymidine kinase as case study, in *Rational Approaches to Drug*

Design: 13th European Symposium on Quantitative Structure-Activity Relationship (eds H.-D. Höltje and W. Sippl), Prous Science, Barcelona-Philadelphia, pp. 92–96.

5. Hetenyi, C. and Van Der Spoel, D. (2002) Efficient docking of peptides to proteins without prior knowledge of the binding site. *Protein Science*, **11**, 1729–37.

6. Minke, W.E., Diller, D.J., Hol, W.G., and Verlinde, C.L. (1999) The role of waters in docking strategies with incremental flexibility for carbohydrate derivatives: heat-labile enterotoxin, a multivalent test case. *Journal of Medicinal Chemistry*, **42**, 1778–88.

7. de Graaf, C., Pospisil, P., Pos, W. *et al.* (2005) Binding mode prediction of cytochrome p450 and thymidine kinase protein-ligand complexes by consideration of water and rescoring in automated docking. *Journal of Medicinal Chemistry*, **48**, 2308–18.

8. Rarey, M., Kramer, B., and Lengauer, T. (1999) The particle concept: placing discrete water molecules during protein-ligand docking predictions. *Proteins*, **34**, 17–28.

9. Culver, K.W., Ram, Z., Wallbridge, S. *et al.* (1992) In vivo gene transfer with retroviral vector-producer cells for treatment of experimental brain tumors. *Science*, **256**, 1550–52.

10. Bonini, C., Ferrari, G., Verzeletti, S. *et al.* (1997) HSV-TK gene transfer into donor lymphocytes for control of allogeneic graft-versus-leukemia. *Science*, **276**, 1719–24.

11. Elion, G.B., Furman, P.A., Fyfe, J.A. *et al.* (1977) Selectivity of action of an antiherpetic agent, 9-(2-hydroxyethoxymethyl) guanine. *Proceedings of the National Academy of Sciences of the United States of America*, **74**, 5716–20.

12. Keller, P.M., Fyfe, J.A., Beauchamp, L. *et al.* (1981) Enzymatic phosphorylation of acyclic nucleoside analogs and correlations with antiherpetic activities. *Biochemical Pharmacology*, **30**, 3071–77.

13. Wild, K., Bohner, T., Aubry, A. *et al.* (1995) The 3-dimensional structure of thymidine kinase from herpes simplex virus type 1. *FEBS Letters*, **368**, 289–92.

14. Champness, J.N., Bennett, M.S., Wien, F. *et al.* (1998) Exploring the active site of herpes simplex virus type-1 thymidine kinase by X-ray crystallography of complexes with aciclovir and other ligands. *Proteins*, **32**, 350–61.

15. Bennett, M.S., Wien, F., Champness, J.N. *et al.* (1999) Structure to 1.9 A resolution of a complex with herpes simplex virus type-1 thymidine kinase of a novel, non-substrate inhibitor: X-ray crystallographic comparison with binding of aciclovir. *FEBS Letters*, **443**, 121–25.

16. AutoDock. http://www.autodock.scripps.edu/.

17. AMBER. http://www.amber.scripps.edu/.

18. FlexX. http://www.biosolveit.de/FlexX/.

19. ACD. http://www.mdli.com/products/experiment/available_chem_dir/index.jsp.

20. DOCK. http://www.dock.compbio.ucsf.edu/.

21. Wild, K., Bohner, T., Folkers, G., and Schulz, G.E. (1997) The structures of thymidine kinase from herpes simplex virus type 1 in complex with substrates and a substrate analogue. *Protein Science*, **6**, 2097–106.

22. Prota, A., Vogt, J., Pilger, B. *et al.* (2000) Kinetics and crystal structure of the wild-type and the engineered Y101F mutant of herpes simplex virus type 1 thymidine kinase interacting with (North)-methanocarba-thymidine. *Biochemistry*, **39**, 9597–603.

23. Fairbanks, L.D., Bofill, M., Ruckemann, K., and Simmonds, H.A. (1995) Importance of ribonucleotide availability to proliferating T-lymphocytes from healthy humans. Disproportionate expansion of pyrimidine pools and contrasting effects of de novo synthesis inhibitors. *The Journal of Biological Chemistry*, **270**, 29682–89.

24. Chen, S.F., Perrella, F.W., Behrens, D.L., and Papp, L.M. (1992) Inhibition of dihydroorotate dehydrogenase activity by brequinar sodium. *Cancer Research*, **52**, 3521–27.

25. Williamson, R.A., Yea, C.M., Robson, P.A. *et al.* (1995) Dihydroorotate dehydrogenase is a high affinity binding protein for A77 1726 and mediator of a range of biological effects of the immunomodulatory compound. *The Journal of Biological Chemistry*, **270**, 22467–72.

26. Liu, S., Neidhardt, E.A., Grossman, T.H. *et al.* (2000) Structures of human dihydroorotate dehydrogenase in complex with antiproliferative agents. *Structure with Folding & Design*, **8**, 25–33.

27. McLean, J.E., Neidhardt, E.A., Grossman, T.H., and Hedstrom, L. (2001) Multiple inhibitor analysis of the brequinar and leflunomide binding sites on human dihydroorotate dehydrogenase. *Biochemistry*, **40**, 2194–200.

28. Baumgartner, R., Walloschek, M., Kralik, M. *et al.* (2006) Dual binding mode of a novel series of DHODH inhibitors. *Journal of Medicinal Chemistry*, **49**, 1239–47.

29. Pospisil, P., Ballmer, P., Scapozza, L., and Folkers, G. (2003) Tautomerism in computer-aided drug design. *Journal of Receptors and Signal Transduction*, **23**, 361–71.

30. Kubinyi, H. (2003) Drug research: myths, hype and reality. *Nature Reviews Drug Discovery*, **2**, 665–68.

31. Fersht, A. (1999) The pH dependence of enzyme catalysis, *Structure and Mechanism in Protein Science*, 2nd Printing edn, W.H. Freeman and Company, New York, pp. 169–90.

32. Strazewski, P. (1988) Mispair formation in DNA can involve rare tautomeric forms in the template. *Nucleic Acids Research*, **16**, 9377–98.

33. Lutz, W.K. (1990) Endogenous genotoxic agents and processes as a basis of spontaneous carcinogenesis. *Mutation Research*, **238**, 287–95.

34. Nederkoorn, P.H.J., Vernooijs, P., Denkelder, G.M.D.O. *et al.* (1994) A new model for the agonistic binding-site on the histamine H-2-receptor – the catalytic triad in serine proteases as a model for the binding-site of histamine H-2-receptor agonists. *Journal of Molecular Graphics*, **12**, 242–56.

35. Boehm, H.-J., Klebe, G., and Kubinyi, H. (1996) *Wirkstoffdesign*, Spektrum Akademischer Verlag GmbH, Heidelberg, Berlin, Oxford.

36. Todorov, N.P., Monthoux, P.H., and Alberts, I.L. (2006) The influence of variations of ligand protonation and tautomerism on protein-ligand recognition and binding energy landscape. *Journal of Chemical Information and Modeling*, **46**, 1134–42.

37. Brandstetter, H., Grams, F., Glitz, D. *et al.* (2001) The 1.8-angstrom crystal structure of a matrix metalloproteinase 8-barbiturate inhibitor complex reveals a previously unobserved mechanism for collagenase substrate recognition. *The Journal of Biological Chemistry*, **276**, 17405–12.

38. Trepalin, S.V., Skorenko, A.V., Balakin, K.V. *et al.* (2003) Advanced exact structure searching in large databases of chemical compounds. *Journal of Chemical Information and Computer Sciences*, **43**, 852–60.

39. Kubinyi, H. (2003) In Virtual screening – problems and success stories, *4th European Workshop in Drug Design*, Siena.

Further Reading

Williamson, R.A., Yea, C.M., Robson, P.A. *et al.* (1996) Dihydroorotate dehydrogenase is a target for the biological effects of leflunomide. *Transplantation Proceedings*, **28**, 3088–91.

7
Chemogenomic Approaches to Rational Drug Design

Until the recent sequencing of the human genome [1, 2], drug discovery had been a multidisciplinary effort to optimize properties of ligands (potency, selectivity, pharmacokinetics) toward a single macromolecular target. It is estimated that, out of the 20 000–25 000 human genes [3] supposed to code for about 3000 druggable targets [4], only a subset of the pharmacological space (about 800 proteins) has currently been investigated by the pharmaceutical industry [5] (Figure 7.1). Remarkably, chemistry showed a parallel boost with the miniaturization and parallelization of compound synthesis, such that over 10 million nonredundant chemical structures cover the actual chemical space, out of which about 1000 are currently approved as drugs. Therefore, only a small fraction of compounds describing the current chemical space has been tested on a fraction of the entire target space. Chemogenomics is the new interdisciplinary field that attempts to fully match target and ligand space, and ultimately identify all ligands of all targets [6]. Various definitions of overlapping fields (chemical genetics, chemical genomics) have been proposed (Table 7.1). We herein consider a broad definition of chemogenomics encompassing chemoproteomics, namely, the study of small-molecular-weight drug candidates on gene/protein function. From the definition of the field, one easily understands that chemogenomics will be at the interface of chemistry, biology and consequently, informatics, since data mining is required to extract reliable information. Furthermore, methodologies at the interface of chemistry and biology (medicinal chemistry), chemistry and informatics (chemoinformatics) and biology and informatics (bioinformatics) will also play a major role in bringing these major disciplines together. Chemogenomics requires at least the following three components, each necessitating hard experimental work: (i) a compound library, (ii) a representative biological system (target library, single cell and whole organism) and (iii) a reliable readout (e.g. gene/protein expression, high-throughput binding or functional assay). The present review will only focus on predictive chemogenomics using *in silico* approaches to derive information from the simultaneous biological evaluation of multiple compounds on multiple targets. By definition, analyzing chemogenomic data is a never-ending learning process aimed at completing a

Molecular Modeling. Basic Principles and Applications. 3rd Edition
H.-D. Höltje, W. Sippl, D. Rognan, and G. Folkers
Copyright © 2008 Wiley-VCH Verlag GmbH & Co. KGaA, Weinheim
ISBN: 978-3-527-31568-0

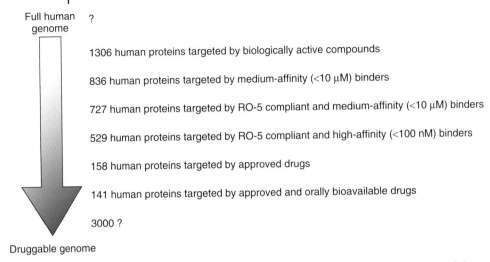

Full human genome ?

1306 human proteins targeted by biologically active compounds

836 human proteins targeted by medium-affinity (<10 μM) binders

727 human proteins targeted by RO-5 compliant and medium-affinity (<10 μM) binders

529 human proteins targeted by RO-5 compliant and high-affinity (<100 nM) binders

158 human proteins targeted by approved drugs

141 human proteins targeted by approved and orally bioavailable drugs

3000 ?

Druggable genome

Figure 7.1 Identification of the druggable human from the full human genome. RO-5 compliant compounds are molecules passing the Lipinski's rule-of-five filters.

Table 7.1 Definition of chemogenomics and related disciplines.

Term	Definition
Chemical genetics	Effect of chemical probes on gene function
Chemical genomics	Effect of target-specific ligands on gene–protein function
Chemogenomics	Effect of target-specific drug candidates on gene–protein function
Chemoproteomics	Effect of target-specific drug candidates on protein function
Genomics	Study of an organism's entire genome
Metabolomics	Study of small-molecule metabolites to be found within a biological sample, such as a single organism
Proteomics	Large-scale study of proteins, particularly their structures and functions
Transcriptomics	Study of all messenger RNA (mRNA) molecules, or 'transcripts', produced in one or a population of cells

two-dimensional matrix (Figure 7.2) where targets/genes are usually reported as columns and compounds as rows, and where reported values are usually binding affinities (K_i, IC_{50}) or functional effects (e.g. EC_{50}). This matrix is sparse as, so far, all possible compounds have not been tested on all possible genes/proteins. Predictive chemogenomics will thus attempt to fill existing gaps by predicting compounds–genes/proteins relationships. *In silico*

Targets

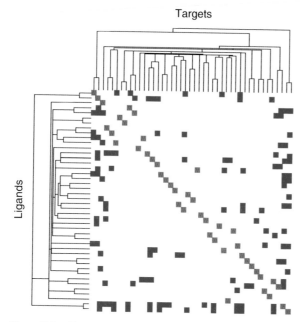

Ligands

Figure 7.2 Chemogenomic binding affinity matrix. Red squares indicate the presence of a ligand–target relationship (binding event) and green squares indicate the absence of any relationship (no binding event). For most target–ligands pairs, such relationships are unknown and predictable.

approaches to predict such data (target selectivity for various ligands, ligand selectivity for various targets) will span ligand-based approaches (comparison of known ligands to predict their most likely targets), target–based approaches (comparison of targets or ligand-binding sites to predict their most likely ligands) or ultimately target-ligand-based approaches (using experimental and predicted binding affinity matrices).

7.1
Description of Ligand and Target Spaces

Basic assumptions of any chemogenomics-based approach are twofold: (i) compounds sharing some chemical similarity should also share targets and (ii) targets sharing similar ligands should share similar patterns (binding sites). Filling the full theoretical chemogenomic matrix thus implies that data on 'unliganded' targets should be gathered from the closest 'liganded' neighboring targets and that data on 'untargeted' ligands should be gathered from the closest 'targeted' ligands. The question is how to measure distances between two ligands and that between two targets.

7.1.1
Ligand Space

To efficiently navigate in ligand space, one needs to first describe the compound using appropriate properties (descriptors) and then use a master equation to measure the distance between two compounds (similarity matrix).

Descriptors are usually classified according to their dimensionality ranging from one-dimensional (1D) to three-dimensional (3D) properties [7] (Figure 7.3, Table 7.2) 1D descriptors are easy and fast to compute. They describe global properties (e.g. molecular weight, atom and bond counts), which can be derived from the chemical formulae and which are used in combination not only to predict physicochemical properties (e.g. solubility) but also to discriminate two sets of compounds (e.g. drugs vs. nondrugs [8], ligands from various target families [9]) by linear or nonlinear QSAR methods. To fasten comparisons, 1D linear representations of compounds are used. The most popular of this kind of simplified string is the 'Simplified Molecular Input Line Entry System' (SMILES) [10] (Figure 7.4).

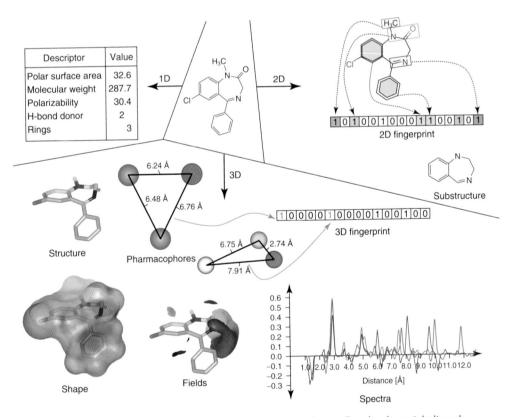

Figure 7.3 Examples of molecular descriptors for small-molecular-weight ligands.

Table 7.2 Ligand descriptors.

Dimension	Nature	Examples
One	Global	Molecular weight, atom and bound counts (e.g. number of H-bond donors, number of rings), polar surface area, polarizability, log P
Two	Topological	Topological and connectivity indices, fragments, substructures (e.g. maximum common substructures), topological fingerprints (e.g. structural keys)
Three	Conformational	*n*-points pharmacophore, shape, field, spectra, fingerprints

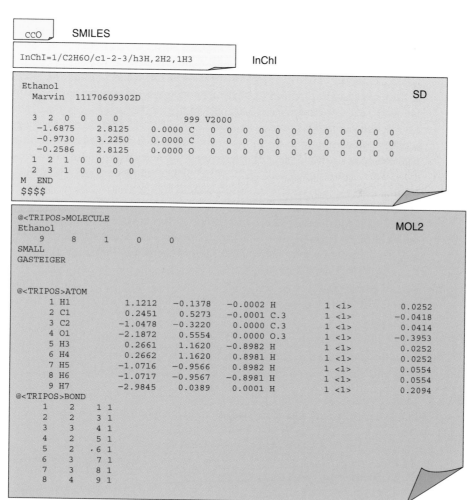

Figure 7.4 Common file formats that can be used for simple 1D queries (e.g. SMILES and InChI), 2D queries (e.g. SD) and 3D queries (e.g. MOL2).

Most ligand descriptors fall in the family of 2D topological descriptors where the connectivity table (list of bonds between ligand atoms) is parsed to encode both atomic and bond properties. The most intuitive way to represent this kind of information is the 2D sketch of the structure (Figure 7.3), which enables browsing of a ligand library for compounds sharing a particular 2D motif (fragment, substructure). Graph-based methods that transform the 2D structure into a molecular graph (atoms being the nodes) are relatively popular for substructure search and clustering chemical compounds into subfamilies [11], but present a noticeable disadvantage of being computationally slow. Fingerprint-based methods are much faster, where the occurrence and absence of predefined structural events (atoms, fragments, rings, substructures, 2D pharmacophores) are encoded into bit strings (sequence of '0' and '1' digits) called *fingerprints*, which are easy to derive, handle and compare. Although receptor–ligand recognition is a 3D event, 2D fingerprints are generally better suited than true 3D fingerprints for similarity searches. The latter descriptors encode conformation–specific properties (atomic coordinates, 3D pharmacophores, shapes, potentials, fields, spectra; Table 7.2) and therefore, usually necessitate a common alignment of molecules to be compared in the same 3D Cartesian space (particularly if grid-based fields or potentials have to be compared), and a relevant sampling of conformational space accessible to each ligand. To avoid the alignment step, which may cause false positives on a virtual screen, 3D information can be translated into a bit string that stores the occurrence of all possible pharmacophore tuplets (doublets, triplets, quadruplets) with their corresponding features (e.g. H-bond acceptor, positively ionizable atom, etc.) and interfeature distances. Hence, comparing bit strings is much easier than comparing structures. Most similarity searches prefer a binary representation of 2D or 3D properties to derive simple similarity indices, the most popular being the Tanimoto coefficient (Equation 1):

$$T_c = \frac{c}{a + b + c},\qquad(1)$$

where

a = count of bits on in compound A but not in compound B,
b = count of bits on in compound B but not in compound A,
c = count of the bits on in both compound A and compound B.

The Tanimoto coefficient will thus range from zero for two completely dissimilar structures to one for two identical compounds.

7.1.2
Target Space

Proteins are commonly classified according to their structures (Table 7.3). The full amino acid sequence is the first interesting information (Figure 7.5),

Table 7.3 Structural classification of proteins.

Dimension	Classification scheme	Databases
1D	By sequence	UniProt [13], Pfam [14]
	By patterns	PRINTS [12], PROSITE [15]
2D	By secondary structure, fold	SCOP [16], CATH [17]
3D	By atomic coordinates	PDB [18], MODBASE [19]
	By binding site	BindingMOAD [20], sc-PDB [21]

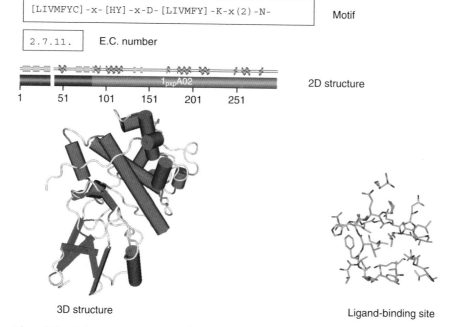

>P24941|CDK2_HUMAN Cell division protein kinase 2 - Homo sapiens (Human).
MENFQKVEKIGEGTYGVVYKARNKLTGEVVALKKIRLDTETEGVPSTAIREISLLKELNHPNIVK
LLDVIHTENKLYLVFEFLHQDLKKFMDASALTGIPLPLIKSYLFQLLQGLAFCHSHRVLHRDLKP
QNLLINTEGAIKLADFGLARAFGVPVRTYTHEVVTLWYRAPEILLGCKYYSTAVDIWSLGCIFAE
MVTRRALFPGDSEIDQLFRIFRTLGTPDEVVWPGVTSMPDYKPSFPKWARQDFSKVVPPLDEDGR
SLLSQMLHYDPNKRISAKAALAHPFFQDVTKPVPHLRL

Sequence

[LIVMFYC]-x-[HY]-x-D-[LIVMFY]-K-x(2)-N-

Motif

2.7.11.

E.C. number

2D structure

3D structure

Ligand-binding site

Figure 7.5 Various representations of a protein using 1D–3D properties.

which enables a reliable clustering of targets by family (e.g. G protein-coupled receptors (GPCRs), kinases). However, sequence lengths may considerably vary within a protein family (e.g., sequence lengths of human GPCRs range from 290 to 6200 residues) such that analyzing similarities and differences first requires an alignment of amino acid sequences, which can be tricky in case of large insertions or deletions. Therefore, one may focus on specific motifs [12] that are a collection of continuous residues specific to a protein family. To take into account the structural organization of the target, it may be of interest to look at the 2D structure (mapping of α-helices, β-sheets, coils and random structures) or, even better, at the 3D structure (atomic coordinates provided either experimentally by X-ray diffraction or nuclear magnetic resonance (NMR)) and/or the corresponding fold. In chemogenomics-related approaches, one usually focuses on the ligand-binding site where structural similarities among related targets are usually much higher than when considering the full 1D sequence or 3D structure.

Targets may also be classified according to their pharmacological profile (binding affinity for a panel of ligands), which means according to the nature of ligands they recognize [5]. Of course, there is a considerable overlap between sequence-based and ligand-based classifications since ligands generally bind to a subset of the protein universe. However, relationships across protein subfamilies are particularly interesting in drug design for predicting or modifying the pharmacological profile of a drug.

7.1.3
Protein–Ligand Space

It is possible to directly navigate in the protein–ligand space by browsing full matrices in which either affinity or structural information is stored. Experimental evaluation of x compounds on y targets (e.g. *in vitro* binding affinity assay) leads to a matrix of xy numbers (e.g. IC50 values) that can be used to predict the affinity of a new compound to an existing target by multivariate linear regression [22], measure a structure-activity relationships (SAR) distance between two targets [23] and predict a global pharmacological profile [24]. A clear advantage of this approach is that it relies on true binding affinity values and that experimentally derived descriptors will usually outperform computed descriptors. A clear drawback is the enormous amount of data required to derive true information so that similar approaches are not realistic in an academic environment. Therefore, one might substitute experimental affinity values with predicted ones derived either from docking (see Chapter 5) or from 3D QSAR (see Chapter 3) approaches [25, 26], although extrapolation is limited here by a small protein space. Since binding free energy is extremely difficult to predict, replacing affinity by molecular interaction descriptors is possible. Of particular interest are structural interaction fingerprints (IFPs) [27] that convert atomic coordinates of a protein–ligand complex into a bit string that features, for each residue of a binding site, those molecular interactions (e.g. H-bond, aromatic interaction, hydrophobic contact) that are developed by a cocrystallized or docked ligand. Comparing a series of complexes between *n* ligands and a single protein or one ligand and *n* related proteins is then performed as for ligands (cf. Section 7.1.1) by computing distances between 1D IFPs (Figure 7.6).

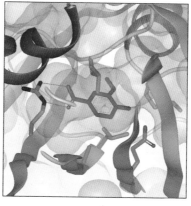

↓ A

M111 | R115 | L126 | Y211 | L213 | E300 | List of residues
10000000000100100000011000001000000000000100   fingerprint

↓ B

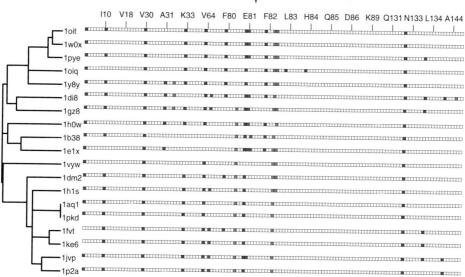

Figure 7.6 Deriving (A) and comparing protein–ligand complexes (B) by molecular interaction fingerprints. Digits '0' and '1' are replaced by color-coded squares for the ease of comparison (blue: hydrophobic interactions, green: aromatic interactions, red: hydrogen bonds).

7.2
Ligand-based Chemogenomic Approaches

7.2.1
Annotating Ligand Libraries

The basic paradigm underlying ligand-based chemogenomic approaches is that molecules sharing enough similarity to existing ligands for which a target profile is known have enhanced probability of sharing the same biological profile (Figure 7.7). It is therefore very important to annotate chemical libraries with biological information (targets, *in vitro* affinity data, absorption, distribution, metabolism, excretion (ADME) properties). Over recent years, there has been a huge effort mainly from small biotech companies to compile such data by an exhaustive survey of literature and patent data (Table 7.4). Since chemogenomic approaches usually focus on target families, most of these archives are related to the pharmaceutically most important target families (GPCRs, kinases, nuclear hormone receptors (NHRs), proteases, phosphodiesterases).

A nice example has been provided by Novartis scientists [28] who linked chemical space to target space by merging fields from separate chemical and biological databases to provide a unified and searchable chemogenomic database. Over 110 000 pharmaceutical ligands were gathered from the Measurement Devices Ltd Drug Data Report (MDDR) (Table 7.3). Annotation of targets was based on existing classifications for enzymes and receptors. Linking MDDR 'activity keys' to the target classification scheme enabled the annotation of 53 000 compounds totaling 799 different activity keys and related targets. Since the target's sequence is linkable to the ligand, sequence-based similarity searches of ligands for protein homolog of liganded targets are therefore feasible. Annotated reference ligands for a particular GPCR were used as starting points to recover either new receptor ligands or ligands of receptors close to the reference GPCR. Interestingly, the efficiency of the virtual screening approach was dependent on the phylogenetic distance between the reference and the query targets. Another straightforward application of biologically annotated compound libraries is the design of target-directed combinatorial libraries [29] focusing on chemotypes preferred by a family of targets.

Natural products also cover a very interesting chemical space of biological relevance because of the evolutionary pressure put on these compounds to bind, usually through highly specific mechanisms, to particular targets. The chemical space spanned by biologically annotated natural products was recently described as a structural and hierarchical scaffold tree [30] that could be browsed to design natural product-oriented chemical libraries.

Biologically annotated compound libraries are a direct source of potentially new biological mechanisms to correct a phenotype. Root *et al.* designed a library of 2036 biologically active compounds covering 169 different biochemical mechanisms, which was shown to be structurally diverse and able to provide

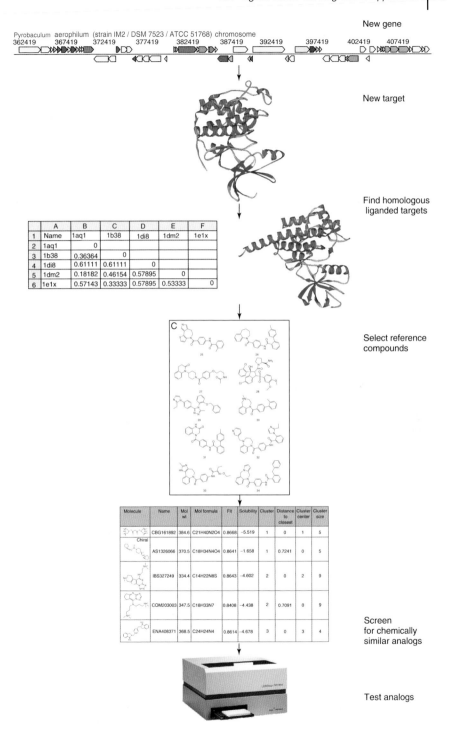

Figure 7.7 Structure-activity relationship homology flowchart.

Table 7.4 Biologically annotated compound libraries.

Database	Description	Web site
AurSCOPE	Target family-oriented knowledge database containing pharmacological and pharmacokinetical data for 160 000 GPCR ligands and 77 000 kinase inhibitors	http://www.aureus-pharma.com
Bioprint	Biological profile (*in vitro* and clinical data) of 2400 small-molecular-weight drugs and druglike compounds	http://www.cerep.fr/
ChemBank	Storage of 50 000 compounds and related biological properties in 441 high-throughput screening and small-molecule microarray assays	http://chembank.broad.harvard.edu/
ChemBioBase	Target centric ligand databases (GPCRs, kinases, phosphodiesterases)	http://www.jubilantbiosys.com/
Kinase knowledgebase	kinase structure-activity and chemical synthesis data	http://www.eidogen-sertanty.com/
MDL Drug Data Report	132 000 biologically relevant compounds and well-defined derivatives	http://www.mdli.com/
MedChem database	650 000 compounds with biological and pharmacological information	http://www.gvkbio.com
StARLITe	Highly curated target–compound SAR relationships	http://www.inpharmatica.co.uk/
Wombat	154 236 entries over 307 700 biological activities on 1320 unique targets.	http://sunsetmolecular.com/

85 hits in a cell viability and proliferation assay [31]. Among the 85 hits, 27 were considered to be active by new biochemical mechanisms.

7.2.2
Privileged Structures

The term 'privileged structure' was first coined by Evans *et al.* [32], who noticed the promiscuity of the 1,4-benzodiazepine scaffold for various targets (Figure 7.8). A privileged structure is defined as a substructure or scaffold

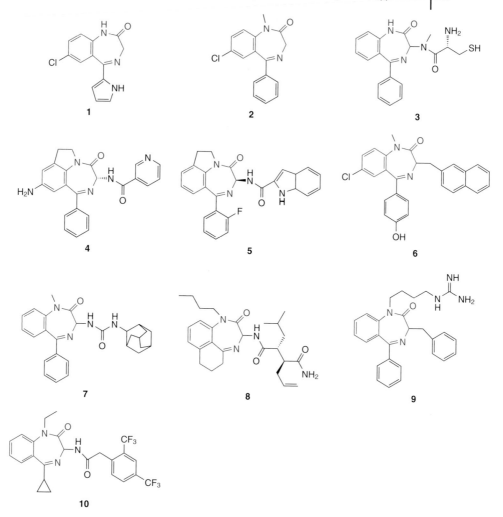

Figure 7.8 Permissivity of the 3H-1,4-benzodiazepin-2-one scaffold (in blue) across various targets. (**1**) Ro-5-3335, HIV-1 Tat inhibitor; (**2**) Diazepam, gamma amino butyric acid-A (GABA-A) receptor ligand; (**3**) 231023: farnesyltransferase inhibitor; (**4**) CI-1044, phosphodiesterase 4 inhibitor; (**5**) Pranazepide, cholecystokinin (CCK) receptor antagonist; (**6**) BZ-423, F1F0 ATPase inhibitor; (**7**) 171 644, oxytocin receptor antagonist, (**8**) 309 060: $\beta-\gamma$ secretase inhibitor; (**9**) 278 588: Stat5 agonist; (**10**) 276 345: KV$_s$ channel blocker.

exhibiting strong preferences for a particular area of the target space (e.g. GPCRs) and suitable to orient the design of targeted compound libraries [33]. In fact, a recent and deeper analysis of MDDR druglike ligands show that privilege appears only upon a certain level of chemical functionality of the scaffold [34]. For example, the biphenyl substructure is not a privileged

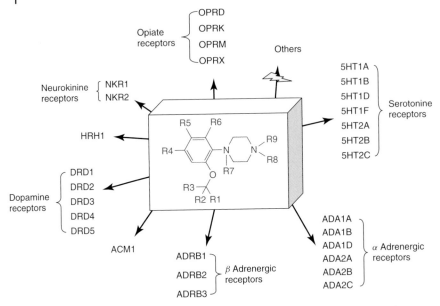

Figure 7.9 Human GPCRs targeted by the orthoalkoxy-N-phenylpiperazine privileged structure (http://bioinfo-pharma. u-strasbg.fr/hGPCRLig/). Remarkably, no other protein ever cocrystallized with druglike compounds is able to recognize this substructure (http://bioinfo-pharma. u-strasbg.fr/scPDB/).

structure but a simple protein-binding motif since it occurs in a wide array of protein ligands with no particular preference for a certain target family. However, extending the biphenyl motif to a 2-tetrazolo-biphenyl dramatically enhance the specificity of the latter substructure for GPCRs [34]. Remarkably, many substructures apparently have corresponding binding sites in unrelated target families (e.g. GPCRs, kinases, ion channels, proteases, NHRs). Only a few of them (e.g. Figure 7.9) are really selective for a certain target family [34]. One main reason for this exquisite specificity is that specific binding sites for peculiar substructure (Figure 7.10) have been conserved along the evolution of target subfamilies [35, 36]. Family-specific privileged structures are of prime importance to design targeted libraries and enhance hit rates when a protein from the targeted family is screened experimentally.

7.2.3
Ligand-based *In silico* Screening

The main target families can be distinguished by a simple look at physicochemical properties (molecular weight, log P, polar surface area, H-bond donor and acceptor counts) of their cognate ligands [9]. One can

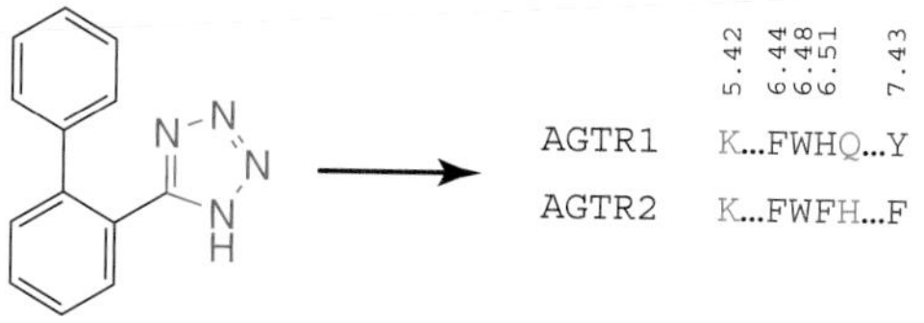

Figure 7.10 Mapping substructure elements of the 2-tetrazolobiphenyl scaffold (biphenyl: blue; tetrazole: red) to conserved hotspots of angiotensin II receptor subtypes (blue: aromatic patch; red: polar patch) to which it specifically binds by complementarity of physicochemical properties [36].

thus easily imagine that more sophisticated descriptors (cf. 7.1.1) can be used to predict a global target profile for any given compound, provided the targets to be predicted are sufficiently well described by existing ligands. Ligand-based *in silico* approaches to target fishing have begun to appear in the literature [37–43]. They all share three basic components: (i) a set of reference compounds from which 2D (scaffold, substructure and fingerprints) or 3D descriptors (pharmacophore) are stored in a database; (ii) a screening procedure using either QSAR, machine learning (Bayesian classification, support vector machines) or pharmacophore searches; and (iii) a screening collection to identify new molecules that are more likely to share the same target or target profile than reference compounds using above-mentioned descriptors (Figure 7.11).

Mestres *et al.* [38, 39] have annotated a library of molecules targeting NHR. Using a hierarchical classification for 2000 ligands and 25 receptors, chemogenomic links bridging ligand to target space can be easily recovered to distinguish selective scaffolds from promiscuous scaffolds. Using Shannon entropy descriptors (SHED) based on the distribution of atom-centered feature pairs, any compound collection can be screened to identify hits presenting SHED distances beyond a defined threshold and likely to share the same NHR profile.

Novartis successfully applied a machine learning algorithm using Bayesian statistics to predict target profiles from extended connectivity fingerprints of compounds from the biologically annotated Wombat database [40]. For each activity class (target), a separate Bayesian model was trained to distinguish known actives from known inactives. Prediction of the most likely targets of compounds in the test set is done by calculating the probability of each test compound to become a ligand for each of the targets. On average, the correct target was found in 77% of the instances when training with Wombat compounds and testing MDDR molecules from 10 different activity classes [40]. A significant improvement in the predictions was observed when considering, instead of a series of individual probabilities, the global profile of all training compounds in which all target-associated probabilities are concatenated into a 'Bayes affinity fingerprint' [41]. Other 2D and 3D descriptors have been assessed for the same application. 2D descriptors were found to be more

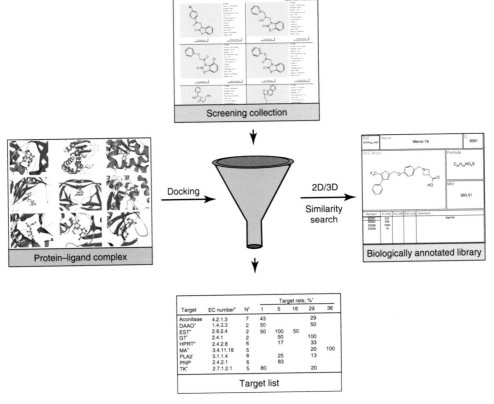

Figure 7.11 *In silico* target fishing approaches.

predictive with regard to correct target prediction than a pure 3D pharma-cophoric approach for test compounds that are structurally similar to those in the training set. For singletons (compounds exhibiting no strong similarity to molecules of the training set), the 3D descriptor was more predictive.

In all these approaches, one must first automatically categorize compounds from the training set according to their molecular target without checking whether each compound really binds to its target, where it binds (which binding site) and how it binds (agonist or antagonist for receptor ligands). Therefore, there is a risk of training a machine learning algorithm with incorrect data and generating false rules. To overcome this drawback, more accurate but slower strategies are available. Among them, a promising approach is to derive 3D pharmacophores from protein–ligand complexes for which experimentally determined atomic coordinates and pharmacological activities exist [37]. The target-annotated pharmacophore database can be browsed to identify target(s) of new compounds by a classical pharmacophore search. The advantage of the method is that it relies on the higher quality of the reference dataset, but is nevertheless, limited by the pharmacophore generation step and the still limited chemical diversity observed among PDB ligands [21]. For example, membrane receptors (e.g. GPCRs, ion channels) cannot be predicted by this approach, as the crystallographic data are very sparse for these protein families, although homology model-based pharmacophores may be theoretically derived.

7.3
Target-based Chemogenomic Approaches

Controlling the selectivity of ligands toward related targets from the same family is crucial information in early drug discovery stages. There is therefore, a growing interest in comparing all targets from the same family, especially those for which there is enough structural data (X-ray or NMR structures) to enable a proteomewide comparative modeling of targets of still unknown structure (e.g. protein kinases). Target-based chemogenomic approaches can be classified into two categories depending on whether the amino acid sequence or the 3D structure of targets is compared.

7.3.1
Sequence-based Comparisons

Sequence-based approaches are intended to be used for any kind of target family provided a multiple alignment of all targets to be compared is possible. They are generally used for target families where a lack of high-resolution structural data hampers target comparison. GPCRs constitute an ideal framework for sequence-based comparisons [36, 44–46] because it is a very important target family for drug design and only one member of this family (bovine rhodopsin) has been crystallized to date [47]. After aligning all sequences, key residues that are considered to map the binding site of most nonpeptide ligands can be extracted and concatenated into an ungapped sequence of a few residues (about 30) which can be later used to derive a distance matrix based on sequence identity [36], sequence similarity [46] or physicochemical property-based bit string [45] (Figure 7.12). An exhaustive cavity-based clustering of 372 human GPCRs has been recently proposed using such a strategy [36]. Interestingly, it perfectly reproduces the full sequence-based tree, suggesting that only a few residues are really important when comparing targets across a family. This simplification enables a much simpler analysis of features (binding site regions) that are responsible for selective or permissive ligand binding by simply looking at residue conservation [36, 44]. There are several potential applications of cavity-based trees in drug discovery. A simple application consists in target hopping, which means discovering receptor ligands for a particular receptor by first considering the known ligands of closely related receptors. For example, CRTH2 (chemoattractant

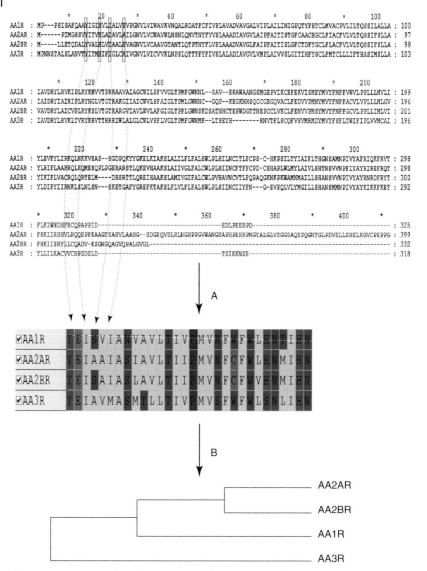

```
           *        20        *        40        *        60        *        80        *       100
AA1R  : MP---PSISAFQAAYIGIEVLIALVSVPGNVLVIWAVKVNQALRDATFCFIVSLAVADVAVGALVIPLAILINIGPQTYFHTCLMVACFVLILTQSSILALLA : 100
AA2AR : M------FIMGSSVYITVELAIAVLAILGNVLVCWAVWLNSNLQNVTNYFVVSLAAADIAVGVLAIPFAITISTGFCAACHGCLFIACFVLVLTQSSIFSLLA :  97
AA2BR : M-----LLETQDAINYALSLVIIAALSVAGNVLVCAAVGTANTLQTPTNYFLVSLAAADVAVGLFAIPFAITISLGFCTDFYGCLFLACFVLVLTQSSIFSLLA :  98
AA3R  : MPNNSTALSLANVTYITMEIPIGLCAIVGNVLVICVVKLNPSLQTTTFYFIVSLALADIAVGVLVMPLAIVVSLGITIHFYSCLFMTCLLLIFTHASIMSLLA : 103

           *       120        *       140        *       160        *       180        *       200
AA1R  : IAVDRYLRVKIPLRYKMVVTPRRAAVAIAGCWILSFVVGLTPMFGWNNL--SAV--ERAWAANGSMGEPVIKCEFEKVISMEYMVYFNFFVWVLPPLLLMVLI : 199
AA2AR : IAIDRYIAIRIPLRYNGLVTGTRAKGIIAICWVLSFAIGLTPMLGWNNC--GQP--KEGKNHSQGCGEGQVACLFEDVVDMNYMVYFNFFACVLVPLLLMLGV : 196
AA2BR : VAVDRYLAICVPLRYKSLVTGTRARGVIAVLWVLAFGIGLTPFLGWNSKDSATNNCTEFWDGTTNESCCLVKCLFENVVPMSYMVYFNFFGCVLPPLLIMLVI : 201
AA3R  : IAVDRYLRVKLTVRYKRVTTHRRIWLALGLCWLVSFLVGLTPMFGWNMK--LTSEYH-------RNVTFLSCQFVSVMRMDYMVYFSFLTWIFIPLVVMCAI : 196

           *       220        *       240        *       260        *       280        *       300
AA1R  : YLEVFYLIRKQLNKKVSAS--SGDPQKYYGKELKIAKSLALILFLFALSWLPLHILNCITLFCPS-C-HKPSILTYIAIFLTHGNSAMNPIVYAFRIQKFRVT : 298
AA2AR : YLRIFLAARRQLKQMESQPLPGERARBTLQKEVHAAKSLAIIVGLFALCWLPLHIINCFTFFCPD-CSHAPLWLMYLAIVLSHTNSVVNPFIYAYRIREFRQT : 298
AA2BR : YIKIFLVACRQLQRTELM----DHSRTTLQREIHAAKSLAMIVGIFALCWLPVHAVNCVTLFQPAQGKNKPKWAMNMAILLSHANSVVNPIVYAYRNRDFRYT : 300
AA3R  : YLDIFYIIRNKLSLNLSN---SKETGAFYGREFKTAKSLFLVLFLFALSWLPLSIINCIIYFN---G-EVPQLVLYMGILLSHANSMMNPIVYAYKIKKFKET : 292

           *       320        *       340        *       360        *       380        *       400       *
AA1R  : FLKIWNDHFRCQPAPPID------------------------------------EDLPEERPD------------------------------------------ : 325
AA2AR : FRKIIRSHVLRQQEPFKAAGTSARVLAAHG--SDGEQVSLRLNGHPPGVWANGSAPHPERRPNGYALGLVSGGSAQESQGNTGLPDVELLSHELKGVCPEPPG : 399
AA2BR : FHKIISRYLLCQADV-KSGNGQAGVQPALGVGL---------------------------------------------------------------------- : 332
AA3R  : YLLILKACVVCHPSDSLD--------------------------TSIEKNSE------------------------------------------------------ : 318
```

A

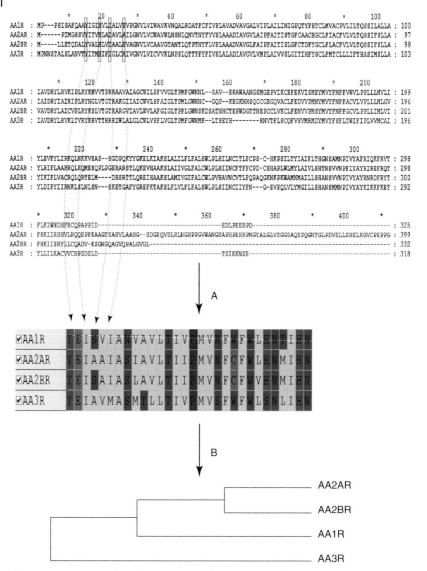

B

AA2AR

AA2BR

AA1R

AA3R

Figure 7.12 Sequence-based comparison of targets exemplified by human adenosine receptors [36]. (A) Selection of key cavity-lining residues; (B) Clustering according to residue conservation.

receptor-homologous molecule expressed on TH2 cells) antagonists could have been identified from existing Angiotensin II type 1 receptor antagonists [45]. In addition, the design of targeted libraries towards a particular area of the cavity-based tree is facilitated by addressing those residues responsible for selectivity or promiscuity [45, 46].

7.3.2
Structure-based Comparisons

Structure-based comparisons are only possible for target families where there are enough good structural templates (X-ray structures) to afford the homology modeling of other related targets. In general, only ligand-binding sites [20, 21] are compared since the basic aim of such comparisons is to understand the selectivity or permissivity features of related targets of known ligands.

7.3.2.1 **Comparing Molecular Fields**
The first possible strategy is to compare computed molecular interaction fields (MIFs) from the cavities to be compared [48–50]. Starting from a structural alignment of all targets, interaction energies generated by rolling several probe atoms (e.g. sp3 carbon atom) at each point of 3D grid encompassing the ligand-binding site are then concatenated into an MIF vector which can be placed in a global matrix where rows describe targets, and columns describe interaction energies at a given 3D grid point (Figure 7.13). Comparing of the MIFs and clustering of the cognate targets can be done either by analyzing the matrix by principal component analysis [48, 49] or by calculating a field difference which will be used as a descriptor to generate a hierarchical clustering of targets [50]. A clear issue with this approach, which is also seen in 3D QSAR methods (see Section 2.6), is that the comparison is highly dependent on the structural alignment, the grid resolution and the choice of the probe atoms. Moreover, it cannot be applied to targets of different families. However, it has been successfully applied to protein kinases [48, 50], serine proteases [50], matrix metalloproteinases [49] and NHRs [50] to pinpoint cavity regions or subpockets explaining either selective or promiscuous ligand binding, and thus to guide the design of compound libraries towards the desirable selectivity pattern.

7.3.2.2 **Comparing 3D Structures**
To avoid the previously reported structural alignment bias, 3D atomic protein coordinates can be directly compared to measure the distance between two targets (Table 7.5). Global structural alignment methods (GASH [51], DaliLite [52] and CE [53]) usually count the number of structurally equivalent residues by comparing overlapping fragments. Such methods, however, do not work very well for discontinuous sequences (active sites) and for proteins exhibiting different folds.

A second approach is to identify predefined structural motifs or templates (e.g. Ser–His–Asp catalytic triad in serine proteases) and align a query to a reference protein by matching templates [54, 55]. However, numerous proteins (e.g. kinases, GPCRs, ion channels) may share a binding site for a unique ligand (adenosine triphosphate: ATP) without sharing any structural template similarity. Most recent approaches to generate structural alignment

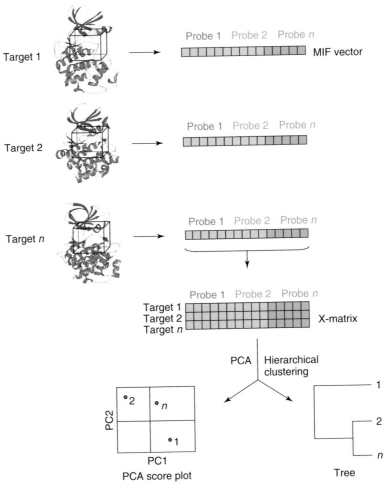

Figure 7.13 Molecular interaction field (MIF)-based clustering of targets.

describe proteins by physicochemical properties at representative locations. Molecular surfaces can be easily discretized in either chemically labelled sparse points [62] or graphs [63], and thus aligned to maximize surface overlap with any reference (Table 7.5). A database of protein surfaces (eF-site) has successfully been browsed to predict the function of a hypothetical archaeon protein (MJ0226) by detection of a mononucleotide binding site [63]. Surface-based comparisons are, however, relatively slow and thus incompatible with proteomewide comparisons. Recent and faster methods (e.g. SuMo [57], Cavbase [58], SiteEngine [59], SitesBase [60] and CPASS [61]) have been developed over the last three years. They all have to represent an active site of interest by pseudocenters (dummy atoms located along or close to every side

Table 7.5 Available programs for comparing protein structure or active sites.

Name	Method	Web site	References
TESS	Distance-based matching to predefined structural templates	http://www.ebi.ac.uk/thornton-srv/databases/CSA/	54
ASSAM	Matching spatially conserved patterns by a subgraph isomorphism algorithm	–	55
eF-site	Matching triangulated surfaces with electrostatic potential information	http://ef-site.hgc.jp/eF-site/	56
SuMo	Matching graphs of triangles from stereochemical groups	http://sumo-pbil.ibcp.fr/cgi-bin/sumo-welcome	57
Cavbase	Matching pseudocenters by a clique detection algorithm	–	58
SiteEngine	Matching triangles of physicochemical properties by geometric hashing	http://bioinfo3d.cs.tau.ac.il/SiteEngine	59
SitesBase	Matching triplets of atoms by geometric hashing	http://www.modelling.leeds.ac.uk/sb/	60
CPASS	Maximizing a rmsd-weighted BLOSUM62 scoring function	http://bionmr-c1.unl.edu/	61

chain of interest) encoding physicochemical properties (H-bonding capacity, aromaticity, hydrophobicity, charge) of their cognate residues, pseudocenters being linked together by edges, and thus defining a molecular graph. Alignment is operated by detection of maximal common subgraphs (clique detection) [64] or geometric hashing [65] from defined pseudocenters. Local similarity at ligand-binding subpockets can thus be detected for proteins with totally different folds and catalytic activities. Predicted, similar binding sites can even be linked together in a global network to better position a protein in the target space [66].

A nice example of binding site similarities for distant proteins has been exemplified by Weber *et al.* [67], who detected crossreactivity of arylsulfonamide-based (cyclooxygenase-2) COX-2 inhibitors withhuman carbonic anhydrase (HCA) based on the similarity of COX-2 and HCA binding pockets. A problem with these matching techniques is that the computed similarity score (usually dependent on the number of atom or pseudocenter or triangle matches) is not always easy to interpret, notably for active sites of different dimensions because large active sites will have a tendency to present more matches than small ones even if the latter are more similar. Therefore, normalized distance metrics similar to those used for comparing ligands (see Section 7.1.1) are needed. A promising approach is proposed by Surgand *et al.* [68], who discretized an active site by a dimensionless 80-triangle sphere and projected various topological and physicochemical descriptors from Cα atoms of cavity-lining residues to the center of the sphere. The distance between two active sites is thus simply computed by summing up the normalized differences in descriptor space between each triangle of the sphere (Figure 7.14).

The current speed of such comparisons enables the definition of all-against-all similarity matrices [58–60] and opens the door to various applications: (i) functional analysis and classification of ligand-binding sites, (ii) predicting potential ligands and (iii) anticipating side effects caused by targeting a peculiar protein.

An alternative approach for comparing ligand-binding sites is to evaluate the similarity of potential ligand-binding envelopes for known X-ray structure of apo or holoproteins [69]. A first draft of the human pocketome, a collection of all possible ligand-binding envelopes for a set of 943 crystallized human proteins, has been recently proposed [69] and clustered by envelope similarity. Interestingly, the envelope tree only partially matches alternative trees based on the amino acid sequence of the target proteins or on ligand-bound similarities [69].

Another recently proposed approach for comparing proteins of the same family is to look at packing defects [70] localized at so-called 'dehydrons' (backbone heavy atoms with unsatisfied H-bonding partners), which are good indicators of protein capacity to interact with potential ligands and can be predicted form the amino acid sequence. Packing distances between 32 PDB-reported kinases were shown to be almost identical to the pharmacological distance between these kinases estimated from an experimental affinity matrix derived for 17 inhibitors [71], and to efficiently guide the structure-based design of selective inhibitors for various enzymes specifically designed to target packing defects [72].

7.4
Target-Ligand-based Chemogenomic Approaches

7.4.1
Chemical Annotation of Target Binding Sites

Numerous biologically annotated chemical libraries can be browsed (see Table 7.4) to link chemical to target spaces and focus ligand-based design to target families [40–42]. However, if the information about the binding

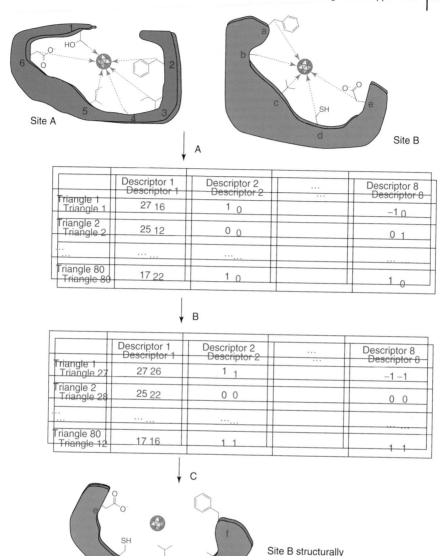

Figure 7.14 Sphere-based protein alignment and comparison [68]. (A) Projection on a triangulated sphere of cavity descriptors (distance from sphere center to Cβ, orientation of side chain vs. sphere, size of the side chain, H-bond donor and acceptor counts, aliphatic character, aromatic character, charge) from Cβ atoms to the sphere center; (B) computing a distance between two sites by measuring a normalized distance in descriptor space after exhaustive rotation/translation of one sphere (site B) to optimize the similarity score and (C) alignment of site B to site A by selection of the sphere rotation/translation giving the lowest distance.

site is missing, there is a potential risk in comparing compounds sharing the same target but not the same binding site (e.g. orthosteric and allosteric ligands). It is, therefore, important to rigorously annotate protein sequences and/or binding site by the chemotype of the ligands they can recognize. The SMID (Small Molecule Interaction Database) is an interesting initiative to annotate protein amino acid sequences by domain-specific ligands [73]. A total of 6300 ligands covering 230 000 experimentally observed domain or small-molecule interactions have been stored in a relational database which can be browsed to predict the most likely ligand of proteins of unknown 3D structures by comparison of their domains using a reverse position-specific BLAST procedure [74]. Ligand-annotated binding sites from the PDB are annotated in several databases [21], but only two of them (BindingMOAD, sc-PDB; Table 7.3) consider the ligand from a pharmacological point of view and are therefore of interest for chemogenomic approaches. Such databases can be used to prioritize either ligands or molecular scaffolds for designing targeted compound libraries covering a well-defined target space (Figure 7.15).

7.4.2
Two-dimensional Searches

In order to browse and predict protein–ligand complexes, one needs to set up simple descriptors for both ligands and proteins from knowledge databases (Table 7.4) and concatenate them into a single protein–ligand description.

The easiest way to encode this information is to start form experimental binding affinity matrices [22–24] and define appropriate QSAR–QSPR models to predict the affinity of new compounds for registered targets or the full virtual profile by general neighborhood behavior modeling [24].

Another approach has been recently proposed for deorphaning GPCRs, in which a ligand atom connectivity fingerprint is merged to a sequence-based target fingerprint if a high-affinity complex ($pK_i > 7$) is reported in the Psychoactive Drug Screening Program (PDSP) database (http://pdsp.med.unc.edu/). A machine learning algorithm was trained from 5319 nonredundant known complexes and applied to a set of 1 911 415 virtual complexes (55 orphan receptors and 34 753 druglike compounds form the NCI database) to predict the most likely associations [75]. Out-of-sample validations (finding the receptors of a promiscuous ligand and the ligands of a single target) were in general agreement with literature data and some predictions still awaiting experimental validations have been made.

7.4.3
Three-dimensional Searches

A straightforward way to predict putative targets of ligands is to dock each of the ligands of the compound library into each of the active site of the

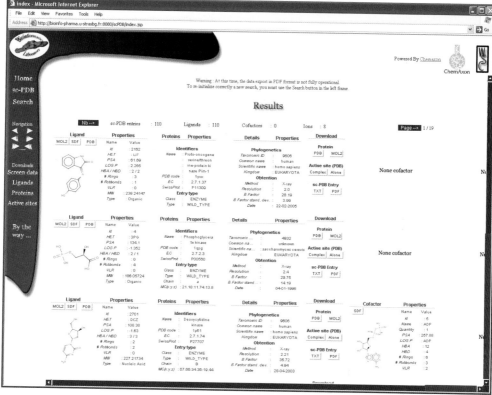

Figure 7.15 Querying the sc-PDB chemogenomic database (http://bioinfo-pharma.u-strasbg.fr/scPDB) for rule-of-five compliant small-molecular-weight fragments (MW < 300, clogP < 3, H-bond donor count < 3, H-bond acceptor count < 6) cocrystallized with protein kinases.

target library. This strategy has been validated by several groups and has been proved capable of recovering the known ligands of known targets and also predicting their off-targets, and thus some potential side effects [76, 77]. Up till now, there is only a single successful target fishing application described in the literature that utilizes a molecular inverse docking approach [78]. Hence, inverse docking requires first, a high-quality 3D dataset of binding sites, whose automated setup is quite difficult, and second, an accurate scoring function to properly rank targets. A problem is that energy-based scoring functions are not very good at quantifying very heterogeneous protein–ligand complexes by decreasing binding free energies [79] and that alternative ways of scoring are requested for efficient target selection. Among the most promising methods is the computation of IFPs between a protein and its ligand. Practically, IFPs are simple bit strings that convert 3D information about protein–ligand interactions into simple 1D bit vector representations (Figure 7.6) that can be quickly compared by the use of traditional metrics (e.g. Tanimoto coefficient, Euclidean distance). Usage of IFPs has shown several promising features: (i) enhancing the quality of pose prediction in docking experiments [80, 81]; (ii) clustering protein–ligand interactions for a panel of related inhibitors according to the diversity of their interaction with a target subfamily [81, 82] and (iii) assisting target-biased library design [83].

However, 3D-based docking-independent methods may also constitute an interesting approach to predict protein–ligand complex. A significant problem is to encode protein and ligand properties with similar descriptors such that one partner can be retrieved by using the second one as a query. A promising solution is proposed with the CoLiBRI (complementary ligands based on receptor information) method [84] in which both ligand and active site atoms are described by the same vector of molecular descriptors (TAE-RECON) derived from shape and electronic properties of isolated atoms. Therefore, it is possible to directly correlate chemical similarities between active site and their ligands by mapping patterns of active sites onto patterns of their complementary ligands and vice versa. When applied to a test data set of 800 high-resolution Protein Data Bank (PDB) complexes, the complementary ligand was ranked among the top 1% of a large library in 90% of tested active sites. Accuracy dropped significantly for active sites very different from those in the test set but still allowed application as a prefiltering step for removing the most improbable ligands [84].

7.5
Concluding Remarks

Chemogenomic approaches to rational drug discovery have been exploding in the last few years as high-throughput data (structure, binding affinity, functional effects) has become available for both targets and ligands of pharmaceutical interest. Numerous ways to link those data have been proposed, focusing on either ligand or target neighborhood. A clear data organization and storage is necessary to foster such applications and these have begun to emerge for the most interesting target families (kinases, GPCRs, NHRs). In the near feature, an earlier and better control of ligand selectivity can be anticipated by using chemogenomic data. This does not mean that more selective ligands are going to be designed, but simply that the observed selectivity profile of the compound will be compatible with a therapeutic usage. In addition, novel genomic targets could be better addressed after locating them in the target space and exploiting the associated chemical information.

References

1. Venter, J.C., Adams, M.D., Myers, E.W. *et al.* (2001) The sequence of the human genome. *Science*, **291**, 1304–51.

2. Lander, E.S., Linton, L.M., Birren, B. *et al.* (2001) Initial sequencing and analysis of the human genome. *Nature*, **409**, 860–921.

3. International Human Genome Sequencing Consortium. (2004) Finishing the euchromatic sequence of the human genome. *Nature*, **431**, 931–45.

4. Russ, A.P. and Lampel, S. (2005) The druggable genome: an update. *Drug Discovery Today*, **10**, 1607–10.

5. Paolini, G.V., Shapland, R.H., van Hoorn, W.P. *et al.* (2006) Global mapping of pharmacological space. *Nature Biotechnology*, **24**, 805–15.

6. Caron, P.R., Mullican, M.D., Mashal, R.D. *et al.* (2001) Chemogenomic approaches to drug discovery. *Current Opinion in Chemical Biology*, **5**, 464–70.

7. Bender, A. and Glen, R.C. (2004) Molecular similarity: a key technique in molecular informatics. *Organic & Biomolecular Chemistry*, **2**, 3204–18.

8. Sadowski, J. and Kubinyi, H. (1998) A scoring scheme for discriminating between drugs and nondrugs. *Journal of Medicinal Chemistry*, **41**, 3325–29.

9. Morphy, R. (2006) The influence of target family and functional activity on the physicochemical properties of pre-clinical compounds. *Journal of Medicinal Chemistry*, **49**, 2969–78.

10. Weininger, D. (1988) SMILES 1. Introduction and encoding rules. *Journal of Chemical Information and Computer Sciences*, **28**, 31–36.

11. Raymond, J.W., Blankley, C.J., and Willett, P. (2003) Comparison of chemical clustering methods using graph-and fingerprint-based similarity measures. *Journal of Molecular Graphics & Modelling*, **21**, 421–33.

12. Attwood, T.K., Bradley, P., Flower, D.R. *et al.* (2003) PRINTS and its automatic supplement, prePRINTS. *Nucleic Acids Research*, **31**, 400–2.

13. Wu, C.H., Apweiler, R., Bairoch, A. *et al.* (2006) The Universal Protein Resource (UniProt): an expanding universe of protein information. *Nucleic Acids Research*, **34**, D187–91.

14. Finn, R.D., Mistry, J., Schuster-Bockler, B. *et al.* (2006) Pfam: clans, web tools and services. *Nucleic Acids Research*, **34**, D247–51.

15. Hulo, N., Bairoch, A., Bulliard, V. *et al.* (2006) The PROSITE database. *Nucleic Acids Research*, **34**, D227–30.

16. Casbon, J. and Saqi, M.A. (2005) S4: structure-based sequence alignments of SCOP superfamilies. *Nucleic Acids Research*, **33**, D219–22.

17. Reeves, G.A., Dallman, T.J., Redfern, O.C. *et al.* (2006) Structural diversity of domain superfamilies in the CATH database. *Journal of Molecular Biology*, **360**, 725–41.

18. Berman, H.M., Westbrook, J., Feng, Z. *et al.* (2000) The protein data bank. *Nucleic Acids Research*, **28**, 235–42.

19. Pieper, U., Eswar, N., Braberg, H. *et al.* (2006) MODBASE: a database of annotated comparative protein structure models and associated resources. *Nucleic Acids Research*, **34**, D291–95.

20. Hu, L., Benson, M.L., Smith, R.D. *et al.* (2005) Binding MOAD (mother of all databases). *Proteins*, **60**, 333–40.

21. Kellenberger, E., Muller, P., Schalon, C. *et al.* (2006) sc-PDB: an annotated database of druggable binding sites from the Protein Data Bank. *Journal of Chemical Information and Modeling*, **46**, 717–27.

22. Kauvar, L.M., Higgins, D.L., Villar, H.O. *et al.* (1995) Predicting ligand binding to proteins by affinity fingerprinting. *Chemistry & Biology*, **2**, 107–18.

23. Vieth, M., Higgs, R.E., Robertson, D.H. *et al.* (2004) Kinomics-structural biology and chemogenomics of kinase inhibitors and targets. *Biochimica et Biophysica Acta*, **1697**, 243–57.

24. Krejsa, C.M., Horvath, D., Rogalski, S.L. *et al.* (2003) Predicting ADME properties and side effects: the BioPrint approach. *Current Opinion in Drug Discovery & Development*, **6**, 470–80.

25. Fukunishi, Y., Kubota, S., and Nakamura, H. (2006) Noise reduction method for molecular interaction energy: application to in silico drug

screening and in silico target protein screening. *Journal of Chemical Information and Modeling*, **46**, 2071–84.

26. Matter, H. and Schwab, W. (1999) Affinity and selectivity of matrix metalloproteinase inhibitors: a chemometrical study from the perspective of ligands and proteins. *Journal of Medicinal Chemistry*, **42**, 4506–23.

27. Singh, J., Deng, Z., Narale, G., and Chuaqui, C. (2006) Structural interaction fingerprints: a new approach to organizing, mining, analyzing, and designing protein-small molecule complexes. *Chemical Biology & Drug Design*, **67**, 5–12.

28. Schuffenhauer, A., Floersheim, P., Acklin, P., and Jacoby, E. (2003) Similarity metrics for ligands reflecting the similarity of the target proteins. *Journal of Chemical Information and Computer Sciences*, **43**, 391–405.

29. Savchuk, N.P., Balakin, K.V., and Tkachenko, S.E. (2004) Exploring the chemogenomic knowledge space with annotated chemical libraries. *Current Opinion in Chemical Biology*, **8**, 412–17.

30. Koch, M.A., Schuffenhauer, A., Scheck, M. *et al.* (2005) Charting biologically relevant chemical space: a structural classification of natural products (SCONP). *Proceedings of the National Academy of Sciences of the United States of America*, **102**, 17272–77.

31. Root, D.E., Flaherty, S.P., Kelley, B.P., and Stockwell, B.R. (2003) Biological mechanism profiling using an annotated compound library. *Chemistry & Biology*, **10**, 881–92.

32. Evans, B.E., Rittle, K.E., Bock, M.G. *et al.* (1988) Methods for drug discovery: development of potent, selective, orally effective cholecystokinin antagonists. *Journal of Medicinal Chemistry*, **31**, 2235–46.

33. Klabunde, T. and Hessler, G. (2002) Drug design strategies for targeting G-protein-coupled receptors. *Chembiochem*, **3**, 928–44.

34. Schnur, D.M., Hermsmeier, M.A., and Tebben, A.J. (2006) Are target-family-privileged substructures truly privileged? *Journal of Medicinal Chemistry*, **49**, 2000–9.

35. Bondensgaard, K., Ankersen, M., Thogersen, H. *et al.* (2004) Recognition of privileged structures by G-protein coupled receptors. *Journal of Medicinal Chemistry*, **47**, 888–99.

36. Surgand, J.S., Rodrigo, J., Kellenberger, E., and Rognan, D. (2006) A chemogenomic analysis of the transmembrane binding cavity of human G-protein-coupled receptors. *Proteins*, **62**, 509–38.

37. Steindl, T.M., Schuster, D., Laggner, C., and Langer, T. (2006) Parallel screening: a novel concept in pharmacophore modeling and virtual screening. *Journal of Chemical Information and Modeling*, **46**, 2146–57.

38. Cases, M., Garcia-Serna, R., Hettne, K. *et al.* (2005) Chemical and biological profiling of an annotated compound library directed to the nuclear receptor family. *Current Topics in Medicinal Chemistry*, **5**, 763–72.

39. Mestres, J., Martin-Couce, L., Gregori-Puigjane, E. *et al.* (2006) Ligand-based approach to in silico pharmacology: nuclear receptor profiling. *Journal of Chemical Information and Modeling*, **46**, 2725–36.

40. Nidhi, Glick, M., Davies, J.W., and Jenkins, J.L. (2006) Prediction of biological targets for compounds using multiple-category Bayesian models trained on chemogenomics databases. *Journal of Chemical Information and Modeling*, **46**, 1124–33.

41. Bender, A., Jenkins, J.L., Glick, M. *et al.* (2006) "Bayes affinity fingerprints" improve retrieval rates in virtual screening and define orthogonal bioactivity space: when are multitarget drugs a feasible concept? *Journal of Chemical Information and Modeling*, **46**, 2725–36.

42. Nettles, J.H., Jenkins, J.L., Bender, A. *et al.* (2006) Bridging chemical and biological space: target fishing using 2D and 3D molecular descriptors. *Journal of Medicinal Chemistry*, **49**, 6802–10.

43. Bhavani, S., Nagargadde, A., Thawani, A. *et al.* (2006) Substructure-based support vector machine classifiers for prediction of adverse effects in diverse classes of drugs. *Journal of Chemical Information and Modeling*, **46**, 2445–56.

44. Crossley, R. (2004) The design of screening libraries targeted at G-protein coupled receptors. *Current Topics in Medicinal Chemistry*, **4**, 581–88.

45. Frimurer, T.M., Ulven, T., Elling, C.E. *et al.* (2005) A physicogenetic method to assign ligand-binding relationships between 7TM receptors. *Bioorganic & Medicinal Chemistry Letters*, **15**, 3707–12.

46. Kratochwil, N.A., Malherbe, P., Lindemann, L. *et al.* (2005) An automated system for the analysis of G protein-coupled receptor transmembrane binding pockets: alignment, receptor-based pharmacophores, and their application. *Journal of Chemical Information and Modeling*, **45**, 1324–36.

47. Palczewski, K., Kumasaka, T., Hori, T. *et al.* (2000) Crystal structure of rhodopsin: a G protein-coupled receptor. *Science*, **289**, 739–45.

48. Naumann, T. and Matter, H. (2002) Structural classification of protein kinases using 3D molecular interaction field analysis of their ligand binding sites: target family landscapes. *Journal of Medicinal Chemistry*, **45**, 2366–78.

49. Pirard, B. and Matter, H. (2006) Matrix metalloproteinase target family landscape: a chemometrical approach to ligand selectivity based on protein binding site analysis. *Journal of Medicinal Chemistry*, **49**, 51–69.

50. Hoppe, C., Steinbeck, C., and Wohlfahrt, G. (2006) Classification and comparison of ligand-binding sites derived from grid-mapped knowledge-based potentials. *Journal of Molecular Graphics & Modelling*, **24**, 328–40.

51. Standley, D.M., Toh, H., and Nakamura, H. (2005) GASH: an improved algorithm for maximizing the number of equivalent residues between two protein structures. *BMC Bioinformatics*, **6**, 221.

52. Holm, L. and Park, J. (2000) DaliLite workbench for protein structure comparison. *Bioinformatics*, **16**, 566–67.

53. Shindyalov, I.N. and Bourne, P.E. (1998) Protein structure alignment by incremental combinatorial extension (CE) of the optimal path. *Protein Engineering*, **11**, 739–47.

54. Wallace, A.C., Borkakoti, N., and Thornton, J.M. (1997) TESS: a geometric hashing algorithm for deriving 3D coordinate templates for searching structural databases. Application to enzyme active sites. *Protein Science*, **6**, 2308–23.

55. Artymiuk, P.J., Poirrette, A.R., Grindley, H.M. *et al.* (1994) A graph-theoretic approach to the identification of three-dimensional patterns of amino acid side-chains in protein structures. *Journal of Molecular Biology*, **243**, 327–44.

56. Kinoshita, K., Furui, J., and Nakamura, H. (2002) Identification of protein functions from a molecular surface database, eF-site. *Journal of Structural and Functional Genomics*, **2**, 9–22.

57. Jambon, M., Imberty, A., Deleage, G., and Geourjon, C. (2003) A new bioinformatic approach to detect common 3D sites in protein structures. *Proteins*, **52**, 137–45.

58. Schmitt, S., Kuhn, D., and Klebe, G. (2002) A new method to detect related function among proteins independent of sequence and fold homology. *Journal of Molecular Biology*, **323**, 387–406.

59. Shulman-Peleg, A., Nussinov, R., and Wolfson, H.J. (2004) Recognition of functional sites in protein structures.

Journal of Molecular Biology, **339**, 607–33.

60. Gold, N.D. and Jackson, R.M. (2006) Fold independent structural comparisons of protein-ligand binding sites for exploring functional relationships. *Journal of Molecular Biology*, **355**, 1112–24.

61. Powers, R., Copeland, J.C., Germer, K. et al. (2006) Comparison of protein active site structures for functional annotation of proteins and drug design. *Proteins*, **65**, 124–35.

62. Rosen, M., Lin, S.L., Wolfson, H., and Nussinov, R. (1998) Molecular shape comparisons in searches for active sites and functional similarity. *Protein Engineering*, **11**, 263–77.

63. Kinoshita, K. and Nakamura, H. (2003) Identification of protein biochemical functions by similarity search using the molecular surface database eF-site. *Protein Science*, **12**, 1589–95.

64. Gardiner, E.J., Artymiuk, P.J., and Willett, P. (1997) Clique-detection algorithms for matching three-dimensional molecular structures. *Journal of Molecular Graphics & Modelling*, **15**, 245–53.

65. Nussinov, R. and Wolfson, H.J. (1991) Efficient detection of three-dimensional structural motifs in biological macromolecules by computer vision techniques. *Proceedings of the National Academy of Sciences of the United States of America*, **88**, 10495–99.

66. Zhang, Z. and Grigorov, M.G. (2006) Similarity networks of protein binding sites. *Proteins*, **62**, 470–78.

67. Weber, A., Casini, A., Heine, A. et al. (2004) Unexpected nanomolar inhibition of carbonic anhydrase by COX-2-selective celecoxib: new pharmacological opportunities due to related binding site recognition. *Journal of Medicinal Chemistry*, **47**, 550–57.

68. Surgand, J.S. (2006) *Développement de Nouvelles Méthodes Bioinformatiques Pour L'étude des récepteurs couplés aux protéines G*, Thèse de l'Université Louis Pasteur – Strasbourg I, France.

69. An, J., Totrov, M., and Abagyan, R. (2005) Pocketome via comprehensive identification and classification of ligand binding envelopes. *Molecular & Cellular Proteomics*, **4**, 752–61.

70. Fernandez, A., Rogale, K., Scott, R., and Scheraga, H.A. (2004) Inhibitor design by wrapping packing defects in HIV-1 proteins. *Proceedings of the National Academy of Sciences of the United States of America*, **101**, 11640–45.

71. Fernandez, A. and Maddipati, S. (2006) A priori inference of cross reactivity for drug-targeted kinases. *Journal of Medicinal Chemistry*, **49**, 3092–100.

72. Fernandez, A. (2005) Incomplete protein packing as a selectivity filter in drug design. *Structure*, **13**, 1829–36.

73. Snyder, K.A., Feldman, H.J., Dumontier, M. et al. (2006) Domain-based small molecule binding site annotation. *BMC Bioinformatics*, **7**, 152.

74. Feldman, H.J., Snyder, K.A., Ticoll, A. et al. (2006) A complete small molecule dataset from the protein data bank. *FEBS Letters*, **580**, 1649–53.

75. Bock, J.R. and Gough, D.A. (2005) Virtual screen for ligands of orphan G protein-coupled receptors. *Journal of Chemical Information and Modeling*, **45**, 1402–14.

76. Chen, Y.Z. and Zhi, D.G. (2001) Ligand-protein inverse docking and its potential use in the computer search of protein targets of a small molecule. *Proteins*, **43**, 217–26.

77. Paul, N., Kellenberger, E., Bret, G. et al. (2004) Recovering the true targets of specific ligands by virtual screening of the protein data bank. *Proteins*, **54**, 671–80.

78. Muller, P., Lena, G., Boilard, E. et al. (2006) In silico guided target identification of a scaffold-focused library: 1,3,5-triazepan-2,6-diones as novel phospholipase A2 inhibitors. *Journal of Medicinal Chemistry*, **49**, 6768–78.

79. Ferrara, P., Gohlke, H., Price, D.J. et al. (2004) 3rd assessing scoring functions for protein-ligand

interactions. *Journal of Medicinal Chemistry*, **47**, 3032–47.

80. Deng, Z., Chuaqui, C., and Singh, J. (2004) Structural interaction fingerprint (SIFt): a novel method for analyzing three-dimensional protein-ligand binding interactions. *Journal of Medicinal Chemistry*, **47**, 337–44.

81. Marcou, G. and Rognan, D. (2007) Optimizing scaffold docking by use of molecular interaction fingerprints. *Journal of Chemical Information and Modeling*, **47**, 195–207.

82. Chuaqui, C., Deng, Z., and Singh, J. (2005) Interaction profiles of protein kinase-inhibitor complexes and their application to virtual screening. *Journal of Medicinal Chemistry*, **48**, 121–33.

83. Deng, Z., Chuaqui, C., and Singh, J. (2006) Knowledge-based design of target- focused libraries using protein-ligand interaction constraints. *Journal of Medicinal Chemistry*, **49**, 490–500.

84. Oloff, S., Zhang, S., Sukumar, N. *et al.* (2006) Chemometric analysis of ligand receptor complementarity: identifying complementary ligands based on receptor information (CoLiBRI). *Journal of Chemical Information and Modeling*, **46**, 844–51.

8

A Case Study for Protein Modeling: the Nuclear Hormone Receptor CAR as an Example for Comparative Modeling and the Analysis of Protein-Ligand Complexes

8.1
The Biochemical and Pharmacological Description of the Problem

8.1.1
Nuclear Hormone Receptor Superfamily

Nuclear hormone receptors (NHRs) are key elements in the intracellular signal transduction of metazoans. Forty eight members of the nuclear receptor (NR) superfamily can be found in humans. By responding to a large variety of hormonal and metabolic signals, all NRs act as ligand-activated transcription factors, thus playing a crucial role in the regulation of gene expression. Moreover, NRs are targeted by other signaling cascades and integrate diverse signal transduction pathways involving them in numerous physiological processes comprising development, differentiation, homeostasis and reproduction [1].

Usually, NRs bind as homo- (Type I) or heterodimers (Type II) to their respective DNA response elements. Activation of gene expression requires coactivators and other protein factors that are recruited to the promoter-bound NR that serves as nucleation site for a large multiprotein complex [2]. Unliganded (inactive) NRs are complexed to corepressors such as the silencing mediator of retinoid and thyroid receptors (SMRT), which recruit histone deacetylases and chromatin remodeling proteins, rendering the promoter transcriptionally silent [3]. The coactivators, for example, the steroid receptor coactivator (SRC-1), bind to the NR via a nuclear receptor interaction domain (NRID), an amphipathic helix containing a conserved amino acid motif (LxxLL). The coactivator surface overlaps with the binding site for corepressors. However, inactive, antagonist-bound NRs show a different architecture of the NRID that prevents coactivator recruitment and promotes corepressor binding [4].

8.1.2
Molecular Architecture and Activation Mechanisms of Nuclear Hormone Receptors

NRs share a conserved structural and functional organization and are grouped in one superfamily (Figure 8.1). Altogether four distinct domains

Molecular Modeling. Basic Principles and Applications. 3rd Edition
H.-D. Höltje, W. Sippl, D. Rognan, and G. Folkers
Copyright © 2008 Wiley-VCH Verlag GmbH & Co. KGaA, Weinheim
ISBN: 978-3-527-31568-0

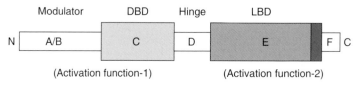

Figure 8.1   Schematic representation of the nuclear receptor architecture.

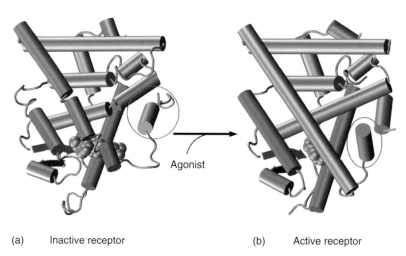

(a)   Inactive receptor   (b)   Active receptor

Figure 8.2   Ligand binding domain (LBD) of a nuclear hormone receptor (estrogen receptor α). (a) The inactive receptor is shown with a bound antagonist (4-hydroxytamoxifen, colored orange). (b) After binding an agonist (estradiol, colored orange) the AF-2 domain (green) is folded into an α-helix and forms the binding site for the co activator (magenta).

have been identified, which comprise the N-terminal region A/B, a conserved DNA binding domain (DBD, region C), a linker region D and a ligand binding domain (LBD) (region E) [5]. The 3D architecture of the DBD as well as the LBD is known from a variety of crystal structures that have been resolved over the last few years [6–8]. Owing to the size and properties of NRs, it was so far not possible to get crystal structures for the whole receptor proteins. From the X-ray structures, it is known that the LBD shows a conserved folding pattern consisting of 12–14 helices arranged in a three-layered helix sandwich and a β-sheet composed of 2–5 strands [9]. As an example, the LBD of the estrogen receptor (ERα) is shown in Figure 8.2. Ligands bind in a mainly hydrophobic binding pocket located between the outer layers of the helix sandwich. Size and shape of the ligand binding pockets vary dramatically between different NRs ranging from 220 Å^3 (ERR3) to 1300 Å^3 (PPARγ) but all are more or less hydrophobic [10]. Besides the ligand binding pocket, the LBD also carries the ligand-dependent activation function 2 (AF-2) on the C-terminal helix 12

(H12). The position and conformation of this helix, which is modulated by the binding of agonist and antagonist, is responsible for the binding of coactivator and corepressor, respectively. Agonist binding induces H12 to cover and seal the binding pocket. The emerging hydrophobic surface composed of helices H3, H4 and H12 enables coactivator binding via specific aliphatic amino acid residues. Binding of antagonists results in transformation of H12 into a disordered conformation disrupting the coactivator binding site, thus enabling corepressor recruitment (see Figure 8.2) [11]. The function or the endogenous ligands are yet unknown for some of the 48 NRs. Therefore, these NRs are designated as *orphan receptors* [12].

8.1.3
The Human Constitutive Active Androstan Receptor (CAR)

Constitutive adrostane receptor (CAR) was discovered in 1994 [13] and was originally found to exhibit an intrinsic basal activity in cell-based reporter assays. This was in contrast to other NRs known so far [14]. CAR is active in the constitutive state, where it is localized in the cytoplasm in a complex with accessory proteins, including heat shock protein 90 (HSP90) and constitutive adrostane receptor cytoplasmic retention protein (CCRP) (Figure 8.3). As an NR with intrinsic basal activity, the ligand-independent gene expression must be repressed in order to acquire responsiveness to activating compounds. Usually, classical NRs reside permanently at the nucleus, thus agonists or antagonists act directly through binding to the chromatin associated receptor. The classical mechanism of CAR activation involves direct binding of an agonist to the receptor's hydrophobic ligand binding pocket, which leads to nuclear translocation and heterodimerization of CAR with another nuclear receptor family member (i.e. the retinoid X receptor α (RXR)) (Figure 8.3) [15].

CAR belongs to the subfamily NR1I of the NR superfamily including the related vitamin D receptor (VDR) and pregnane X receptor (PXR). CAR is part of the metabolic defense in humans. In conjunction with the closely related PXR, CAR regulates the expression of metabolizing enzymes upon xenobiotic stress (e.g. the cytochromes *CYP2B* and *CYP3A*) [16]. For more information about the biochemical details, the reader is referred to the literature [15].

8.1.4
CAR Ligands

Only a few ligands of CAR are known so far. Assignment of a ligand to a certain category (i.e. agonists or antagonists) is often problematic. Depending on the cell line used for assays, some compounds can turn out to be agonists, antagonists or inverse agonists. The varying expression

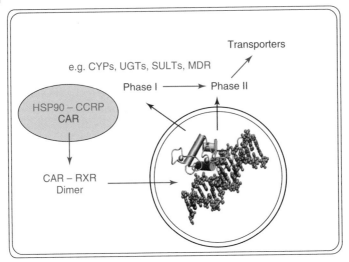

Figure 8.3 Schematic representation of the CAR system: CAR is active in the constitutive state, where it is localized in the cytoplasm in a complex with heat shock protein 90 (HSP90) and CAR cytoplasmic retention protein (CCRP). The activated CAR binds with RXR as dimer to DNA response elements where the complex is responsible for the expression of proteins involved in biotransformation (e.g. CYP, cytochrom P450; UGT, glucuronyltransferases; SULT, sulfotransferases; MDR, multidrug resistance protein).

pattern of NR coactivators in different cell lines may lead to differential response of NR modulators [17]. As an example, clotrimazole has been reported as agonist in distinct cell lines (HEK293 cells) [18], whereas in another cell line (CV-1) clotrimazole behaves as an inverse agonist [19]. Besides several steroid derivatives, some nonsteroid ligands are also known to activate human CAR, for example 6-(4-chlorophenyl)imidazo[2,1-b][1,3]thiazole-5-carbaldehyde O-(3,4-dichlorobenzyl)-oxime (CITCO) or tri-*p*-methylphenylphosphate (TMPP) [20].

To understand the interaction of known CAR ligands and to develop novel potent and selective ligands, the 3D structure of the receptor protein is a prerequisite. However, when the CAR project was initiated, no crystal structure of the receptor protein was available, and the molecular determinants for the CAR-ligand specificity remained obscure, partly for that reason.

8.2
Comparative Modeling of the Human Nuclear Hormone Receptor CAR

In terms of the present drug design project, it is important to note the following points:

(1) No X-ray structure of the receptor protein was available at the beginning of the project that could help to understand the constitutive activity of CAR. This means that comparative modeling was the only solution in order to get structural information. The modeling approach was closely connected with site-directed mutagenesis experiments in order to analyze the CAR activation mechanism on the one site and to check the reliability of the homology model on the other site.

(2) Only a few CAR ligands were known and therefore no detailed structure-activity relationships for CAR ligands could be established. The ligand recognition of CAR is completely different compared to the related PXR and VDR. This means that a generated CAR protein model should also be able to explain ligand binding in order to develop novel CAR ligands.

8.2.1
Choosing Appropriate Template Structures

The selection of suitable template structures is a critical step in the homology modeling process since the choice of a wrong or inappropriate template may result in an inadequate model. Therefore, the starting point of every modeling approach is the identification of proteins that qualify as feasible templates by both, a sequence related to that in the target as well as the availability of structural data (as already discussed in detail in Section 4.3). In case of CAR, sequences of the closely related VDR and PXR receptors (same NR subfamily), the crystal structures of which were available, show between 40 and 50% sequence identity within the LBD, respectively. To evaluate the homology modeling procedure using only a single template structure, homology models for PXR and VDR were generated and compared with the available PXR and VDR X-ray structures [21–25]. Thus, the VDR homology model was based on the PXR template structure, and vice versa. The sequence identity between PXR and VDR in the LBD is 37%, which is expected to be high enough to yield reliable models. However, sequence identity between template and target protein is only one criteria that has to be considered. In contrast to the common topology of NRs, VDR and PXR contain an additional domain inserted between helices H1 and H3. In case of the PXR, this domain consists of a helical segment and two β-strands. The insertion domain in the X-ray structure of the VDR contains two additional helices and an artificial loop segment. A further deviation from the common topology in PXR occurs in the region of helices H6 and H8. The generated homology models were refined in order to remove unfavorable steric contacts resulting from the homology modeling procedure.

The superimposition of the minimized VDR and PXR models and their corresponding crystal structures revealed large deviations, indicating that the strategy using one single template structure did not yield a reliable receptor model. In addition, molecular dynamics (MD) simulations were carried out for the derived PRX and VDR models. The simulations produced major discrepancies between the two starting structures of the models. Both models produced an unacceptably large deviation during the 5-nanosecond simulation time. From this, ligand binding pockets resulted in both models that would never be able to accommodate the known ligands and thus were worthless for any further docking or drug design procedure. Therefore, in the second step a CAR model based on two template structures was generated by manually selecting the coordinates from VDR and PXR. As we were interested in analyzing the constitutive activity, the CAR model was generated in the active state. For this reason, the X-ray structures of PXR and VDR in complex with an agonist were considered to model the CAR protein. The multiple sequence alignment of CAR with the related VDR and PXR receptors was carried out using the well-known program CLUSTALW [26] and is shown in Figure 8.4. The individual colors in the alignment indicate which part of the PXR and VDR structure was taken for the generation of the CAR model.

```
CAR   106      LSKEQEELIRTLLGAHTRHMGTMFEQFVQFRPP--------------------
PXR   142       GLTEEQRMMIRELMDAQMKTFDTTFSHFKNFRLPGV|REEAAKWSQVRKDLCSLK
VDR   120      LRPKLSEEQQRIIAILLDAHHKTYDPTYSDFCQFRPP--------------------


CAR   139      ----------AHLFIHHQP-LPTLAPVLPLVTHFADINTFMVLQVIKFTKDLPVFRSL
PXR   211      VSLQLRGEDGSVWNYKPPA-DSGGKEIFSLLPHMADMSTYMFKGIISFAKVISYFRDL
VDR   157      ----------VRVNDGGGS|VTLELSQLSMLPHLADLVSYSIQKVIGFAKMIPGFRDL


CAR   186      PIEDQISLLKGAAVEICHIVLNTTFCLQTQNFLCG--PLRYTIEDGARVGFQVEFLEL
PXR   268      PIEDQISLLKGAAFELCQLRFNTVFNAETGTWECG--RLSYCLEDTAG-GFQQLLLEP
VDR   255      TSEDQIVLLKSSAIEVIMLRSNESFTMDDMSWTCGNQDYKYRVSDVTKAGHSLELIEP


CAR   242      LFHFHGTLRKLQLQEPEYVLLAAMALFSPDRPGVTQRDEIDQLQEEMALTLQSYIKGQ
PXR   323      MLKFHYMLKKLQLHEEEYVLMQAISLFSPDRPGVLQHRVVDQLQEQFAITLKSYIECN
VDR   313      LIKFQVGLKKLNLHEEEHVLLMAICIVSPDRPGVQDAALIEAIQDRLSNTLQTYIRCR


CAR   300      QRRPRDRFLYAKLLGLLAELRSINEAYGYQIQHI---QGLS-AMMPLLQEICS
PXR   381      RPQPAHRFLFLKIMAMLTELRSINAQHTQRLLRI---QDIHPFATPLMQELFGI
VDR   371      HPPP|--LLYAKMIQKLADLRSLNEEHSKQYRCLSFQPECSMKLTPLVLEVFG
```

Figure 8.4 Sequence alignment of the two template structures PXR and VDR and the target sequence of CAR. Vertical lines indicate missing segments. Residues within the CAR sequence are colored depending on the origin of the structural information used (PXR: red, VDR: green).

8.2.2
Homology Modeling of the Human CAR

The CAR model was generated taking the two template structures mentioned earlier, and the Homology module within INSIGHT II [27]. Coordinates for most amino acids were borrowed from the PXR structure (see Figure 8.4 for details). To obtain the common NR fold, coordinates for H6 and H7 were taken from the VDR X-ray structure 1 DB1. Additionally, coordinates for helices H10 and H11 and the C-terminal H12 were adopted from the VDR template. Compared to PXR, VDR shares a significantly higher sequence identity with CAR within helices H10 and H11. Smaller amino acid side chains on the H12 of CAR are believed to enable closer attachment of H12 on the LBD in CAR than observed in PXR [28]. This was also found in the VDR structure, and thus coordinates for H12 were adopted from this receptor. For the H1–H3 region (29 amino acids) the application of loop search approaches or *de novo* construction methods failed to determine a reliable conformation. Thus the protein backbone of the H1–H3 region was completely adopted from the VDR receptor. This procedure was facilitated by the identical number of amino acids in the corresponding segment of VDR.

The program SCWRL was applied to model the side-chain conformations [29]. The prediction of the numerous side-chain conformations is, by far, a more complex problem than the prediction of the backbone conformation of a homologous protein. On the basis of a backbone-dependent rotamer library, SCWRL adds side chains to a protein backbone trying to avoid side chain–backbone and side chain–sidechain clashes. To incorporate side-chain conformations of conserved amino acids directly from the template, amino acids can be excluded from the assignment procedure. The accuracy of SCWRL for NRs was tested in advance on the available crystal structures of VDR and PXR. For both receptors, most side chains were correctly assigned. Thus SCWRL seemed to be an appropriate approach to assign the side chains for NRs. To enhance the quality of the side-chain assignment, conformations of conserved amino acids among NRs were directly adopted from the template structures.

NRs bind coactivator or corepressor proteins at particular binding epitopes that are needed for regulating the activation/inactivation process. In order to simulate the influence of coactivator or corepressor binding, another CAR model was generated. This CAR model included the SRC-1 coactivator fragment. To model the CAR–SRC-1 complex, the coordinates for the coactivator (amino acids 682–696) were completely adopted from the crystal structure of PXR bound to SRC-1 (PDB code: 1NRL) [30, 31].

8.2.3
Setting up the System for the Molecular Dynamics Simulations

As mentioned in Section 4.4, the refinement of models derived from comparative modeling studies is a must. Loop and side-chain conformations of the derived protein model represent only one possible conformation. In order to detect the energetically stable 3D-structures of a system, searching the conformational space is usually carried out by MD simulations.

Therefore, the individual CAR models were refined by minimization procedures and subsequent MD simulations using the GROMACS software package [32]. This program includes the GROMOS96 force field, which has been developed for the simulations of protein structures [33]. To mimic an aqueous environment, the models were placed in a solvent box. Water molecules were represented using the simplified SPC model [32]. Long-range electrostatic interactions were considered by applying the particle mesh Ewald method [34]. In this approach, the long-range electrostatic interactions observed in protein structures are considered in a more sophisticated way. An application of the particle mesh Ewald method for calculation of Coulomb interactions requires the system to be neutral; therefore, sodium ions were added to yield an uncharged system. Additionally, sodium and chlorine ions were added to yield physiological conditions, because an explicit consideration of ions has been shown to have beneficial effects on secondary structure stability.

The resulting minimized structures were then used as input for subsequent MD simulations. Periodic boundary conditions were applied to simulate a bulk fluid. The system was kept at constant temperature of 310 K using a Berendsen thermostat with a coupling time of 0.1 picosecond [35]. Further, constant pressure was maintained by coupling to an external bath. Bonds involving hydrogen atoms were constrained to their equilibrium length using the Lincs algorithm [36]. Equilibration runs with decreasing constraints on backbone atoms ($1000-100$ kJ mol^{-1}) were carried out, first following free MD simulations of 2.25 nanoseconds. The long simulation time was chosen to ensure that the receptor structures reached an equilibrated state.

8.3
Analysis of the Models that Emerged from MD Simulations

8.3.1
Atomic Fluctuations

The MD simulations of both CAR models (CAR and CAR–SRC1) showed that the overall fold of the NR remained stable. The rmsd (root mean square deviation) did not exceed a value of 2.5 Å within the backbone region (see Figure 8.5). The low deviation is the result of the compact architecture of the three-layered helix sandwich that allowed only limited motions of the individual domains. The AF-2 domain assumes the active conformation, forming a lid over the ligand binding cavity that is significantly smaller (630 Å^3 before and 480 Å^3 after MD simulation) than the binding cavity of the PXR receptor (1294 Å^3). As observed for PXR and VDR, the CAR-ligand binding pocket is

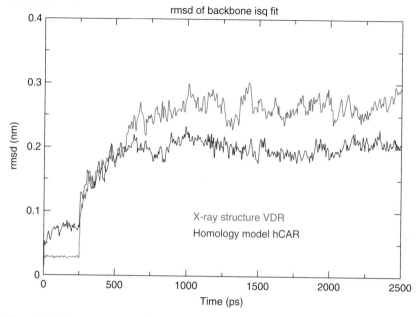

Figure 8.5 Root mean square deviation of backbone atoms of the CAR homology model (blue) and the VDR X-ray structure (green) during the MD simulation.

mainly composed of hydrophobic residues with only a few contributing polar residues.

In order to analyze the stereochemical quality of the simulated homology models, representative MD frames were selected. For this purpose, the resulting trajectories were clustered using the program NMRCLUST [37] by superimposition of all frames on backbone atoms and subsequently grouping them into clusters of similar conformations. For each cluster, a representative frame was selected and analyzed using the program PROCHECK [38]. The Ramachandran plots for CAR and CAR–SRC-1 models obtained with PROCHECK assessed 88.4 and 86.4%, respectively, of the $\phi - \psi$ torsion angles as being within the favorable region. These values were found to be in agreement with values from known protein crystal structures. In conclusion, the detailed structure validation showed that the models were in agreement with stereochemical parameters observed for high-resolution X-ray structures. However, it should be kept in mind that the stereochemical accuracy is by no means a sufficient criteria for a reliable homology model! It is very important to use further modeling methods and all experimental information that can be accessed concerning the target and related protein structures.

The analysis of the MD trajectories showed that in both models the AF-2 domain remained closely attached to the LBD during the entire simulation. The superimposition of the CAR and CAR–SRC-1 model after the MD simulation on the backbone atoms is shown in Figure 8.6. It must be stated that a large movement of the AF-2 domain cannot be expected at the time range of the chosen MD simulation (2.5 nanoseconds). However, the high stability of the LBD–AF-2 interaction was in close agreement with the known experimental data [39, 40]. To get an impression of the reliability of the CAR MD simulation, the available X-ray structures of PXR and VDR were analyzed using the same MD simulation setup as for CAR. The backbone regions of VDR and PXR showed comparable atomic fluctuations as observed for the CAR models (Figure 8.5), which can be regarded as further support for the chosen strategy.

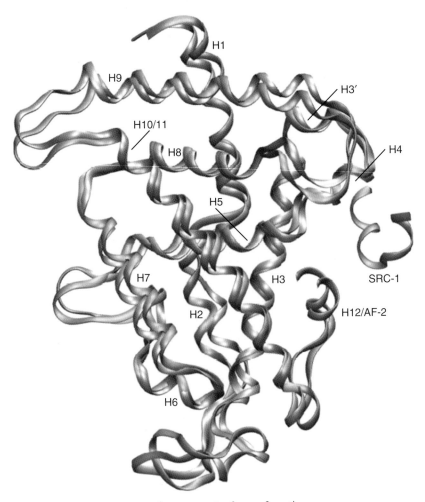

Figure 8.6 Superimposition of representative frames from the MD simulations of CAR and CAR–SRC-1 (color code: CAR: gray; CAR–SRC-1: orange). Structural differences between both models are observed mainly in the loop regions.

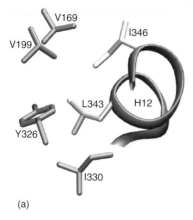

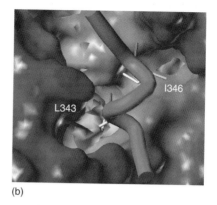

(a)

(b)

Figure 8.7 (a)Amino acid residues located at the AF-2–LBD interface. The AF-2 (H12) helix is shown as ribbon with amino acid whereas the amino acid residues of the ligand binding pocket are colored cyan. (b) The favorable hydrophobic interaction regions (colored yellow) of the AF-2 truncated LBD (calculated with the GRID methyl probe, contour level −2.5 kcal mol−1) are displayed. The MOLCAD surface of the LBD is shown, colored according to the cavity depth.

8.3.2
AF-2 Interaction Domain

The next step was a detailed analysis of the interaction of the AF-2 helix with the residues forming the binding pocket. In both CAR models, strong interactions between the AF-2 helix and the residues of the LBD were observed. Leu343 and Ile346 located at one site of the AF-2 helix were found to be in close contact with hydrophobic residues of the LBD (Val199, Tyr326, Ile330; see Figure 8.7). To rationalize the binding of the hydrophobic side chains from the AF-2 domain and the residues of the LBD, the optimum interaction site was computed. The calculation was carried out using the well-known program GRID (see Sections 2.5 and 4.6.2 for details). An AF-2-truncated CAR model was constructed and the GRID interaction fields were subsequently calculated for the AF-2 interaction site. To detect favorable hydrophobic and van der Waals (vdW) interactions, the hydrophobic DRY and the aliphatic methyl probes were selected. The obtained contour maps were then viewed superimposed on the structure of the CAR model (Figure 8.7). Two main favorable interaction regions for the methyl probe were detected at the LBD–AF-2 interface. The location and size of the calculated contour map was in perfect agreement with the position of the hydrophobic residues Leu343 and Ile346 located on the AF-2 helix. A closer inspection of the AF-2–LBD interaction domain showed that several aromatic residues of the LBD interact with the AF-2 residues. Tyr326, especially, is in close proximity to the aliphatic residues of the AF-2 domain. Tyr326 is surrounded by a cluster of aromatic or hydrophobic residues (Val199,

His203, Phe234, Phe238, Ile330) that seem to fix the location of Tyr326 side chain partly shown in Figure 8.7. Each of the described interactions between AF-2 and the LBD was also observed in the CAR–SRC-1 homology model. However, at the end of the 2.5-nanosecond simulation the AF-2 helix was positioned closer to the LBD as observed for the CAR model, without SRC-1 (Figure 8.6).

In order to validate homology models, experimental data such as site-directed mutagenesis data are very helpful. Therefore, we first analyzed the literature for available experimental data. It was reported that replacement of either Leu343 or Ile330 by Ala reduced the basal activity significantly [39]. This was in accordance with our observation that Leu343 is the only amino acid from AF-2 that permanently shows vdW interactions with the LBD in the MD simulation. Upon mutation to alanine, the contact to Tyr326 is disrupted and the remaining interactions between Ile346 and LBD are not sufficient to keep AF-2 anchored to the active conformation. Mutation of Ile330 to alanine not only reduced the hydrophobic surface area and the number of potential interaction partners for Leu343, but also destabilized the position of Tyr326, because Ile330 is one of the amino acids that restrain the side chain of Tyr326. This stabilization is reduced by mutation to alanine, resulting in an increased flexibility for Tyr326.

8.3.3
Deciphering the Structural Basis for Constitutive Activity of Human CAR

The AF-2 domain located at the C-terminal end is the key element for the activation of NRs including CAR. Agonists and antagonists induce conformational changes of AF-2 that subsequently results in formation of a complex with coactivator or corepressor proteins, respectively. In the case of an agonist-occupied binding site, AF-2 covers the ligand binding pocket like a lid. According to the generated homology model, AF-2 residues Leu342, Leu343, Ile346, Cys347 and Ser348 were found to contribute to the basal activity of CAR owing to their stabilizing interactions with helices H11, H5 and H4. In contrast to other NRs, CAR is constitutively active. Of several unique residues involved in these interactions, Tyr326 was found to be of special importance. A cluster of hydrophobic and aromatic residues around Tyr326 fix the side chain to enable vdW interactions with the AF-2 domain (Figure 8.7). Phe238, which is located in close proximity, was found to prevent rotation of the Tyr326 side chain, thus blocking the position of Tyr326. Val199, His203, Phe234 and Ile330 are positioned above and below the plane of the Tyr326 side chain, thereby also fixing the position. Assisted by its surrounding residues, Tyr326 emerged as a central interaction partner for AF-2 that keeps it closely attached to the LBD. This interaction pattern was thought to be unique among related NRs, thus providing a convincing explanation for the constitutive activity of CAR. To test the derived hypothesis, the Ala326 mutant of the human CAR was expressed and tested. As proposed, the mutated receptor had lost its basal activity (Figure 8.8) but could be activated by activators such as TMPP, which are able to directly interact with the AF-2 domain [41]. On the basis of these data, we suggested that vdW interactions between LBD and AF-2 are key elements for the constitutive activity of CAR. This hypothesis was further supported by analyzing the crystal structures of other NRs, that is, the murine liver receptor homolog 1 (LRH-1) and the human estrogen-related receptor 3 (ERR-3) [42]. Both receptors are constitutively active and show a comparable stabilization of the AF-2 helix by hydrophobic residues of the LBD, as observed for the CAR model. Also, in the agonist-bound structures of PXR and VDR, vdW interactions between the bound agonist and the AF-2 domain can be observed [23, 24]. These are believed to maintain AF-2 attached to the LBD, thus enabling coactivator binding. The CAR homology model showed that the single residue Tyr326 makes similar vdW contacts to AF-2. Tyr326 thereby accommodates a corresponding position relative to AF-2 as the agonists do in PXR and VDR structures (Figure 8.9) Therefore, we suggest that the constitutive activity of CAR results from a 'molecular mimicry' (i.e. Tyr326) of a bound agonist. In active NRs the AF-2, domain adopts a position that, together with residues of the LBD, allows the formation of a hydrophobic binding pocket for a coactivators. The AF-2 domain in CAR is able to form this hydrophobic groove even in the absence of any bound agonist.

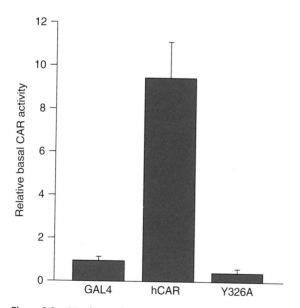

Figure 8.8 Site-directed mutagenesis results: relative activity obtained for the free medium (GAL), the CAR wild type and the Tyr326Ala mutant.

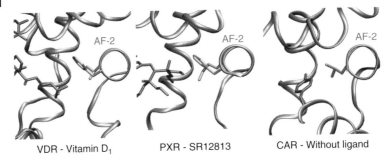

VDR - Vitamin D₁   PXR - SR12813   CAR - Without ligand

Figure 8.9 Molecular mimicry: Interaction between the helix 12/AF-2 (green) and the LBD domain (purple) for VDR (a), PXR (b) and CAR (c). Ligands for PXR (SR12813) and VDR (vitamin D3) as well as the corresponding amino acid Tyr326 in CAR are colored brown.

8.3.4
Coactivator Binding

Activation of NRs requires the binding of coactivators such as SRC-1 in case of CAR. The known crystallographic and experimental data revealed that NRs possess specific interaction patterns for the binding of coactivators [43] that seem to be essential for their function. Thus, reproducing the interactions between LBD and coactivator was thought to be prerequisite for a reliable homology model. In the known NR crystal structures, the interaction domain of the coactivator shows an α-helical segment containing the conserved LxxLL motif. This motif interacts with residues located in a hydrophobic pocket in the region of the AF-2 domain. A closer look at the CAR–SRC-1 model reveals that this hydrophobic groove is formed by 11 amino acid residues. The electric dipole of the interacting helix of the SRC-1 protein is known to be stabilized by two conserved amino acids interacting with its N- and C-terminal residues that form a 'charge clamp'. This conserved interaction pattern was also observed in the CAR–SRC-1 model where these residues are represented by Lys177 (H3) and Glu345 (AF-2) (Figure 8.10). Lys177 forms a hydrogen bond with the backbone carbonyl group on SRC-1, whereas Glu345 interacts with the backbone amide groups of Ile689 and Leu690. In order to analyze the stability of the receptor–coactivator interaction, we decided to carry out MD simulations on the CAR–SRC-1 complex using the same setup as described in Section 8.2.3. The SRC-1 peptide that was used to simulate the coactivator binding remained bound throughout the simulation time of 5 nanoseconds. All hydrogen bonds that were detected in the homology model were still present at the end of the MD simulation, indicating a highly stable complex between SRC-1 and the active CAR. Thus, the homology model was able to deliver a structural explanation why SRC-1 is strongly bound by CAR in its active form.

In conclusion, the important lesson to learn was that the automatically derived homology models using a single template resulted in unacceptable

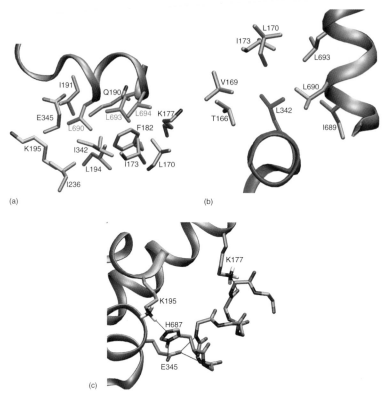

Figure 8.10 Interactions between LBD and SRC-1: (a) Several residues from H3, H3', H4 and AF-2 (carbon atoms in orange) form a hydrophobic groove to which SRC-1 (cyan) can bind. Leucines from the LxxLL motif on SRC-1 are shown explicitly. (b) Binding of SRC-1 to the LBD fixes Leu342 from AF-2 (green) in a hydrophobic pocket formed by several amino acids from LBD (orange) and SRC-1 (cyan). (c) The helix dipole of SRC-1 is stabilized by Lys177 (H3) and Glu345 (AF-2) located on the LBD (orange), forming the so-called 'charge clamp'. Lys177 interacts via a hydrogen bond with the backbone carbonyl group of Leu693, whereas Glu345 interacts with the backbone nitrogen atoms of Ile689 and Leu690.

receptor models, whereas the manually generated model based on two templates yielded a reliable model in accordance with the known biochemical data.

8.4
Analysis of CAR Mutants

8.4.1
Identifying Important Amino Acids for CAR Activation

On the basis of the CAR homology model (and the MD simulations), the molecular determinants for the ligand specificity, as well as the importance of amino acid residues forming the ligand binding pocket, should be analyzed in the next step. In order to select the critical residues for a mutation analysis, the CAR–SRC-1 homology model was investigated for residues lining the binding pocket and the AF-2 interface. In this way, 22 amino acids were manually identified, for which alanine scanning mutagenesis experiments were carried out. The activities and ligand specificities of these mutants were tested in a mammalian activity assay. In addition, the yeast-two-hybrid assay was applied to study the interactions between the mutated LBDs and the steroid receptor coactivator-1 (SRC-1) (for methodological details on the biochemical experiments, see the literature) [41].

First, the basal activities of the 22 CAR mutants were analyzed and compared with the wild-type data. The generated CAR–SRC-1 model suggested that even without a bound ligand, helix H12 (AF-2) adopts the active conformation owing to hydrophobic interactions of Leu343 and Ile346 from AF-2 with residues Val199, Tyr326 and Ile330 from the LBD. In addition, a hydrogen bond that stabilized the aromatic side chain in its position was observed between Asn165 and Tyr326. The central role of residues Asn165, Val199, Tyr326, Ile330 and Leu343 for basal activity was confirmed by a dramatic decrease in activity and SRC-1 interaction by their mutation to alanine (see Figures 8.11 and 8.12).

Interestingly, the basal activity (white bars in Figure 8.11) of 16 mutants was decreased by more than 90% compared to the wild-type value. The interpretation was that mutation of aromatic (Phe161, Phe234, Phe238, Tyr326) or hydrophobic (Cys202, Ile164, Met168, Ile330, Ile333, Met339) residues that protrude into the ligand binding pocket to alanine, creates more space and reorganizes the surrounding binding pocket residues. This reorganization decreases the basal activity in most cases via the central residue Tyr326, the position of which is crucial for the stabilization of helix H12 in active conformation. Three mutants (Phe129Ala, Phe217Ala and Tyr224Ala) also had low basal activity and they could not be activated in mammalian cells, or could not elicit a SRC-1 response in yeast like the other 19 mutants did. Residues Phe217 and Tyr224 form a wall in the ligand binding pocket (Figure 8.12), and their change to alanine is likely to disrupt protein folding locally. Finally, the remaining six mutants (His203Ala, Leu206Ala, Thr209Ala, Leu242Ala, Phe243Ala and Gln329Ala) retained 30–80% of the wild-type activity. These residues were located on the top and back of the binding pocket, (Figure 8.12) away from the central residue Tyr326. Therefore, their larger distance from Tyr326 was consistent with the finding that their mutations did not profoundly attenuate the basal activity of CAR.

Taking all these complex experimental data together, a visual analysis of the model indicated that the presence of extensive contacts between helices H11 and H12 may be fundamental to hold CAR in the active state. The observed interactions contributed to helix H12 (AF-2) acquiring the active position, because mutation of each residue to alanine caused marked decreases in the basal activity. Thus, we were able to explain the activation profile for all generated receptor mutants on a structural level.

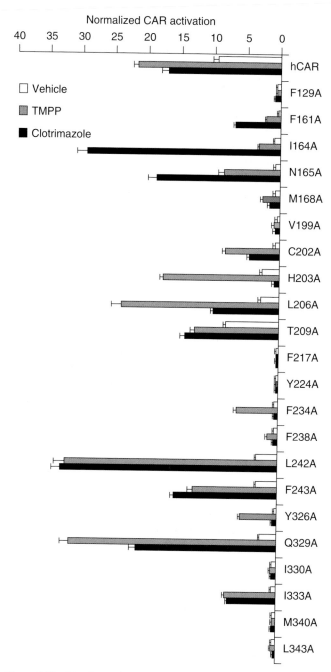

Figure 8.11 Site-directed mutagenesis data for human CAR. The reporter activities were measured in HEK293 cells for activators and normalized to hCAR without activator (set at 1.0).

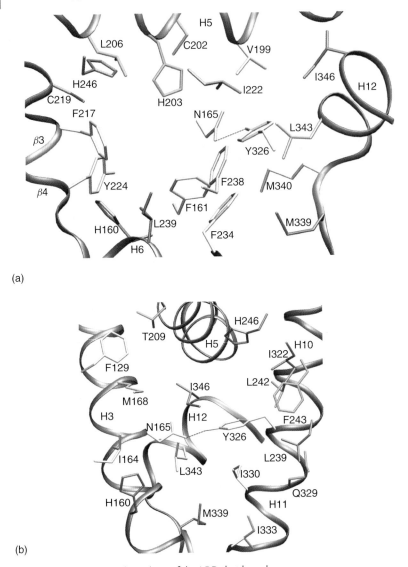

Figure 8.12 Amino acid residues of the LBD that have been mutated are shown in black, other residues are kept gray. The hydrogen bond between N165 and Y326 is indicated. (a) Side view of the binding pocket. (b) Rotation about 90°.

8.4.2
MD Simulations of Selected CAR Mutants

In order to prove the consistency between the experimental and theoretical data, a few receptor mutants were modeled and analyzed by means of MD simulation. As this is more time consuming than simple visual inspection, we focused on receptor mutants, where the mutated amino acid was not directly in contact with the AF-2 domain. For most of the CAR mutants, it was obvious from a visual inspection of the homology model why the mutated receptor showed a dramatic decrease of constitutive activity (i.e. for amino acids that maintain the interaction between the LBD and the AF-2 domain), or why practically no reduction was observed (i.e. for amino acids located on the top and back of the binding pocket far away from Tyr326). Two mutants (Phe234Ala and Phe238Ala) where the modified amino acid residues are part of the LBD were considered [44]. During the simulation of the two CAR mutants – for which the same setup was used as for the other CAR models – the Phe238Ala mutation provoked a rotation of the side chain of Tyr326, resulting in a more buried conformation of Tyr326 (Figure 8.13). As a result, the interactions of Tyr326 with Leu343 and Ile346 were disrupted. The hydrogen bond between Asn165 and Tyr326 was also lost. The mutation Phe234Ala forced H7 toward the β-sheet by more than 3 Å, and as a result helices H10/11 were pushed in the same direction. These movements resulted in a pronounced reorientation of several residues within the binding pocket (Figure 8.13). Tyr326 now pointed more deeply in the ligand binding cavity and the interaction with AF-2 was partly lost. In addition the conformation of Phe238 changed, resulting in loss of the stabilizing effects on Tyr326.

The CAR activity can also be modulated by residues more distantly located from the LBD–AF-2 interface: the mutation Phe234Ala resulted in a modification of the overall shape of the ligand binding pocket that caused a displacement of helix H10/11 and a subsequent reorientation of Tyr326 (Figure 8.13). Thus, the experimentally observed loss of basal activity of these

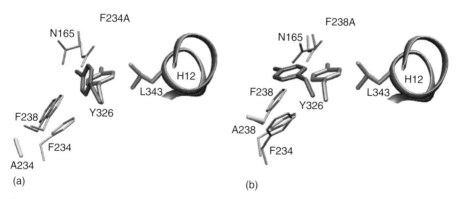

(a) (b)

Figure 8.13 (a) Mutation of Phe234 to alanine (yellow) results in side-chain reorientation of several residues. Because of the displacement of H10/11, Tyr326 now points into the ligand binding pocket (mutated receptor is colored magenta).

(b) Mutation of Phe238 to alanine (yellow): During the MD simulations Tyr326 changed its position pointing now into the ligand binding pocket (mutated receptor is colored magenta). The contact between Tyr326 and AF-2 is lost.

two mutants can be explained by the missing stabilizing interactions between Tyr326 and the AF-2 domain.

8.5
Modeling of CAR-Ligand Complexes

In order to analyze the further activation of CAR upon agonist binding, a ligand docking study was carried out using the CAR homology model. The CAR–SRC-1 structure as described in Section 8.2.1 and the well-known docking program GOLD (Cambridge Crystallographic Data Centre, Cambridge, UK), which is based on a genetic search algorithm, were selected. The implemented scoring function GoldScore was chosen and for each ligand 30 different docking poses were calculated. First, we decided to analyze the two known CAR activators, clotrimazole and TMPP. The challenge for any docking program in case of NRs is the mainly hydrophobic ligand binding pocket including only a few polar residues able to form directed interactions. The obtained docking results were analyzed by visual inspection and grouped into clusters of similar conformations according to their rmsd values. For both activators, two clusters were obtained that differed only slightly from each other. Clotrimazole was found to bind deeply in the ligand binding pocket (Figure 8.14). Interestingly, no direct contact between clotrimazole and the AF-2 helix was observed. The aromatic side chains of the ligand mainly interact with aromatic residues of the binding pocket (Phe112, Phe161, Phe234 and Tyr326). In order to evaluate whether the predicted receptor–ligand complexes are energetically feasible, the two CAR activator complexes were subjected to MD simulations. During the simulation time of 2.5 nanoseconds, the position of clotrimazole did not change significantly (rmsd values below 1.5 Å). Only Phe161 moved slightly toward the interface between LBD and AF-2, establishing vdW interactions with Met339 from the AF-2 domain.

Tri-*p*-methylphenylphosphate showed a different binding mode (Figure 8.15). It is located much closer to the interface between LBD and AF-2 compared to clotrimazole. One of the methylphenyl groups directly interacts with Leu343 and Ile346 from AF-2. The two remaining methylphenyl groups point into the ligand binding pocket, both interacting with Phe161 and pushing it deeper into the pocket. As a result, the distance to Met339 was increased compared to the ligand-free receptor. Additional vdW interactions between the methylphenyl groups and the two aromatic residues Phe234 and Tyr326 could be observed. The phosphate group formed strong hydrogen bonds to both Asn165 and Tyr326 that remained stable during the MD simulation. In agreement with the simulation of clotrimazole, the distance between Tyr326 and Leu343 (AF-2) was also decreased in the simulation of TMPP.

In conclusion, both investigated agonists interact with amino acids surrounding Tyr326 leading to a stabilization of this side chain. As a result, the distance between Tyr326 and Leu343 of AF-2 is decreased in the MD

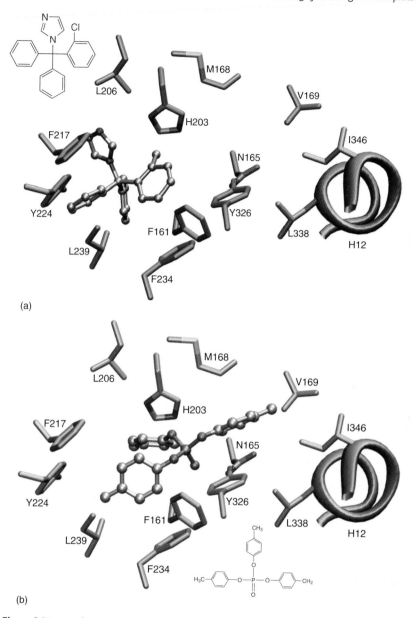

(a)

(b)

Figure 8.14 Binding modes of CAR activators (a) clotrimazole (carbon atoms in gray) is positioned deeply in the LBD without any contact to the AF-2 helix (orange). Main interactions are observed with aromatic amino acids (Phe112, Phe161, Phe234 and Tyr326). (b) TMPP (carbon atoms in gray) is located close to the AF-2—LBD interface establishing vdW interactions with Leu343 and Ile346 from AF-2. Additionally, hydrogen bonds with residues Asn165 and Tyr326 are observed.

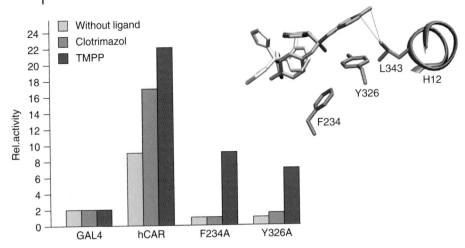

Figure 8.15 Effect of activators on human CAR mutants. TMPP (colored magenta) is able to activate both the Tyr326Ala as well as the Phe234Ala CAR mutant because it is able to directly interact with the AF-2 H12 helix.

simulations compared to the empty receptor. We suggest that the induced stabilization of the LBD–AF-2 interaction facilitates SRC-1 binding. Although the adopted binding mode of clotrimazole and TMPP is quite different, both ligands induced comparable structural changes at the AF-2/LBD interface.

The function of CAR as a xenosensor requires recognition of a diverse set of ligands. Thus, the ability of the binding site to adapt to a variety of ligands is essential. Upon agonist binding, the ligand binding pocket is able to expand up to 80%. Increase in the size of the cavity has been also reported for the related PXR complexed with hyperforin [23–25]. In the MD simulations of CAR, two regions of moderate flexibility were detected upon ligand binding, which are responsible for the flexibility of the binding pocket. In contrast to PXR, structural adaptions took place within parts of the β-sheet and residues located at the interface LBD–AF-2. The smaller ligand spectrum of CAR, compared to PXR, might thus be due to the significantly smaller ligand binding pocket and the limited flexibility of regions located therein.

8.6
The CAR X-ray Structure Comes into Play

8.6.1
How Accurate is the Generated CAR Model?

When working with homology models, the question that often arises is, how reliable is a theoretical model and can it be used for drug design studies?

Luckily, during the course of the work, the X-ray structures of human (PDB code: 1XV9, 1XVP) and mouse CAR (PDB code: 1XLS, 1XNX) in complex with two agonists were reported [45, 46]. The available 3D data for human CAR enabled us to evaluate the quality of our homology model and the proposed activation mechanism.

To investigate how well the modeled structure matches the X-ray data, the CAR homology model and crystal structures were superimposed on their backbone atoms [47] (Figure 8.16). The overall arrangement of helices and loops in the model was found to be in excellent agreement with the corresponding elements in the X-ray structures. Solely, helices H3 and H10/11 showed an additional turn in the homology model. The most striking difference was observed in the region connecting H2 and H3. The X-ray structures possess an additional helix (H2') here, whereas our CAR model shows a flexible loop. Also the orientation of the H2–H3 loop was found to be different compared to the corresponding element in the X-ray structures.

To assess the model accuracy, the rmsd values between superimposed model and crystallographic structures were calculated. The values of rmsd for the backbone atoms were found to vary between 3.4 and 3.8 Å (for the individual chains of the X-ray structures), suggesting a suboptimal modeling quality. However, from a visual inspection, a good overall agreement of secondary structural elements of the homology model and the X-ray structures is observed. In fact, high rmsd values originated mainly from large deviations

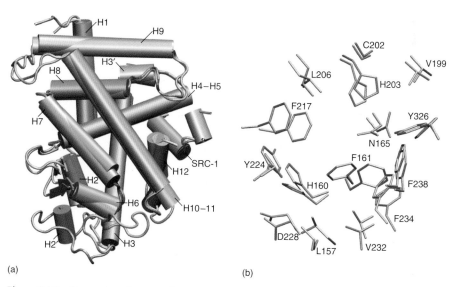

(a)

(b)

Figure 8.16 Comparison between the docking poses of CITCO (a) and 5β-pregnanedione (b) and the positions observed in the X-ray structures. Carbon atoms of the cocrystallized ligands are colored white. Docking poses are shown in pink. Helix 12 (AF-2) is shown as green ribbon.

in the H2–H3 region. Excluding this segment (amino acids 139–153) from the measurement, the rmsd dropped significantly to 1.8 Å. Since the H2'–H3 loop is located at the protein surface and the residues located therein, neither contribute to the formation of the binding pocket nor to the dimerization interface, it was suggested that the false prediction in this region has no major influence on the general reliability of the CAR model. This was also supported by the closely related VDR: it contains a significantly larger H2–H3 region that does not affect the main receptor function [48].

To further evaluate the accuracy of the CAR model, the side-chain orientations of the amino acids forming the binding pocket in the model and the X-ray structures were compared. For this purpose, 29 amino acids contributing to the accessible surface of the ligand binding pocket were analyzed. CAR model and X-ray structures were superimposed and the rmsd was calculated for all heavy atoms (between 1.5 and 1.9 Å, depending on the chosen monomer of the X-ray structure). Figure 8.16 shows the superimposed binding pockets of the homology model and the X-ray structure 1XVP, illustrating the high accuracy of the CAR model. For comparison, the rmsd values for the individual chains of the CAR X-ray structures lie in the range between 0.4 and 0.6 Å.

From the modeling results yielded, it can be stated that CAR models based on a single template (i.e. PXR) generally resulted in an overall unfavorable architecture owing to the structural deviations of PXR from the common NR topology. Models based on two templates in which the VDR X-ray structure is used to model the problematic regions yielded more reliable structures. An additional, important feature of the CAR model was the high accuracy of the side-chain orientations of the residues forming the binding pocket. Especially, if homology models are used for docking studies or structure-based drug design, highly accurate binding pockets are a prerequisite to obtain reliable results.

8.6.2
Docking Studies Using the CAR X-ray Structure

In addition to the already carried out docking of clotrimazole and TMPP, it was tested whether a docking program can correctly reproduce the binding mode of the two cocrystallized agonists 5β-pregnane-dione and CITCO. The conformation obtained by the docking (using the same setup as for clotrimazole and TMPP, see Section 8.5) was found to be close to the experimentally observed position indicated by an rmsd of 1.48 Å for CITCO and 0.65 Å for 5β-pregnanedione, respectively. The position of CITCO docked in the homology model was almost identical to that in the crystal structures showing an rmsd of 1.97 Å (Figure 8.17a).

The experimentally determined position of 5β-pregnanedione could not be correctly reproduced using the CAR homology model (rmsd 4.0 Å). This was

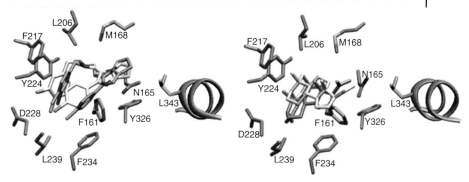

Figure 8.17 Structural features observed in the CAR X-ray structure which were postulated to be essential for maintaining the constitutive activity: the hydrophobic barrier (marked cyan), the 'helix X' (marked green), and the charge clamp (marked red). The AF-2 helix is displayed as green ribbon.

due to a different side-chain conformation of Tyr224 in the homology model as compared to the crystal structure. The conformation in the model prevented a correct placement of the ligand. To analyze whether the presence of β-pregnanedione would change the Tyr224 conformation, an MD simulation (2.25 nanoseconds) of the CAR model in complex with the docked ligand was carried out. The simulation showed a movement of Tyr224, thus leading to a orientation of 5β-pregnanedione that is consistent with the crystal structure (rmsd 2.0 Å) (Figure 8.17b). Reorientation of side chains upon ligand binding has also been observed during the MD simulation of the CAR model complexed with either clotrimazole or TMPP. Two amino acid residues of the LBD emerged as highly flexible: Phe161 as well as Tyr224 were found to adopt different conformations during the simulation. This conformational flexibility of side chains facing the binding pocket seems to be responsible for the adaption to structurally diverse ligands.

8.6.3
The Basis for Constitutive Activity Revisited

On the basis of the CAR crystal structures, a hypothesis for the structural basis of the constitutive activity has been deduced by the authors [45]. They state that CAR basal activity is mainly achieved by three structural features (Figure 8.18): first, a hydrophobic barrier formed by residues F161, N165, Phe234 and Tyr326 is interacting with helix H12 keeping it in the active conformation. Second, the missing C-terminal extension of H12 – observed in other NRs – is considered to allow the formation of a salt bridge between K195 on helix 5 and the C-terminal free carboxylate further stabilizing the active position of H12. Third, an additional helix (termed as *helix X*) located

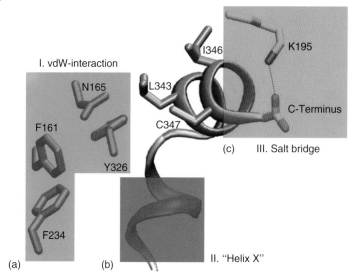

Figure 8.18 Left: Superimposition of CAR–SRC-1 X-ray structure (1XVP.pdb blue) and the CAR–SRC-1 homology model (green) on backbone atoms (α-helices are depicted by cylinders, β-sheets by arrows, random coil by tube). Right: Comparison of side-chain orientations of residues belonging to the ligand binding pocket (same color scheme).

between the helices H11 and H12 is supposed to orient H12 in its active position and additionally interact with the hydrophobic barrier.

The homology model was first analyzed to see it is in agreement with the features stated for the X-ray structure. Furthermore, MD simulations of the CAR X-ray structures were used to evaluate the structural features proposed to contribute to the CAR basal activity. A hydrophobic barrier could also be observed in the homology model including Tyr326 as central feature. Additionally, other vdW interactions between LBD and helix H12 were observed in the model involving the amino acids Val199, Ile330 and Ile346. Further, amino acids that contribute to the interaction of the LBD with H12 can be observed in the X-ray structures as well in the homology model. Thus, the postulated hydrophobic barrier was consistent with the observation in the CAR model that several aromatic and hydrophobic residues (especially, Tyr326) maintain the interaction with the AF-2 helix. For most of these residues, the influence on constitutive activity was already confirmed by the site-directed mutagenesis data [39, 41].

The salt bridge between Lys195 and Ser348 detected in the CAR X-ray structure (Figure 8.18) could not be detected in the homology model. Instead, Lys195 made a hydrogen bond to His687 located on the SRC-1 peptide, whereas the C terminus (Ser348) interacts with Gln331 on H11. The importance of Gln331 for the constitutive activity was already demonstrated by site-directed mutagenesis [41]. Mutation of Gln331 to alanine resulted in a decrease of the basal activity by about 70%. To elucidate the stability of the salt bridge in the CAR X-ray structure, MD simulations (5.25 nanoseconds) of the CAR–SRC-1 crystal structure were carried out. During the simulation, the salt bridge between Lys195 and the C terminus (Ser348) was found to be instable. A generated distance plot clearly indicated an early separation of the two amino acids. Instead, similar to the CAR–SRC-1 homology model, Lys195 is involved in a hydrogen bond with His687 located on the SRC-1 peptide (Figure 8.10c). These findings together with the site-directed mutagenesis data supported the hypothesis that the salt bridge between Lys195 and Ser348 is an artifact of the crystallization procedure and not an essential feature maintaining the constitutive activity of CAR.

Since CAR has been crystallized only in complex with agonists and not in unliganded form, it was not clear whether the so-named *helix X* is an essential feature for maintaining the basal activity, or rather formed upon agonist binding. However, visual inspection of a variety of other NR X-ray structures revealed that the 'helix X' motif can also be observed among other receptors. Altogether 26 X-ray structures in the Protein Databank showed a comparable helical element (Table 8.1). In the case of VDR, which shows no basal activity, all crystal structures complexed with an agonist possess this 'helix X'. The same holds true for the retinoid acid-related orphan receptor α and β (RORα and $-\beta$) (Figure 8.19). Another important observation was the fact that the thyroid receptor (TR) shows variability in the H11–H12 region. Among the 13 X-ray structures inspected, seven showed a 'helix X'-like element whereas six do not possess an helical segment. Another example for the observation of the 'helix X' is the glucocorticoid receptor (GR, 1P93). When GR is cocrystallized with dexamethasone, two of the four monomers in the PDB structure show an 'helix X', whereas the others do not!

In conclusion, these observations stressed the question whether the occurrence of a single turn helix (i.e. helix X) can serve as an explanation for the basal activity of CAR. According to the common annotation of helices in nuclear receptors, we suggested that 'helix X' should be rather termed helix 11' as it has been done at first in case of RORβ.

Table 8.1 Occurrence of the 'helix X' motif in NR X-ray structures (PDB code).

CAR	VDR	TRα/β	GR	RORα/β
1XV9, 1XVP, 1XLS	1DB1, 1IE8, 1IE9, 1RJK, 1RK3, 1RKG, 1RKH, 1S19, 1TXI	1N46, 1NAV, 1NAX, 1NQ0, 1NQ2, 1R6G, 1Y0X	1M2Z, 1P93	1K4W, 1N4H, 1N83, 1NQ7, 1S0X

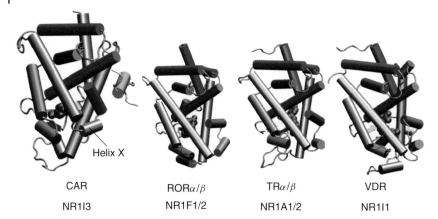

CAR   RORα/β   TRα/β   VDR

NR1I3   NR1F1/2   NR1A1/2   NR1I1

Figure 8.19 Occurrence of the 'helix X' motif in NRs. Besides CAR (displayed PDB entry 1XVP), all VDR X-ray structures crystallized so far contain this helical element (displayed PDB entry 1 DB1). 'Helix X' is colored green and the cocrystallized vitamin D is shown in magenta. Several thyroid hormone receptors (TR α and β) show a 'helix X'-like structure (displayed PDB entry 1N46 with a complexed thyromimetic compound) as also the RORα and β receptors (RORβ is displayed with bound stearate, PDB entry 1K4W).

The important lesson to learn from the analysis of the CAR X-ray structure and other NR structures was that the structural hypothesis for the constitutive activity that we postulated on the basis of our CAR model was verified. In addition, the examination of all available NR X-ray structures showed that a 'helix X'-corresponding element can also be observed in receptors possessing no constitutive activity. The observed structural data did not support the interpretation of 'helix X' as an essential structural feature for CAR basal activity. However, since the crystal structure of ligand-free CAR has not been obtained, future studies will clarify whether helix 11' ('helix X') is a permanent feature or – as we predicted – is rather formed upon agonist binding.

8.7
Virtual Screening for Novel CAR Activators

Constitutive adrostane receptor activators are regarded as potential therapeutics in cholestasis or neonatal jaundice. Currently, only a few nonsteroid, druglike CAR activators (e.g. CITCO) are known, of which some have been already applied or are discussed as potential therapeutics [49]. In order to find new and structurally different human CAR agonists that might be used as new lead compounds and as pharmacological tools to further analyze CAR, a virtual screening experiment was carried out using the CAR homology model and the human X-ray structure. A variety of docking programs and docking parameters were tested and evaluated on known NR X-ray structures to get an impression of the docking accuracy and the enrichment of real active ligands. Such a retrospective analysis on related targets gives the kind of objective measure that can be expected from a virtual screening experiment (for details see Chapter 7). We observed sufficiently high accuracy and enrichment for the analyzed NRs, using the GOLD program and the modified GoldScore as scoring function.

At first a database search was conducted using the LeadQuest database (Tripos Inc., St. Louis, USA) containing about 85 000 molecules. Compounds violating Lipinski's 'rule of five' and steroid derivatives were excluded. A search query, consisting of the Conolly surface of the binding pocket and two sites of hydrophobic character, was defined within the UNITY module of SYBYL (Figure 8.20). The tolerance of the pocket surface was set to 0.5 Å. The selection of the hydrophobic features with a tolerance of 1.0 Å was based on both observed vdW interactions between the CAR homology model and the docked ligands clotrimazole and TMPP, as well as favorable sites of lipophilic interactions within the ligand binding pocket that were evaluated by calculating the lipophilic potential [50]. Active compounds were expected to fit into the features of the search query, whereas inactive molecules would not (Figure 8.20). Altogether 9700 compounds (representing about 11% of the whole database) were found to fit into the spatial environment of the search query, on the basis of the CAR homology model.

The compounds that emerged from the 3D database search were subsequently subjected to a molecular docking procedure carried out using GOLD (CCDC, Cambridge, UK). GoldScore was selected as the scoring function. In order to take the coherence of the GoldScore and the molecular weight into account, the GoldScore value was corrected by dividing the original score over the square root of the number of heavy atoms. The 9700 compounds were docked into the CAR model and X-ray structure. The top-ranked docking

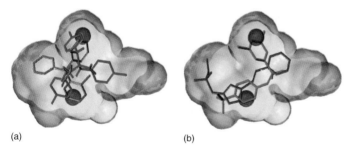

(a)   (b)

Figure 8.20 UNITY search query (a) defined by the pocket volume (yellow) and two hydrophobic features (blue spheres) placed on the basis of the docked ligands Clotrimazole (magenta) and TMPP (green). (b) Conformation of an active (cyan) and inactive (violet) compound within the binding pocket.

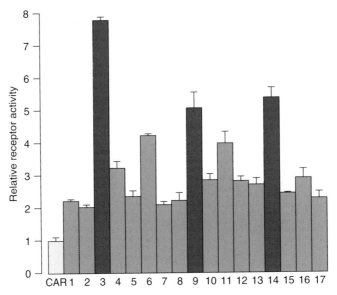

Figure 8.21 Ligand-induced CAR activation. Relative raise in receptor activity upon addition of the indicated agonists. The 17 found virtual screening hits activating CAR by at least twofold are presented. Compounds activating CAR more than fivefold are colored in dark gray. The basal activity of CAR is shown by the white column. The data are expressed as mean $\pm$ SD of at least three independent experiments.

solutions obtained by both strategies were visually analyzed and compared with the docking results for the known activators. In this manner, 66 compounds were selected for further experimental investigation, of which 30 were obtained from the suppliers. In the selected biological test system, compounds activating CAR by at least 100% above the basal activity rate were considered as agonists by the molecular biologist. According to this criterion, 17 compounds were identified as CAR agonists corresponding to a success rate of 57% (Figure 8.21). Most of the compounds that were found, activated CAR by two to fourfold, whereas three compounds were found to induce CAR activation between five and eightfold. The molecules found could be grouped into three chemical families: *N*-substituted and *N, N*-disubstituted sulfonamides, as well as thiazolidin-4-one derivatives that represented completely novel structural scaffolds for CAR activators. The docking poses for the detected hits showed different binding modes compared to the known activators, but were found to be in agreement with the calculated GRID interaction fields for the binding pocket. As an example, the docking pose of one potent sulfonamide is shown in Figure 8.22. The detailed analysis of the docking results would be out of the scope of this book.

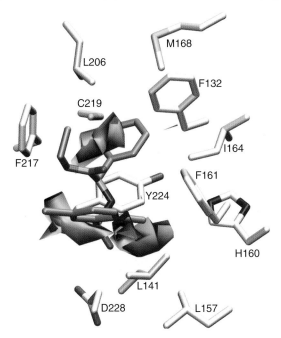

Figure 8.22 Interaction of the most potent sulfonamide derivative (colored cyan) with the residues of the binding pocket. Favorable GRID interaction field for an aromatic probe (shown as yellow contour map, energy level -2.5 kcal mol^{-1}) is in good agreement with the location of the hydrophobic parts of the docked ligand.

8.8
Concluding Remarks

While everyone is aware of the fact that models do not represent reality and that their predictivity is often overestimated, the modeling procedure presented for CAR has two facets of general importance that are worth mentioning.

It has been shown that a careful and critical integration of theoretical and experimental data gave rise to new mechanistic findings. The carefully generated CAR homology model, together with the site-directed mutagenesis study, led to the establishment of the first structural hypothesis for the constitutive activity of human CAR. The detailed comparison between experimental data and the CAR model further supported the reliability of the derived homology model. As we were in the lucky situation that the CAR X-ray structure was published during the course of the work, the theoretical results could be carefully analyzed and compared with the crystallographic data.

The second facet is that the present approach has yielded structurally novel CAR activators. Sulfonamide and thiazolidinone derivatives that were found by the virtual screening are the most potent CAR activators known so far and represent interesting tools for the further pharmacological characterization of CAR. Also here, the theoretical studies provided the basis of working hypotheses which, together with the biological testing, led to novel, interesting and active CAR ligands. The experience that was gained throughout the course of this work will be helpful for future work on the CAR project.

References

1. Mangelsdorf, D.J., Thummel, C., Beato, M. *et al.* (1995) The nuclear receptor superfamily: the second decade. *Cell*, **83**, 835–39.
2. Acevedo, M.L. and Kraus, W.L. (2004) Transcriptional activation by nuclear receptors. *Essays in Biochemistry*, **40**, 73–88.
3. Kraus, W.L. and Wong, J. (2002) Nuclear receptor-dependent transcription with chromatin. Is it all about enzymes? *European Journal of Biochemistry*, **269**, 2275–83.
4. Steinmetz, A.C., Renaud, J.P., and Moras, D. (2001) Binding of ligands and activation of transcription by nuclear receptors. *Annual Review of Biophysics and Biomolecular Structure*, **30**, 329–59.
5. Robinson-Rechavi, M., Carpentier, A., Duffraise, M., and Laudet, V. (2002) How many nuclear hormone receptors are there in the human genome? *Trends in Genetics*, **17**, 554–56.
6. Hard, T., Kellenbach, E., Boelens, R. *et al.* (1990) Solution structure of the glucocorticoid receptor DNA-binding domain. *Science*, **249**, 157–60.
7. Luisi, B.F., Xu, W.X., Otwinowski, Z. *et al.* (1991) Crystallographic analysis of the interaction of the glucocorticoid receptor with DNA. *Nature*, **352**, 497–505.
8. Bourguet, W., Ruff, M., Chambon, P. *et al.* (1995) Crystal structure of the ligand-binding domain of the human nuclear receptor RXR. *Nature*, **375**, 377–82.
9. Wurtz, J.M., Bourguet, W., Renaud, J.P. *et al.* (1996) A canonical structure for the ligand binding domain of nuclear receptors. *Nature Structural Biology*, **3**, 87–94.
10. Nolte, R.T., Wisely, G.B., Westin, S. *et al.* (1998) Ligand binding and co-activator assembly of the peroxisome proliferator-activated receptor. *Nature*, **395**, 137–43.
11. Brzozowski, A.M., Pike, A.C., Dauter, Z. *et al.* (1997) Molecular basis of agonism and antagonism in the estrogen receptor. *Nature*, **389**, 753–58.
12. Evans, R.M. (1988) The steroid and thyroid hormone receptor superfamily. *Science*, **240**, 889–95.
13. Baes, M., Gulick, T., Choi, H.S. *et al.* (1994) A new orphan member of the nuclear hormone receptor superfamily that interacts with a subset of retinoic acid response elements. *Molecular and Cellular Biology*, **14**, 1544–52.
14. Forman, B.M. (2002) A structural model of the constitutive androstane receptor defines novel interactions that mediate ligand-independent activity. *Molecular and Cellular Biology*, **22**, 5270–80.
15. Goodwin, B. and Moore, J.T. (2004) CAR: detailing new models. *Trends in Pharmacological Sciences*, **25**, 437–41.
16. Wei, P., Zhang, J., Egan-Hafley, M. *et al.* (2000) The nuclear receptor CAR mediates specific xenobiotic induction of drug metabolism. *Nature*, **407**, 920–23.
17. Liu, Z., Auboeuf, D., Wong, J. *et al.* (2002) Coactivator/corepressor ratios modulate PRmediated transcription by the selective receptor modulator RU486. *Proceedings of the National Academy of Sciences of the United States of America*, **99**, 7940–44.

18. Honkakoski, P., Palvimo, J.J., Penttila, L. *et al.* (2004) Effects of triaryl phosphates on mouse and human nuclear receptors. *Biochemical Pharmacology*, **67**, 97–106.

19. Moore, L.B., Parks, D.J., Jones, S.A. *et al.* (2000) Orphan nuclear receptors constitutive androstane receptor and pregnane X receptor share xenobiotic and steroid ligands. *Journal of Biological Chemistry*, **275**, 15122–27.

20. Kobayashi, K., Yamanaka, Y., Iwazaki, N. *et al.* (2005) Identification of HMG-CoA reductase inhibitors as activators for human, mouse and rat constitutive androstane receptor. *Drug Metabolism and Disposition*, **33**, 924–29.

21. Rochel, N., Wurtz, J.M., Mitschler, A. *et al.* (2000) The crystal structure of the nuclear receptor for vitamin D bound to its natural ligand. *Molecular Cell*, **5**, 173–79.

22. Tocchini-Valentini, G., Rochel, N., Wurtz, J.M. *et al.* (2001) Crystal structures of the vitamin D receptor complexed to superagonist 20-epi ligands. *Proceedings of the National Academy of Sciences of the United States of America*, **98**, 5491–96.

23. Watkins, R.E., Davis-Searles, P.R., Lambert, M.H., and Redinbo, M.R. (2003) Coactivator binding promotes the specific interaction between ligand and the pregnane X receptor. *Journal of Molecular Biology*, **331**, 815–28.

24. Watkins, R.E., Maglich, J.M., Moore, L.B. *et al.* (2003) 2.1 Å crystal structure of human PXR in complex with the st. john's wort compound hyperforin. *Biochemistry*, **42**, 1430–38.

25. Watkins, R.E., Wisely, G.B., Moore, L.B. *et al.* (2001) The human nuclear xenobiotic receptor PXR: structural determinants of directed promiscuity. *Science*, **292**, 2329–33.

26. Thompson, J.D., Higgins, D.G., and Gibson, T.J. (1994) CLUSTAL W: improving the sensitivity of progressive multiple sequence alignment through sequence weighting, position-specific gap penalties and weight matrix choice. *Nucleic Acids Research*, **22**, 4673–80.

27. Accelrys Inc., San Diego (USA), http://www.accelrys.com.

28. Dussault, I., Lin, M., Hollister, K. *et al.* (2002) A structural model of the constitutive androstane receptor defines novel interactions that mediate ligand-independent activity. *Molecular and Cellular Biology*, **22**, 5270–80.

29. Dunbrack, R.L. Jr. (1999) Comparative modeling of CASP3 targets using PSI-BLAST and SCWRL. *Proteins*, **3**, 81–87.

30. Gampe, R.T. Jr., Montana, V.G., Lambert, M.H. *et al.* (2000) Asymmetry in the PPARgamma/RXRalpha crystal structure reveals the molecular basis of heterodimerization among nuclear receptors. *Molecular Cell*, **5**, 545–55.

31. Xu, H.E., Lambert, M.H., Montana, V.G. *et al.* (1999) Structural determinants of ligand binding selectivity between the peroxisome proliferator-activated receptors. *Proceedings of the National Academy of Sciences of the United States of America*, **98**, 13919–24.

32. Berendsen, H.J.C., van der Spoel, D., and van Drunen, R. (1995) GROMACS: a message-passing parallel molecular dynamics implementation. *Computer Physics Communications*, **91**, 43–56.

33. Scott, W.R.P., Hunenberger, P.H., Tironi, I.G. *et al.* (1999) The GROMOS biomolecular simulation program package. *Journal of Physical Chemistry A*, **103**, 3596–608.

34. Essmann, U., Perera, L., Berkowitz, M.L. *et al.* (1995) A smooth particle-mesh Ewald potential. *Journal of Computational Physics*, **103**, 8577–92.

35. Berendsen, H.J.C., Postma, J.P.M., DiNola, A., and Haak, J.R. (1984) Molecular dynamics with coupling to an external bath. *Journal of Chemical Physics*, **81**, 3684–90.

36. Hess, B., Becker, H., Berendsen, H.J.C., and Fraaije, J.G.E.M. (1997) LINCS: a linear constraint solver for molecular simulations. *Journal of Computational Physics*, **18**, 1463–72.

37. Kelley, L.A., Gardner, S.P., and Sutcliffe, M.J. (1996) An automated approach for clustering an ensemble of NMR-derived protein structures into conformationally related subfamilies. *Protein Engineering*, **9**, 1063–65.

38. Laskowski, R.A., MacArthur, M.W., Moss, D.S., and Thornton, J.M. (1993) PROCHECK: a program to check the stereochemical quality of protein structures. *Journal of Applied Crystallography*, **26**, 283–91.

39. Frank, C., Molnar, F., Matilainen, M. *et al.* (2004) Agonist-dependent and agonist-independent transactivations of the human constitutive androstane receptor are modulated by specific amino acid pairs. *Journal of Biological Chemistry*, **279**, 33558–66.

40. Andersin, T., Väisänen, S., and Carlberg, C. (2003) The critical role of carboxy-terminal amino acids in ligand-dependent and -independent transactivation of the constitutive androstane receptor. *Molecular Pharmacology*, **17**, 234–46.

41. Jyrkkärinne, J., Windshügel, B., Mäkinen, J. *et al.* (2005) Amino acids important for ligand specificity of the human constitutive androstane receptor. *Journal of Biological Chemistry*, **280**, 5960–71.

42. Sablin, E.P., Krylova, I.N., Fletterick, R.J., and Ingraham, H.A. (2003) Structural basis for ligand-independent activation of the orphan nuclear receptor LRH-1. *Molecular Cell*, **11**, 1575–85.

43. Feng, W., Ribeiro, R.C.J., Wagner, R.L. *et al.* (1998) Hormone dependent coactivator binding to a hydrophobic cleft on nuclear receptors. *Science*, **280**, 1747–49.

44. Windshügel, B., Jyrkkärinne, J., Poso, A. *et al.* (2005) Deciphering the structural basis for the constitutive activity of the human CAR receptor. *Journal of Molecular Modeling*, **11**, 69–79.

45. Suino, K., Peng, L., Reynolds, R. *et al.* (2004) The nuclear xenobiotic receptor CAR: structural determinants of constitutive activation and heterodimerization. *Molecular Cell*, **16**, 893–905.

46. Shan, L., Vincent, J., Brunzelle, J.S. *et al.* (2004) Structure of the murine constitutive androstane receptor complexed to androstenol: a molecular basis for inverse agonism. *Molecular Cell*, **16**, 907–17.

47. Windshügel, B., Jyrkkärinne, J., Vanamo, J. *et al.* (2007) Comparison of homology and X-ray structures of the nuclear receptor CAR: assessing the structural basis of constitutive activity. *Journal of Molecular Graphics & Modelling* **25**, 644–57.

48. Rochel, N., Tocchini-Valentini, G., Egea, P.F. *et al.* (2001) Functional and structural characterization of the insertion region in the ligand binding domain of the vitamin D nuclear receptor. *European Journal of Biochemistry*, **268**, 971–79.

49. Chang, T.K. and Waxman, D.J. (2006) Synthetic drugs and natural products as modulators of constitutive androstane receptor (CAR) and pregnane X receptor (PXR). *Drug Metabolism Reviews*, **38**, 51–73.

50. Ghose, A.K., Viswanadhan, V.N., and Wendoloski, J.J. (1998) Prediction of hydrophobic (lipophilic) properties of small organic molecules using fragmental methods: an analysis of ALOGP and CLOGP methods. *Journal of Physical Chemistry A*, **102**, 3762–72.

Index

Molecular Modeling. Basic Principles and Applications. 3rd *Edition*
H.-D. Höltje, W. Sippl, D. Rognan, and G. Folkers
Copyright © 2008 Wiley-VCH Verlag GmbH & Co. KGaA, Weinheim
ISBN: 978-3-527-31568-0